Diversity of Plants and Fungi

Diversity of Plants and Fungi

Rudolf Schmid

Department of Integrative Biology
University of California, Berkeley

BURGESS INTERNATIONAL GROUP INC
Bellwether Press Division

Front cover drawing by Priscilla Fawcett of *Monotropa* (Indian pipe), a non-photosynthetic, heterotrophic flowering plant (angiosperm) that derives its nutrition from the roots of host trees or from decomposing organic matter. (From *Plant biology,* by K. Norstog & R. W. Long, © 1976 by W. B. Saunders Co., Philadelphia.)

Title page drawing by Priscilla Fawcett of *Rafflesia,* a rare genus of 16 species native to western Malesia. This leafless parasite is found on the stems of a tropical vine and has the largest flowers of all angiosperms (each is up to one meter in diameter). The flowers emit a carrion odor that attracts pollinating flies. For details see Kamarudin Mat Salleh's excellent *Rafflesia: Magnificent flower of Sabah* (Borneo Publishing Co., Kota Kinabalu, Malaysia, 1991). (From *Plant biology,* by K. Norstog & R. W. Long, © 1976 by W. B. Saunders Co., Philadelphia.)

The bacteriophagus virus on the back cover has no relevance to this botany manual.

ISBN 0-8087-7931-1

Printed in the United States of America.
J I H G F E

This book was written in WordStar 5.5 and formated in WordPerfect 5.1 on a 486DX 33-MHz PC. Text was printed on a Hewlett Packard LaserJet III Printer in CG Times and Univers fonts. The final copy was produced camera-ready-copy by Bellwether Press.

Address orders to:

BURGESS INTERNATIONAL GROUP, Inc.
7110 Ohms Lane
Edina, Minnesota 55439-2143
Telephone 612/831-1344
EasyLink 629-106-44
Fax 612/831-3167

Bellwether Press
A Division of BURGESS INTERNATIONAL GROUP, Inc.

To Manuel

Overheard on TV the Other Night

While watching William Faulkner's *The sound and the fury* (book first published 1929, movie version 1959):

Conversation of Quentin Compson (played by Joanne Woodward [born 1930!] in the 1959 movie), rebellious teenager, with pawnbroker, to whom she had pawned her high school textbooks:

Quentin: "What I got to find out I can't find out in books. Besides [Quentin picks up Fuller & Tippo's *College botany,* 1954 revised edition], what good is botany in my life?"

Pawnbroker: "It don't hurt to understand the beauties of nature [Quentin turns to title page of book, with color frontispiece of *Crocus*], long with making a living."

Quentin: "Right at this particular minute I got enough trouble with nature."

Table of Contents

Preface

Biology textbooks nowadays are all so bulky (some weigh three kilograms or so) that students are very reluctant to carry them around. Thus the textbooks are rarely available for use in a teaching lab, despite the fact that they would be most useful to the students for diagrams, photographs, definitions of terms (the glossary), and background information. Naturally, the textbook has no specific lab instructions. On the other hand, lab manuals, whether standalone efforts or ones designed to accompany a textbook, are generally too skimpy. Sometimes this skimpiness involves actual instructions to the students, but invariably the lab manual offers minimal background information and definitions of terms. After all, this information is supposed to be in the textbook. Furthermore, many lab manuals are sparsely illustrated.

In 1949 Adriance Foster of the then Department of Botany, University of California, published his *Practical plant anatomy* (2nd ed., 1st ed. 1942). This work was interesting not only because it was totally devoid of illustrations but particularly because at the end of each detailed chapter on a topic there were lengthy sections entitled "Material for the study of . . ." and "Suggested drawings and notes" (e.g., the sections in the chapter on the stem involve, respectively, 14 pages and one page).

Taking a cue from Foster's anatomy textbook, I developed the present course manual on plant and fungal diversity for use in my courses in the Department of Integrative Biology at UC Berkeley. Highlights of the present version of this manual are:

1. Extensive background information is integrated with instructions for actual lab observations by the student.
2. There are 13 supplements that give background information. Some topics actually fairly closely follow my lecture material.
3. There are 13 lab exercises and five lab exercise supplements that contain a fair amount of background material in addition to the usual instructions for the students.
4. Pull-out pages are provided for suggested lab drawings.
5. There are repeated warnings that students must read the appropriate material ***before*** coming to a lab because there is insufficient time in the lab for them to read the material from scratch. Due to the large amount of background information presented in the manual, a ●●● is used to mark those paragraphs calling for actual observations or diagrams in the lab. That is, most unmarked paragraphs reiterate information presented in lecture. Moreover, in my courses the lab topics are covered in the lecture either before or concurrently with the lab on the same topic.
6. A bullet (●) is used to signify an important concept or technical term. Important terms in this course manual are also in bold italics, and are often defined several times.
7. All technical terms are concisely defined when first used. I also use simpler terms wherever possible, for instance, "stalk" instead of "seta" for the sporophyte of bryophytes, or "cone" instead of "strobilus" for the sporophyte of various vascular plants. However, I have not totally avoided some more unusual non-scientific words as I firmly believe that no one ever got brain damage from using a dictionary.
8. Throughout the manual I use abbreviations for gametophyte, sporophyte, and their related terms, for instance: SPT (sporophyte), SPTs (sporophytes), SPTic (sporophytic), GPT (gametophyte), GPTs (gametophytes), GPTic (gametophytic, microGPT (male gametophyte), megaGPT (female gametophyte), etc.

9. A few references are used, mainly to document unusual facts or tallies for the number of taxa in a group.
10. The end of this manual gives sample exam questions in multiple-choice format. Variants of these questions commonly appear on my exams. In my courses I tell students that all the information needed for the diversity part of the course and to answer successfully the exam questions will be presented in my lectures, labs, and course manual. This is another incentive to read the course manual beforehand.
11. Most of the 13 supplements, the 13 lab exercises, and the five lab exercise supplements contain figures that are referred to by the appropriate number. A reference to "Fig." is for a figure in a lab exercise, for example, "Fig. 6–3" for Lab Exercise 6 or "Fig. LabSup1–1" for Lab Exercise 1 Supplement A; in contrast, a reference to "Fig. Sup" is for a figure in a supplement, for instance, "Fig. Sup6–3" for Supplement 6. The figure captions are purposely made detailed, with some terms defined therein, and with restatements of the essential features of the types of life histories. This should make the figures a useful vehicle for students preparing for exams. As a further learning aid, many figure captions ask the student to locate on the figure where meiosis and syngamy (fertilization) occur in the life history.
12. Concomitantly, the descriptions in the text have detailed references to the figures. The figures are cited whenever it is most advantageous to view them, and the student thus is strongly advised to examine the figures in conjunction with the text.
13. Most of the illustrations are from the fine introductory textbook by Knut Norstog & Robert W. Long (*Plant biology,* 1976); acknowledgement is noted in the captions.

Acknowledgements:

I am very much indebted to my daughter Mena for editorial assistance and for taking the edited copy done in WordStar 5.5 and final formatting it in WordPerfect 5.1. Eric Andersen, Leonard Cohen, Butch Hancock, Ian & Sylvia, Buffy Sainte-Marie, Richard Thompson, among others, provided moral support. I also thank Knut Norstog for allowing me to reproduce the excellent artwork from his & Robert W. Long's fine 1976 textbook, *Plant biology.* The illustrations are by Priscilla Fawcett ("PF"), Jack R. Schroeder, and others.

When I choose a word . . . it means just what I choose it to mean—neither more nor less.—Humpty Dumpty (and, usually, Rudolf Schmid)

Introduction to Material on Plant and Fungal Diversity

NOTE: These remarks pertaining to material on plant and fungal diversity offer guidelines not only for studying your textbook(s) but also for efficiently finishing the lab material. The preface explains how this manual is organized.

ACCOUNTABILITY ON EXAMS FOR TERMS AND NAMES OF PERSONS AND ORGANISMS

1. **Terminology:** Important terms in this course manual are in bold italics and are usually defined when first used. Simpler terms are used wherever possible, and alternate terms are given in parentheses, for example, "***female cone*** (megastrobilus, megasporangiate strobilus, ovule cone, ovulate cone, seed cone)" or "***seed coat*** (testa)." Unless noted to the contrary, you are responsible *only* for such bold italicized terms. Some alternative terminology is given because your textbook(s), your TA, or another instructor may prefer that. Any definitions of terms presented in the lectures or course manual take precedence over similar terms in the textbook(s). Your lecture textbook has a good glossary of scientific terms. Use it! *Note:* A bullet (•) at the left margin in the text signifies an important term or concept.
2. **Names of persons:** Unless noted to the contrary, you are *not* accountable for names of persons incidentally mentioned in the manual. Naturally, if a person is repeatedly mentioned in connection with an elaborate discussion, you can assume that that person is important.
3. **Common and scientific names of plants and fungi:** Similarly, unless noted to the contrary, you are accountable for names (either common or generic, as you choose) of plants and fungi wherever any discussion of the plant or fungus in a lab exercise involves a half page or so in any given block (i.e., in *one* place) of a lab exercise, for example, such names as pine or *Pinus*. You are *not* accountable for any species names, such as the *"aromaticum"* part of *Syzygium aromaticum* (clove) or for any family names. In addition, it will be necessary for you to learn the common or scientific names of some of the important divisions of plants and fungi (see Supplement 1).
4. **Exam questions:** In conclusion, the exam questions will generally use common names and simpler terms, often also giving scientific names as well as alternative terms, for instance: "The cone (strobilus) of pine (*Pinus*) differs from that of horsetail (*Equisetum*) in having [choice of five items]"; and "The green algae (Chlorophyta) differ from the brown algae (Phaeophyta) in having [choice of five items]." The end of the manual gives sample exam questions in multiple-choice format. Some of these questions may well appear in identical or revised form on the examinations.

LAB MATERIALS NEEDED

Unless instructed to the contrary, each student must supply his/her own lab supplies, including the following: single-edge razor blades, dissecting equipment (needles and forceps), microscope slides, microscope slide cover glasses (cover slips), drawing supplies (drawing-type pencil, eraser, and sketch or drawing paper if the pull-out pages for the lab diagrams are not used). Bring these items to the *first* lab period. A small number of microscope slides and microscope cover glasses may be supplied to each student in the first lab period *only*.

LAB DIAGRAMS AND ACCOUNTABILITY OF LAB MATERIAL

Lab diagrams are suggested. Space is provided in the manual for lab diagrams, and the pages for these can be detached and turned in if desired by your TA and/or instructor. Note that stereotyped diagrams of items are wanted, not artistic-type drawings; detailed cellular diagrams are a waste of time and are *not* wanted. The suggested sketches or outline diagrams and their appropriate labels are listed in the individual lab exercises. The extra act of sketching seems to etch more firmly on one's mind the structure observed. Your sketches should also prove useful for reviewing for subsequent quizzes and exams. Label any sketches fully since unlabeled or poorly labeled diagrams are of little value.

Your lab diagrams may be graded, as elaborated by your TA and/or instructor. The grade will be based on neatness and accuracy depicting representative material from a lab. You will *not* be graded for artistic ability.

Warning: Note that lab diagrams are suggested for only selected items. Nevertheless, you are accountable for lab material for which *no* diagrams are suggested, in particular material with an extensive treatment.

SUGGESTIONS TO FACILITATE LABWORK

To make effective use of your time in the lab, note the following:

1. The lab topics will be covered in the lecture either before or concurrently with the lab on the same topic. However, read the suggested readings in the textbook(s) and, *in particular, read the relevant lab exercise in the course manual* ***before coming to the lab.*** The lab exercises contain a large amount of written background information; trying to read all of this material in the lab the first time will be quite counterproductive. If you try to read the material in lab from scratch, your performance will not be up to scratch. In the lab exercises a ●●● is used to mark those paragraphs calling for actual observations or diagrams in the lab; most unmarked paragraphs reiterate information presented in the lecture. Also note that the material for which diagrams are suggested probably is the most important!
2. Bring your copy of the textbook and especially the course manual and any required lab textbook to each lab. Figures in the course manual and the textbooks should be frequently consulted in conjunction with actual examination of the material.
3. If in the beginning of the lab period, an exercise is crowded with persons trying to get material or to look at demonstrations, start at a later point in the lab exercise.
4. Work in pairs, especially to prepare water mount slide preparations of material. This will save considerable time. If you have a particularly good slide that you prepared, call it to the attention of your TA or instructor and pass it on to one of your neighbors when you are finished.

CARE OF MICROSCOPES

The microscope is a delicate and expensive instrument and requires the same care as a camera or a piece of electronic equipment. Lab Exercise 1, Supplement A, on the use of the compound microscope elaborates on the following points.

1. Always carry the microscope with two hands, one hand on the arm of the microscope, the other hand supporting its base.

2. Never lift the microscope by the binocular observation tube, and check that this is locked into place.
3. Do *not* excessively tilt the microscope, or else the oculars will fall out of the binocular tube.
4. Before commencing observations, check that the oculars, eyepieces, and other components of the light path are reasonably clean.
5. If water or reagents are spilled onto the stage or other part of the microscope, clean up the mess immediately. If immersion oil is used, also clean up this.
6. When putting away the microscope, return the light intensity sliding control to zero and turn off the lamp. Set the revolving nosepiece so that the 4X objective is in place. Remove any slide from the microscope stage. Coil up the electrical cord and cover the microscope with its plastic cover. Properly carry the microscope to its cabinet as outlined above.

CARE OF MICROSCOPE SLIDES

Permanent microscope slide preparations are expensive and in some cases are irreplaceable. Please observe these precautions:

1. Be careful with the high power objectives so as not to damage slides.
2. Take only a few slides at a time, that is, *not* the slides for the entire lab. This is especially important in a crowded lab because there are not enough slides for each person to observe at the same time.
3. At the end of the lab, remove any slide from the microscope stage and return it and any other borrowed slides to their proper boxes (some slides may be color coded to facilitate this).

Begin at the beginning . . . and go on till you come to the end: then stop.
—The King of Hearts

SUPPLEMENT **1**

Classification of Organisms Historically Regarded as Plants

I. PERSPECTIVE

The diversity of organisms can be formally expressed by means of a classification system. Although there are official rules for naming organisms, for example, by the *International code of botanical nomenclature* (latest edition 1988) and the comparable codes in zoology and bacteriology, there are, as elaborated below, appreciable differences in opinion about the relationships of organisms and hence on their relative rankings.

The information below, especially that in Parts III through VI and in the synopses at the end of this supplement, *is mainly for reference purposes and thus should not be memorized.* The purpose of the lists following is severalfold: (1) to relate briefly how the five-kingdom system of classification evolved and how the plant, fungal, and other kingdoms differ; (2) to indicate the relative positions in these kingdoms of various groups discussed in this course manual; (3) to convey some information about scientific and common names; and (4) to give you some appreciation of what groups are embraced by terms such as "conifer," "lycopod," "moss," etc.

You are not expected to memorize a host of scientific names of specific fungi (plural, singular *fungus*) or plants, including algae (plural, singular *alga*). During this course you will become familiar with the scientific and/or especially common names (see Part V) of some of the major genera of living organisms, for example, *Pinus* (pine). However, you will *not* be held accountable for the scientific names of species, for instance, the "longaeva" of *Pinus longaeva* (bristlecone pine). In addition, it will be necessary for you to learn the scientific or especially common names of some of the important divisions of plants and fungi. The examination questions will generally use common names and simpler terms, often also giving scientific names as well as alternative terms.

II. HISTORICAL EVOLUTION OF THE FIVE-KINGDOM CLASSIFICATION SYSTEM

All living organisms can be fundamentally divided into two groups representing superkingdoms:

- ***prokaryotes,*** lacking nuclei and other membrane-bounded organelles;
- ***eukaryotes,*** having nuclei and other membrane-bounded organelles.

Fungi have traditionally been regarded as plants, but appreciation of the distinctive mode of nutrition of fungi and their other peculiar features has, in recent years, led to their recognition as a separate kingdom (Fungi or Myceteae). The tables on the next page show the historical evolution of the five-kingdom classification system of organisms and some of the salient characteristics of these kingdoms:

PERSON(S)	ANIMALS	PLANTS	PROTISTS	PROKARYOTES	FUNGI
Ancient Greeks, & historically	yes	yes (incl. fungi, protists, prokaryotes)			
Haeckel 1866	yes	yes	yes (incl. fungi, prokaryotes)		
Copeland 1938	yes	yes	yes (incl. fungi)	yes	
Whittaker 1969	yes	yes	yes	yes	yes
Your textbook	yes	yes	yes	yes	yes
This manual	yes	yes	yes	yes	yes

CHARACTER	ANIMALS	PLANTS	PROTISTS	PROKARYOTES	FUNGI
Nuclei	yes	yes	yes	no	yes
Cell wall	no	yes	no	yes	yes
Wall chemical	no wall	cellulose	no wall	not cellulose	chitin
Chloroplasts	no	yes	no	no	no
Vacuoles	no	yes	no	no	in some
Main nutrition	ingestion	photosynthesis	ingestion	absorption	absorption
Other nutrition	absorption	absorption ingestion		photosynthesis chemosynthesis	ingestion

See Lab Exercise 1, Part III, for details of basic cell structure, Supplement 11, Part II, for enumeration of the major types of biological nutrition, and Supplement 12, Part III, for comments on some of these features contrasting the five kingdoms. *Note:* In the above table I include *all* algae (plural, singular *alga*) among the plants and *all* fungi among the fungi. In contrast, some textbooks, including Raven & Johnson (1992) on basic biology, and Raven et al. (1992) and Stern (1991) on basic botany, include various algae and fungi in the protists. See Part VI below for elaboration.

Robert H. Whittaker in 1959 was the first to emphasize the distinctiveness of the fungi, in particular their characteristic mode of obtaining nutrition by absorption. In 1969 Whittaker formally recognized a fifth kingdom, the "Fungi." Since 1959 there has been a lot of discussion about the four-kingdom versus five-kingdom classification system. A major focus has been the arbitrariness of Protista (or Protoctista) as a formal group. Most works now recognize five kingdoms of organisms. However, some recent textbooks, for instance, Raven et al. (1992; but not Raven & Johnson 1992), divide the prokaryotes into two kingdoms, Archaebacteria and Eubacteria, the latter with the bacteria and cyanobacteria. Some alternative systems have even inflated to 19 kingdoms (Leedale 1974).

III. ENDINGS OF THE MAJOR TAXONOMIC CATEGORIES

The following endings of the major taxonomic categories are accepted in the *International code of botanical nomenclature* (latest edition 1988):

- *Superkingdom:* no specified endings
  - *Kingdom:* no specified endings
    - *Subkingdom:* no specified endings [category rarely used, e.g., by Ray et al. (1983), who use the ending "-onta," which, however, we and Bold et al. (1987) use for superkingdoms]
      - *Division:* "-phyta" ("-mycota" for fungi) [the comparable zoological rank of "phylum" is not recognized in the *Botanical code*]
        - *Subdivision:* "-phytina" ("-mycotina" for fungi)
          - *Class:* "-opsida" ("-mycetes" for fungi, "-phyceae" for algae)
            - *Subclass:* "-idae" ("-mycetidae" for fungi, "-phycidae" for algae)
              - *Order:* "-ales" or "-ae"
                - *Suborder:* "-ineae" or "-ae"

Family: "-aceae" ("-ae" when conserved, which is uncommon, e.g., for the grass family Gramineae or Poaceae, the sunflower family Compositae or Asteraceae, and six other families)
Subfamily: "-oideae"
Tribe: "-eae"
Subtribe: "-inae"
Genus: no specified endings
Subgenus: no specified endings
Species: no specified endings
Subspecies: no specified endings
Variety: no specified endings.

The "sub" categories and those categories below the rank of order will be given very little application in this course manual. The endings just listed refer to modern systems of classification. Endings of names in old systems of classification (but also in many new ones) commonly do not correspond to the ranks given above.

The term ***taxon*** (singular, plural ***taxa***) is a useful designation for reference to a taxonomic group of *any* rank. For example, both the genus *Magnolia* and the order Magnoliales are taxa, but obviously of different rank.

IV. EXAMPLES OF THE CLASSIFICATION OF TWO PLANTS AND A FUNGUS

The classification below uses the above hierarchy to classify two plants and a fungus (alternative names are noted in parentheses):

Superkingdom	Eukaryonta (Eukaryota)	Eukaryonta (Eukaryota)	Eukaryonta (Eukaryota)
Kingdom	Plantae (Phyta)	Plantae (Phyta)	Fungi (Myceteae)
Division	Anthophyta (Magnoliophyta)	Phaeophyta	Basidiomycota (Amastigomycota)
Subdivision	Anthophytina (Magnoliophytina)	Phaeophytina	Basidiomycotina
Class	Dicotyledones (Magnoliopsida, Annonopsida)	Phaeophyceae	Basidiomycetes
Subclass	Dicotyledoneae (Magnoliidae, Annonidae)	Phaeophycidae	Holobasidiomycetidae
Order	Magnoliales (Annonales)	Fucales	Phallales
Suborder	Magnoliineae (Annonineae)	Fucineae	Phallineae
Family	Magnoliaceae	Fucaceae	Phallaceae
Subfamily	Magnolioideae	Fucoideae	Phalloideae
Tribe	Magnolieae	Fucieae	Phalleae
Subtribe	Magnoliinae	Fucinae	Phallinae
Genus	*Magnolia*	*Fucus*	*Phallus*
Species	*grandiflora*	*distichus*	*impudicus*
Species authority	Linnaeus	Linnaeus	Persoon
Common name	magnolia, bull bay	rockweed	stinkhorn

For levels above the rank of family there usually is spirited debate which name should be applied to a taxon. Superorders are recognized for some groups, for example, for angiosperms Magnolianae, Magnoliiflorae, or Annoniflorae.

V. MISCELLANEOUS COMMENTS ON NAMES

The scientific name of a plant, fungus, or other type of organism is a ***binomial,*** that is, a two-word name, one name for the genus, another for the species. For plants the binomial system of nomenclature originated in 1753 with Carolus Linnaeus's (1707–1778) *Species plantarum.*

Binomials should be italicized when printed or underlined when written. The generic name is always capitalized. The species or generic names have fairly standardized meanings. Thus in the examples given above, for *Magnolia grandiflora,* the generic name honors the French botanist Pierre Magnol (1638–1715) whereas the species name means "large-flowered." For *Fucus distichus* the generic name means "seaweed" and comes from the Greek *phykos* whereas the species name means "two-ranked." For *Phallus impudicus* the generic name is obvious whereas the species name means "lewd" or "shameless." The same species name (specific epithet) can be used for different genera, for example, *Nepeta grandiflora* (catnip) and *Magnolia grandiflora.* The generic name, however, can be properly used for only one type of plant or fungus.

In essence, a given scientific name can be used for only one taxon. Incidentally, it is the same with movie and TV stars. The Screen Actors Guild (SAG) prohibits two performers from using identical or similar stage names. Thus a recent report (*The San Francisco chronicle,* 26 May 1992) notes that SAG wants comedian and talk show host Dennis Miller to change his name because there exists a Denny Miller, who was in the 1957–65 *Wagon train* TV series and who also had the title role in the 1959 film *Tarzan, the ape man.* There is also the case of former Miss America and current singer-actress Vanessa Williams and actress Vanessa Williams, who was in the 1991 film *New Jack City.* However, the former, registered as Vanessa L. Williams, though never using the initial, is safe as long as the latter, who registered first, does not formally complain.

Scientific names of all ranks have attached to them the name or names of the person or persons proposing the name, that is, the authority for the name. These authorities have more or less standardized abbreviations. Hence the authorities for the three examples in Part IV are Linnaeus and Christiaan Hendrik Persoon (1761–1836) and the most complete rendering of the names would be *Magnolia grandiflora* L., *Fucus distichus* L., and *Phallus impudicus* Pers. The "L." and "Pers." are abbreviations for Linnaeus and Persoon, and the two Linnean names specifically appeared in his *Species plantarum.* Authorities for scientific names are used only in technical papers and monographs, almost never in elementary and general works.

Unfortunately, the great German mycologist, Karl Wilhelm Gottlieb Leopold Fuckel (1821–1876), whose name is not pronounced the way you may think it is, did not have anything to do with the fungal order Phallales and its most famous member *Phallus.* Fuckel did not describe the stinkhorn *Phallus* (if so, it would be nomenclaturally *Phallus* Fuckel), nor any species in it (e.g., nomenclaturally not *Phallus impudicus* Fuckel). As noted above, Persoon described this species, which is a club fungus (Basidiomycota). Fuckel is eponymized by *Fuckelia, Fuckelina,* and *Neofuckelia,* which are all genera of sac fungi (Ascomycota). Incidentally, Fuckel's name is rarely abbreviated as the authority for scientific names of taxa he did author.

Because scientific names are Latin, they generally receive Latin pronunciation. However, Europeans and Latinos, in particular, usually pronounce the names phonetically. This can lead to interesting results for names such as *Pinus* or *Fucus.*

In routine, informal descriptions botanists frequently use for convenience a variety of common (or vulgar) names rather than the relevant formal, taxonomic names, which are often more cumbrous, ostentatious, and even recondite. Some of the common names listed in the "Detailed classification of organisms" at the end of this supplement thus are important and will be used in this course manual and especially in the lectures. Consequently you should become familiar with them.

VI. CLASSIFICATION OF ORGANISMS TRADITIONALLY REGARDED AS PLANTS

The synopsis at the end of this supplement gives the divisions (and some classes) of organisms recognized in this course manual. A sampler of common names likely to be used in this course manual is also included. Alternative names of some of the divisions and of some of the common names are given in parentheses to guide you through the terminology in various books and papers. You are *not* responsible for any of these alternative names.

The classification system at the end of this supplement differs somewhat from that in basic biology or botany textbooks such as those by Raven & Johnson (1992) and Raven et al. (1992). These authors include certain fungi (divisions 4–7), all the algae (divisions 12–19), and the protozoans in Kingdom Protista. Protista as so defined is a gallimaufry group. Because of this, and for other reasons, we will follow Bold et al. (1987) and include all algae in Kingdom Plantae and all the fungi in an inclusive Kingdom Fungi. However, in contrast to Bold et al. (1987), Raven & Johnson (1992), and Raven et al. (1992), we follow Schofield's (1985) recent book on bryophytes and recognize an inclusive division Bryophyta rather than three separate divisions, as do Bold et al. (1987). In addition, our classificatory treatment of the vascular plants is more elaborate than such treatments by Bold et al. (1987) and Raven et al. (1992) because it follows Taylor's (1981) book on paleobotany and includes a number of exclusively fossil groups.

VII. LITERATURE CITED

A separate bibliography gives references for citations used throughout the course manual. The following are some general references on the structure (morphology) of plants and fungi:

on all plants and fungi: Bold et al. (1987);
on algae: Bold & Wynne (1985);
on bryophytes: Schofield (1985);
on all non-vascular plants: Scagel et al. (1982);
on vascular plants: Gifford & Foster (1989);
on fungi: Alexopoulos & Mims (1979), Hanlin & Ulloa (1988);
on paleobotany: Stewart (1983), Taylor (1981);
excellent basic textbooks on botany: Corner (1964), Kaufman et. al. (1989), Norstog & Long (1976), Raven et al. (1992), Ray et al. (1983), Stern (1991), Weier et al. (1982);
excellent basic textbooks on biology: Raven & Johnson (1992), Starr (1991);
well-illustrated "children's books": Johnson (1982, 1983), Selsam (1986), Watts (1986).

Standard textbooks on anatomy include Esau (1965), Fahn (1990), Foster (1949), and Panshin & de Zeeuw (1980). ***Morphology*** is the study of the form of organisms and their development, especially of external form and reproductive structure, whereas ***anatomy*** is the study of their internal structure.

> A biologist, regardless of his [or her] line of specialization, cannot afford to lose sight of the whole organism if his [or her] goal is the understanding of the organic world.
> —Katherine Esau, *Plant anatomy,* 2nd ed., 1965

Detailed classification of organisms

Superkingdom 1: prokaryotes (*Prokaryonta,* Prokaryota)
  Kingdom 1: monerans (*Monera*)
    Division 1: bacteria (*Bacteria*), 2,600 spp., including 100 spp. archaebacteria
    Division 2: cyanobacteria or blue-green bacteria ("blue-green algae") (*Cyanophyta*), 7,500 spp.
    Division 3: Prochlorophyta (with chlorophylls *a* and *b*—proposed 1977), 2 spp.
Superkingdom 2: eukaryotes (*Eukaryonta,* Eukaryota)
  Kingdom 2: protozoans (*Protista,* Protoctista) [not treated here. Some authors, e.g., Raven et al. (1992) include divisions 4-7, 12–19 in this kingdom]
  Kingdom 3: animals (*Animalia*) [not treated here]
  Kingdom 4: fungi (*Fungi,* Myceteae)
    Division 4: cellular slime molds (*Acrasiomycota*), 65 spp.
    Division 5: plasmodial slime molds (*Myxomycota,* Plasmodiomycota), 450 spp.
    Division 6: chytrids (*Chytridiomycota*), 750 spp.
    Division 7: water molds (*Oomycota*), 475 spp.
    Division 8: bread/etc. molds (*Zygomycota*), 765 spp.
    Division 9: sac fungi (*Ascomycota*), 30,000 spp.
    Division 10: club fungi (*Basidiomycota*), 25,000 spp.
    Form-division 11: imperfect fungi (*Deuteromycota*), 17,000 spp.
    Lichens (not a formal division): lichens, 18,000–20,000 spp.
  Kingdom 5: plants (*Plantae,* Phyta)
    Division 12: diatoms, golden and yellow-brown algae (*Chrysophyta*), 6,900 spp.
    Division 13: dinoflagellates (*Pyrrhophyta*), over 1,000 spp.
    Division 14: cryptomonads (*Cryptophyta*), 100 spp.
    Division 15: red algae (*Rhodophyta*), 5,200 spp.
    Division 16: brown algae (*Phaeophyta*), 2,000 spp.
    Division 17: green algae (*Chlorophyta*), 6,750 spp.
    Division 18: stoneworts (*Charophyta*), 250 spp.
    Division 19: euglenoids (*Euglenophyta*), 450 spp.
    Division 20: bryophytes (*Bryophyta*), over 22,400 spp.
      Class 1: liverworts (hepatics) (*Hepaticae*), 8,000 spp.
      Class 2: hornworts (*Anthocerotae*), 100 spp.
      Class 3: mosses (*Musci*), over 14,300 spp.
    Division 21: Rhyniophyta (s.s.)**
    Division 22: Zosterophyllophyta**
    Division 23: Trimerophytophyta**
    Division 24: whisk ferns, psilophytes (*Psilophyta,* Psilotophyta), 12 spp.
    Division 25: lycopods (lycophytes, club mosses) (*Lycophyta,* Microphyllophyta, Lepidophyta), 1,260 spp.
    Division 26: horsetails (sphenophytes, articulates) (*Sphenophyta,* Arthrophyta, Calamophyta), 29 spp.
    Division 27: ferns (*Pterophyta,* Filicophyta, Polypodiophyta, Pteridophyta s.s.), 9,000 spp.
    Division 28: progymnosperms (*Progymnospermophyta*)**
    Division 29: seed ferns (*Pteridospermophyta*)**
    Division 30: cycads (*Cycadophyta*), 137 spp.
    Division 31: cycadeoids (*Cycadeoidophyta*)**
    Division 32: ginkgophytes (*Ginkgophyta*), 1 sp.
    Division 33: conifers (*Coniferophyta* s.s.), 536 spp.

Class 1: cordaites (*Cordaitopsida*)**
Class 2: conifers and taxads (*Coniferopsida*, Pinopsida), 628 spp.
Division 34: gnetophytes (chlamydosperms) (*Gnetophyta*), 76 spp.
Division 35: flowering plants, angiosperms (*Anthophyta*, Magnoliophyta, Angiospermophyta), some 240,000 spp. (13,725 genera—Mabberley 1987)
Class 1: dicotyledons ("dicots") (*Dicotyledones*, Magnoliopsida), 188,000 spp. (10,907 genera)
Class 2: monocotyledons ("monocots") (*Monocotyledones*, Liliopsida), 52,000 spp. (2,818 genera)

Key (to "Detailed classification of organisms"):
s.s. = sensu stricto (in a narrow sense);
s.l. = sensu lato (in a wide sense);
** = only fossil representatives;
sp./spp. = abbreviations for singular/plural of species.

Other notes (to "Detailed classification of organisms"):
The "detailed" list above is substantially *for reference purposes only*. Many of these groups will *not* be treated in this course manual.
Formal names are given in parentheses, and the preferred ones are italicized. This course manual will generally use the common names.
The numbers for species are for living (***extant***) representatives of a group only; that is, fossil (***extinct***) taxa are excluded. Data are from various sources (see lab exercises following). *Note:* The numbers given above and throughout this manual are for named or described species, that is, "known species diversity" rather than "actual species diversity," which is probably vastly greater (see Ehrlich & Wilson 1991).
Raven et al. (1992), but not Raven & Johnson (1992), divide Kingdom 1, the monerans, into two kingdoms, Archaebacteria and Eubacteria, the latter with the bacteria and cyanobacteria.

Colloquial names of various groups

Fungi = divisions 4–11
Algae = divisions 12–19
Non-vascular plants = divisions 12–20
Land plants = divisions 20–35
Vascular plants = divisions 21–35
Nonseed plants = divisions 12–28
Nonseed vascular plants (pteridophytes) = divisions 21–28
Seed plants = divisions 29–35
First vascular plants (Rhyniophyta s.l.) = divisions 21–23

"What's the use of their having names," the Gnat said, "if they won't answer to them?"
"No use to *them*," said Alice; "but it's useful to the people that name them, I suppose. If not, why do things have names at all?"

Quickie classification of organisms

#Prokaryotes: *nuclei and other organelles absent*
  *Monerans—bacteria, blue-green bacteria, etc. (3 divisions)
#Eukaryotes: *nuclei and other organelles present*
  *Protists—protozoans (many divisions): *cell wall absent; chloroplasts absent*
  *Animals—animals (many divisions): *ditto*
  *Fungi (7 divisions, 1 form-division): *cell wall present; chloroplasts absent*
  *Plants (24 divisions): *cell wall present; chloroplasts present*
    Algae (8 divisions)
    Bryophytes—mosses, liverworts, hornworts (1 division)

NON-VASCULAR PLANTS
VASCULAR PLANTS

Nonseed plants or pteridophytes (vascular cryptogams)—ferns, club mosses, horsetails, etc. (8 divisions, 4 strictly fossil)
Seed plants (phanerogams)—(7 divisions, 2 strictly fossil)
  Gymnosperms—conifers, cycads, *Ginkgo,* etc. (6 divisions)
  Angiosperms—flowering plants (1 division)
    Dicotyledons
    Monocotyledons

Key (to "Quickie classification of organisms"):
  # = superkingdom;
  * = kingdom.

Other notes (to "Quickie classification of organisms"):
"Land plants" = bryophytes + vascular plants.
Supplement 6 contrasts bryophytes and vascular plants, Supplement 7 pteridophytes and seed plants, Supplement 9 gymno- and angiosperms, and Supplement 10 di- and monocotyledons.

SUPPLEMENT 2

Basic Descriptive Terminology in Morphology and Anatomy

I. PERSPECTIVE

Certain terms indicating direction or position are used repeatedly to describe structures of plants and fungi (and animals), in particular structures of the land plants, that is, the bryophytes and vascular plants. ***Morphology*** is the study of the form of organisms and their development, especially of external form and reproductive structure, whereas ***anatomy*** is the study of their internal structure.

II. TERMINOLOGY

See Fig. Sup2–1. The terms marked with an asterisk (*) are the most important and, accordingly, should be memorized:

adventitious,* arising from a source other than the usual one, as adventitious roots;
abaxial,* away from the axis, that is, away from the angle the leaf forms with the stem, thus situated on the side of an organ away from the axis, hence facing the base of the organism;
adaxial,* toward the axis, that is, toward the angle the leaf forms with the stem, thus situated on the side of an organ toward the axis, hence facing the apex of the organism;
dorsal, on or toward the upper surface (or toward the back of animals);
ventral, on or toward the lower surface (or toward the undersurface or front of animals);
inferior, growing or placed below, as an inferior ovary located below the other floral parts;
superior, growing or placed above, as a superior ovary located above the other floral parts;
anterior, before or toward the front;
posterior, after or toward the back;
distal, away from a reference point, which is usually the main part of the body; see also "apical";
proximal, toward a reference point, which is usually the main part of the body; see also "basal";
apical,* at or near the apex, often used in the sense of distal;
basal, at or near the base, often used in the sense of proximal;
lateral,* at the sides, as the lateral meristems, cork and vascular cambium;
intercalary, between other parts, intercalated, thus not at the tips;
median or ***medial,**** in or through the middle;
acropetal, from the base toward the apex;
basipetal, from the apex toward the base;
centrifugal, from the inside toward the outside, as centrifugal force;
centripetal, from the outside toward the inside, as centripetal force;
anticlinal, perpendicular to the surface;

periclinal, parallel to the surface;
radial,* along the radius of an axis; a radial section (abbreviated "r.s.") is cut parallel to the radius;
tangential,* along the tangent of an axis; a tangential section (abbreviated "t.s.") is cut perpendicular to the radius;
transectional, cross sectional, or ***transverse,**** perpendicular to the longitudinal axis; a transection or cross section (abbreviated "x.s.") is cut perpendicular to the long axis;
longisectional or ***longitudinal,**** parallel to the longitudinal axis; a longisection or longitudinal section (abbreviated "l.s.") is cut parallel to the long axis and may be either radial or tangential;
paradermal, parallel to the epidermis, especially of a leaf; a paradermal section (abbreviated "p.s.") is cut parallel to the surface.

Such precise morphological terms are preferable to more colloquial designations. For instance, for flattened organs such as leaves "top" is commonly used as roughly equal to "dorsal" or "adaxial," whereas "bottom" is commonly used as roughly equal to "ventral" or "abaxial." However, because structures may rotate during their development, the "top" becomes "bottom," and vice versa. For example, young female pine cones are upright but at maturity are usually pendant, with the result that the ovules on the "top" sides of the cone scales become seeds on their "bottom" sides as the cone rotates. In contrast, the ovule or seed described as "adaxial" is truly adaxial whether the cone is erect or pendant.

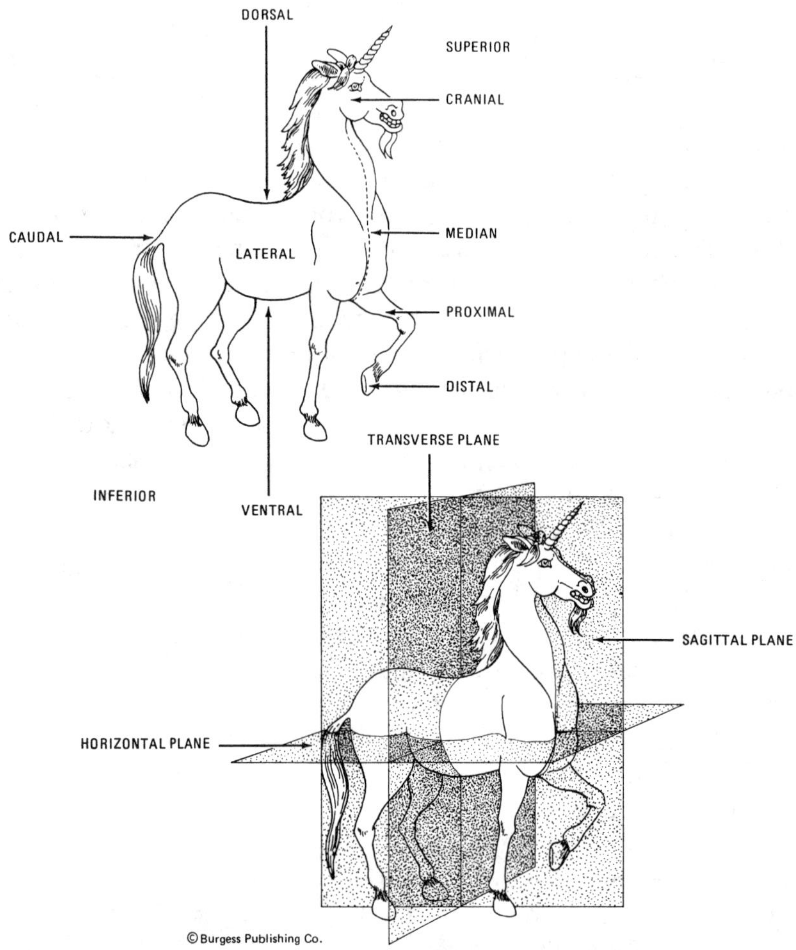

Fig. Sup2-1. Descriptive terms for viewing planes and orientation to body parts. The following are zoological terms not relevant to this botanical course manual: caudal, cranial, horizontal plane, sagittal plane, and transverse plane.

SUPPLEMENT **3**

The Geological Column and the Early Evolution of Life

I. PERSPECTIVE

Most of the discussion of the diversity of plants and fungi deals with living or extant plants. However, considerations of fossil or extinct plants are very important.

II. THE GEOLOGICAL COLUMN

Geological time (see also Fig. Sup3–1) is divided into various subdivisions, as follows (from Taylor 1981:20):

Era	*Period*	*Epoch*	*Began millions years ago*
Cenozoic	Quaternary	Recent	= last 5000 years
		Pleistocene	2.5
	Tertiary	Pliocene	7
		Miocene	26
		Oligocene	38
		Eocene	54
		Paleocene	65
Mesozoic	Cretaceous		136 [130 = earliest angiosperms]
	Jurassic		190
	Triassic		225 [200 = angiosperm precursors]
Paleozoic	Permian		280
	Carboniferous	Pennsylvanian	325
		Mississippian	345
	Devonian		395 [410 = oldest land plants]
	Silurian		430
	Ordovician		500
	Cambrian		570
Precambrian			4600 [= age of earth]

III. SOME IMPORTANT EVENTS MAINLY IN THE PRECAMBRIAN

Some important events (see also Fig. Sup3–1), mainly in the Precambrian, are listed from the oldest to youngest (bya = billion years ago; mya = million years ago):

1. Age of earth: 4.6 bya. This is based on the age of the oldest rocks in northwestern Canada, 3.96 bya.
2. Oldest known chemicals of organic origin: 3.8 bya, from Greenland. Thus the implication is that life was present on earth 3.8 bya.
3. Oldest known fossils, presumed bacteria (= individual cells and filaments): 3.5 bya, from Australia (similar fossils in South Africa, 3.3 bya). Thus considerable diversity of unicellular life already existed on earth at this time and obviously had originated earlier. The first life on earth was prokaryotic.
4. Oldest known cyanobacteria: 2.2 bya, from South Africa (similar fossils in Canada, 1.9 bya).
5. Oldest known eukaryotic cells, of indeterminable nature: 1.4 bya. Thus eukaryotic cells probably originated 1.5–2.0 bya.
6. Oldest known green algae (unicellular): 900 mya, from Australia.
7. Oldest known multicellular organisms (invertebrates): 700 mya.
8. Oldest known land (vascular) plants: 410 mya.
9. Oldest known flowering plants: 130 mya.

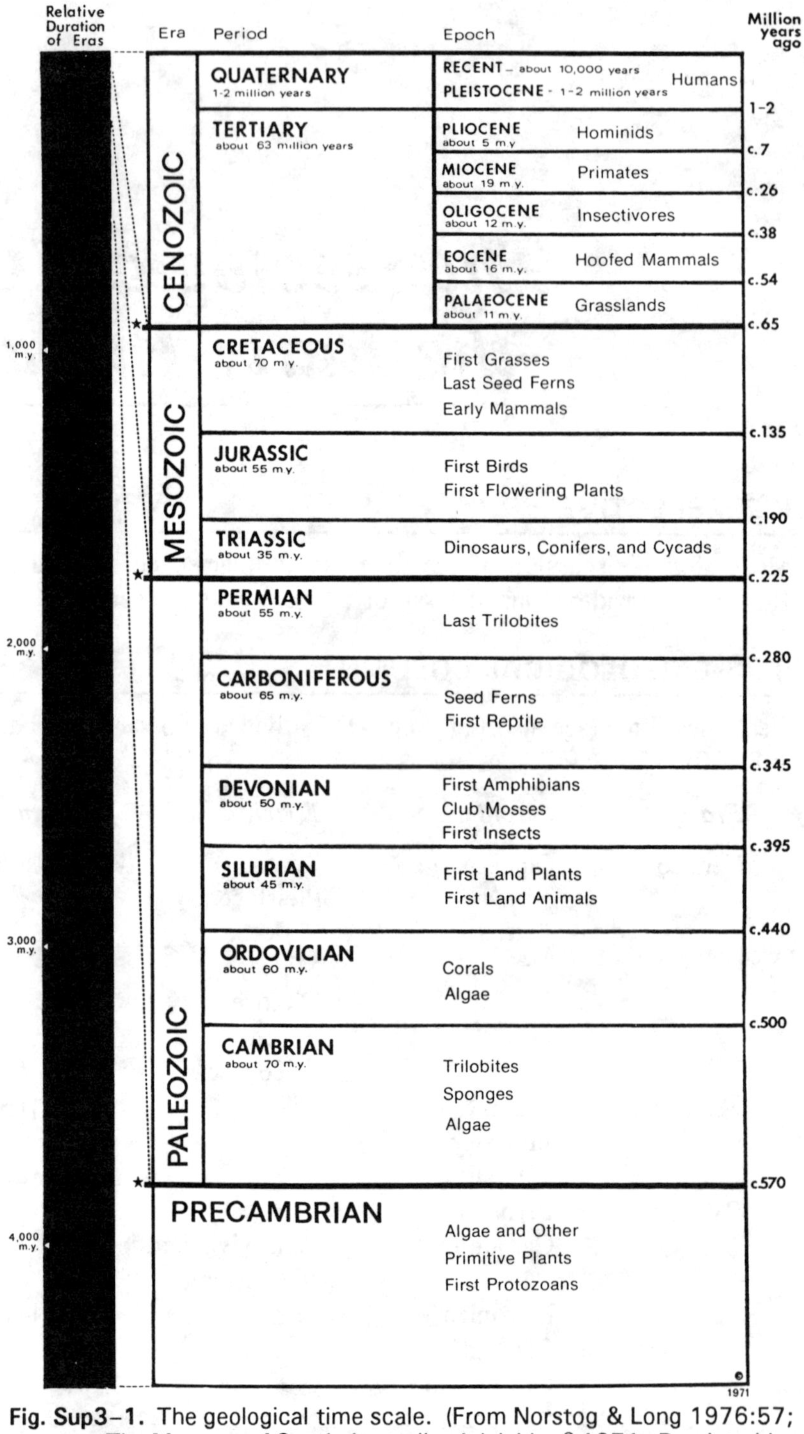

Fig. Sup3–1. The geological time scale. (From Norstog & Long 1976:57; courtesy The Museum of South Australia, Adelaide, © 1971. Reprinted by permission.)

Note: "Oldest known" refers to the oldest record of a fossil that is generally accepted as representative of a group. Many groups have debatable representatives from yet earlier times.

Note: Know the eras and roughly when they began!

SUPPLEMENT **4**

Mitosis, Sexual Reproduction (Meiosis and Syngamy), and Asexual Reproduction

I. PERSPECTIVE

Cells go through a regular sequence of events known as the ***cell cycle,*** which consists of alternating periods of:

- cell growth called interphase;
- cell division involving mitosis or meiosis.

In addition, two or more specialized cells may fuse by syngamy (fertilization) to form a new cell. Sexual reproduction consists of meiosis and syngamy, whereas mitosis is involved in asexual reproduction.

II. CELL DIVISION—MITOSIS VERSUS MEIOSIS

A dividing cell undergoes:

- ***cytoplasmic division*** (cytokinesis), that is, the division of the cytoplasm into equal parts;
- ***nuclear division*** (karyokinesis), that is, the division of the nucleus into equal parts by:
  - ***mitosis,*** the process of nuclear division involving chromosomes that are replicated and distributed equally between the nuclear progeny (i.e., there is *no* change in chromosome number);
  - ***meiosis,*** the process of nuclear division involving two successive nuclear divisions in which the chromosome number is halved and in which genetic ***segregation*** occurs (i.e., the genetic material is divided up differently among the progeny cells).

Cytoplasmic division usually quickly accompanies nuclear division, but some cells may become ***multinucleate*** (with many nuclei) due to delay in cytoplasmic division (e.g., the pine megagametophyte takes a year to become cellular). In plants and fungi a new cell wall (cell plate) eventually forms between the newly divided nuclei.

- ***Interphase*** is the phase between successive mitotic or meiotic divisions when the cell grows, when the genetic material (DNA) is actually replicated, and when the chromosomes are actually duplicated and all "set to go" for the ensuing cell division (mitosis or meiosis).

Only eukaryotic cells undergo mitosis and meiosis and have an interphase because prokaryotic cells lack nuclei.

There are four stages in the single dividing process of mitosis or in each of the two successive dividing processes of meiosis (i.e., meiosis I, meiosis II). Mitosis usually occurs in a day or less (e.g., 2–3 hours in onion root tip cells). Meiosis takes much longer: about two weeks, or even months or years. [Incidentally, in women this is up to about 50 years because cells are in meiotic arrest. Meiosis I begins just before birth, which means that all the eggs a woman will ever have are present at birth. Meiosis II occurs much later, during ovulation. This is the basis for the expression "biological clock" to refer to a woman running out of her eggs, that is, "a mechanism perceived as inexorably marking the passage of one's youth and especially one's ability to bear children" as opposed to the older sense of "an innate mechanism of the body that regulates its rhythmic and periodic cycles, as that of sleeping and waking" (definitions from *Random House Webster's college dictionary*, 1991). Thus at birth a womyn is given a stock of 400–450 eggs that is gradually depleted with the monthly menstrual cycles.]

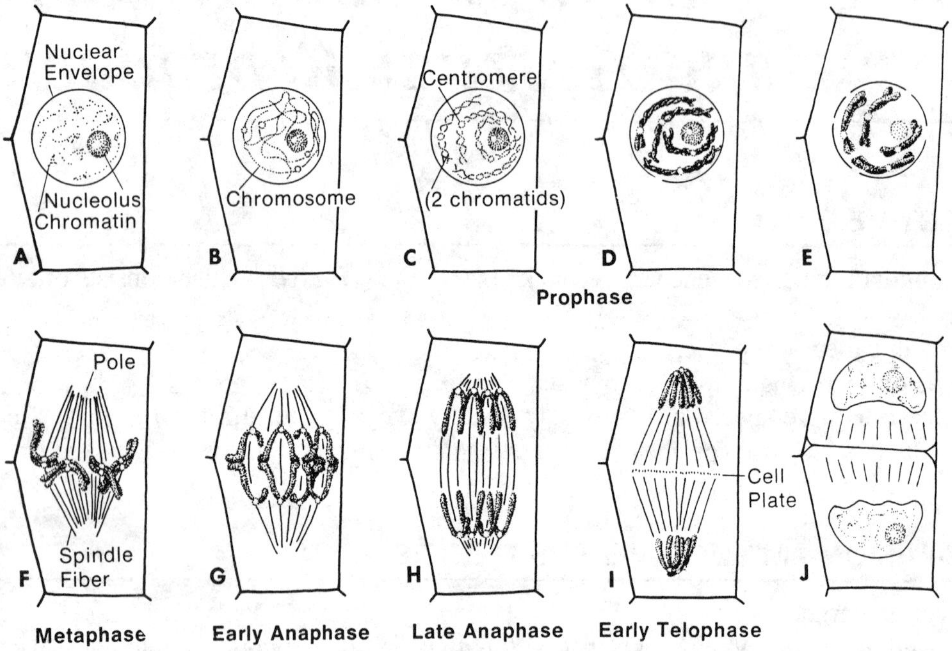

Fig. Sup4–1. Interphase and mitosis in a eukaryotic plant cell. *A*, interphase; *B-E*, prophase; *F*, metaphase; *G, H*, anaphase; *I, J*, telophase. See text (Part II) for details. (From Norstog & Long 1976:87; redrawn from *Botany: An introduction to plant biology*, 5th ed., by T. E. Weier, C. R. Stocking & M. G. Barbour, © 1974 by John Wiley and Sons, Inc., New York. Reprinted by permission of the publisher.)

Mitosis has the following stages (Fig. Sup4–1):

- ***Prophase:*** a rather long phase in which the chromosomes shorten and thicken ("condense") and move to the central plane of the cell (so-called metaphase plate or equator); the membranes of the nucleus and nucleoli break down. Actually, each chromosome is double and consists of two halves (***chromatids***) joined by a ***centromere*** attached to the spindle fibers.
- ***Metaphase:*** a brief phase during which the chromosomes (chromatid pairs) lie in the central plane (metaphase plate) of a ***spindle apparatus*** consisting of small tubelike structures.
- ***Anaphase:*** the very brief phase during which the chromatids move in unison to opposite poles of the spindle apparatus.
- ***Telophase:*** the final long phase during which the chromosomes (the former chromatids) become reorganized into new nuclei. This is basically the reverse of prophase: chromosomes become indistinct; the spindle disappears; the membranes of the nuclei and nucleoli reappear.

Interphase precedes prophase and telophase precedes a new interphase.

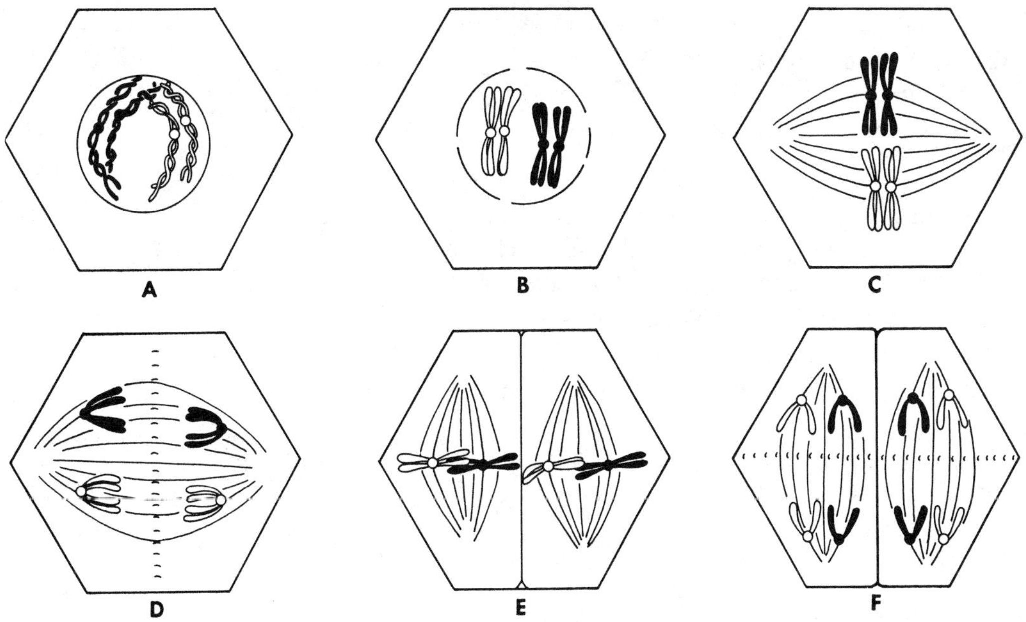

Fig. Sup4–2. The main steps of meiosis in eukaryotic cells. *A-D,* meiosis I (*A, B,* respectively, mid and late prophase; *C,* metaphase; *D,* anaphase); *E, F,* meiosis II (*E,* metaphase; *F,* telophase). Meiosis II somewhat resembles mitosis. See text (Part II) for details. (From Norstog & Long 1976:90.)

Meiosis consists of a series of stages (Fig. Sup4–2) comparable to mitosis but actually double the number of stages. For our purposes, the stages themselves are not that important, but rather what actually results from the process. Meiosis has two divisions:

- ***Meiosis I*** (first meiotic division): Here the chromosomes pair up according to their size, shape, and type of genes; these are ***homologous*** (i.e., comparable) chromosomes and carry equivalent genes. Actually each chromosome at this point consists of two chromatids joined at the centromere. Each set of homologous chromosomes separates in the anaphase and goes to the opposite pole. Meiosis I has been called the "reductional division" because each of the two temporary cells has only half the number of chromosomes.
- ***Meiosis II*** (second meiotic division): Meiosis II more closely resembles mitosis. Here the chromosomes line up again in a metaphase and then their chromatids are pulled to opposite poles. Each resultant cell can be genetically the same (if no crossing over occurs), and so Meiosis II has been called the "equational division."

The essential difference between these two divisions is that in Meiosis I homologous chromosomes separate and move to opposite poles of the cell, whereas in Meiosis II the chromatids of a chromosome (sister chromatids) separate.

Meiosis has resulted in four cells, two of which are genetically different from the other two cells. However, even more variation can occur as a result of ***crossing over*** whereby corresponding segments of homologous chromosomes are exchanged in late prophase of Meiosis I. In this case, all four progeny cells resulting from meiosis are different.

In summary, nuclear divisions may be either mitotic or meiotic. An organism has many more mitotic divisions (mitoses) than meiotic divisions (meioses), and the effects of these differ. Mitosis and meiosis have several significant aspects (Fig. Sup4–3):

MITOSIS:

- 1. The chromosome number is unchanged.
- 2. Cells resulting from mitosis are genetically identical (barring mutations and other rare genetic events).

- 3. Many mitotic divisions may occur in a life history, that is, both in frequency and in location (e.g., to regenerate cells as in the case of wounds).
- 4. Mitosis is *not* a sexual event or sexual process.

MEIOSIS:

- 1. The chromosome number is changed. A cell divides to form typically four cells that each have half the chromosome number. Four cells result because there are two divisions in meiosis, namely, Meiosis I and Meiosis II.
- 2. Cells resulting from meiosis are genetically quite different due to segregation and crossing-over.
- 3. Meiosis occurs only uncommonly in a life history, that is, only at particular times and only in specialized places (e.g., in sex organs).
- 4. Meiosis *is* a sexual event or sexual process.

The points above are numbered for direct comparison with similar points mentioned below for syngamy (fertilization) and asexual reproduction.

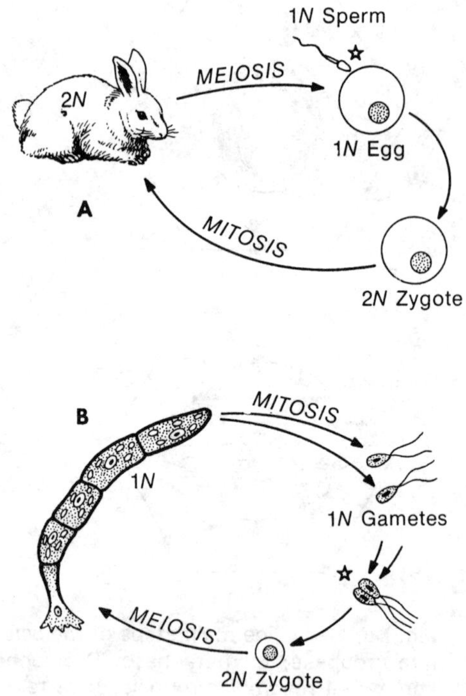

Fig. Sup4–3. Summary of the roles of meiosis, syngamy (fertilization) (at ☆), and mitosis in the life histories of animals and algae (or fungi). *A,* diploid (2*n*) life history; *B,* haploid (1*n*) life history. (From Norstog & Long 1976:89.)

III. CELL FUSION—SYNGAMY (FERTILIZATION)

Sexual reproduction consists of meiosis and syngamy. The fusion of cells is called ***syngamy*** or ***fertilization,*** and as a result of such fusion a ***zygote*** forms. In other words, ***gametes,*** the mature functional reproductive or ***sex cells,*** fuse to form a zygote. With syngamy new genetic diversity occurs because the zygote, the fusion product, is different from the parent cells that underwent meiosis. Usually two cells are involved in syngamy, such as the egg and sperm that fuse in humans and the land plants. However, up to 15 nuclei may fuse in the double fertilization of angiosperms (see Lab Exercise 9, Part III-D), and entire sex organs (***gametangia*** or gamete-producing organs) may fuse in sac fungi. Here a unicellular, but multinucleate antheridium deposits its contents into a unicellular, but multinucleate ascogonium via its trichogyne (Figs. 12–6, 12–7).

Cell fusion (syngamy, fertilization) consists of:

- ***cytoplasmic fusion*** (plasmogamy);
- ***nuclear fusion*** (karyogamy).

Usually these fusions occur very quickly after each other, so that for practical purposes there is only one fusion. In contrast, in some cases there can be appreciable delay between cytoplasmic fusion and nuclear fusion, as much as eight to ten months. We will deal with this fungal procrastination in more detail in Supplement 12 and Lab Exercises 12 and 13.

There are several types of syngamy (fertilization) (Fig. Sup4–4):

- ***isogamy,*** gametes morphologically the same or equal;
- ***anisogamy,*** gametes morphologically the same but of unequal size;

- ***oogamy,*** gametes differentiated as egg (nonmotile) and sperm (motile).

These gametes may either have or lack flagella or cilia (*Note:* These may occur on cells other than gametes).

- A ***flagellum*** (singular, plural ***flagella;*** the word is Latin for "small whip") is a long, hairlike extension protruding from the surface of a cell and functioning to move or propel a cell.
- ***Cilia*** (plural, singular ***cilium***) are flagella that are short and usually more numerous.

Thus, in oogamy a definite sperm swims or otherwise moves toward the egg, which never has a flagellum. In isogamy and anisogamy with their similar gametes, mating strains (types) usually occur and are designated "+" and "−" (Figs. 3-5, 12-2, 12-5, 12-7).

These three types of syngamy sometimes characterize whole groups of organisms. For example, *all* the bryophytes and vascular plants are strictly oogamous, and in these groups sperm are the only flagellate cells. In contrast, *all* three types of syngamy can occur in one genus, for instance, the green alga *Chlamydomonas.*

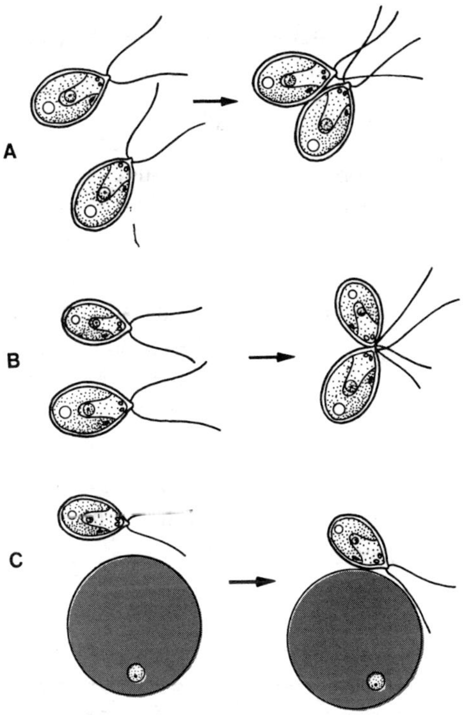

Fig. Sup4-4. The three types of syngamy (fertilization) found in the green algae (Chlorophyta). *A,* isogamy, gametes morphologically the same or equal; *B,* anisogamy, gametes morphologically the same but of unequal size; *C,* oogamy, gametes differentiated as egg (nonmotile) and sperm (motile). (From Norstog & Long 1976:175.)

In summary, syngamy (fertilization) has several significant aspects:

- 1. As with meiosis, the chromosome number is changed. Two cells fuse to give rise to a new cell with double the chromosome number.
- 2. As with meiosis, the cells resulting from syngamy are genetically quite different due to recombination of genetic material.
- 3. Syngamy is a very rare event in a life history. It is much less common than meiosis. However, like meiosis, syngamy occurs only in a specialized place.
- 4. As with meiosis, syngamy *is* a sexual event. However, syngamy involves recombination of genetic material whereas meiosis involves its segregation.

IV. MITOSIS VERSUS MEIOSIS VERSUS SYNGAMY

Mitosis, meiosis, and syngamy (fertilization) are best compared on the basis of their effect on the chromosome number of a cell. The following terminology is useful:

- ***haploid,*** with a single complement (set) of chromosomes, designated "1n";
- ***diploid,*** with two complements (sets) of chromosomes (i.e., both sets of homologous chromosomes), designated "2n";
- ***triploid,*** with three complements (sets) of chromosomes, designated "3n";
- ***polyploid,*** with more than three complements of chromosomes.

Chromosome numbers are usually expressed for 2n numbers and can vary from $2n = 4$ for *Haplopappus gracilis,* a sunflower relative, to $2n = 1262$ for the fern *Ophioglossum.* Most eukaryotes have 10–50 chromosomes per cell; humans have 46 per cell.

Mitosis, meiosis, and syngamy (fertilization) can now be compared (see Fig. Sup4–3 and above for elaboration of points 1 through 4):

Character	*Mitosis*	*Meiosis*	*Syngamy*
1. *Chromosome number*	Unchanged $1n$ cell → 2 $1n$ cells $2n$ cell → 2 $2n$ cells	Changed 1 $2n$ cell → 4 $1n$ cells No meiosis in $1n$ cells, i.e., no $\frac{1}{2}n$ cells!	Changed $1n$ + $1n$ cells → 1 $2n$ cell
Humans:	Cell w/ 46 chromosomes → 2 cells, each with 46 chromosomes	Cell w/ 46 chromosomes → 4 cells, each with 23 chromosomes	Cell w/ 23 chromosomes + cell w/ 23 → cell with 46 chromosomes
2. *Progeny cells*	Genetically the same	Genetically different	Genetically different
3. *In the life history*	Occurs very many times Occurs in many places	Occurs uncommonly Occurs in a definite place	Occurs very rarely Occurs in a definite place
4. *A sexual event?*	No	Yes	Yes

V. ASEXUAL (VEGETATIVE) REPRODUCTION

Organisms reproduce both sexually and asexually.

- ***Asexual reproduction,*** which is also called ***vegetative reproduction,*** "vegetative propagation," or "somatic reproduction," is any reproductive process *not* involving meiosis or the union of nuclei, sex cells, or sex organs.
- ***Sexual reproduction,*** in contrast, is characterized by syngamy (fertilization), that is, the union or fusion of gametes (or gametangia), followed by meiosis at some point in the life history.

Asexual reproduction is common in the life histories of many organisms (plants, fungi, animals, and protists), and in plants, depending on the species, may involve the gametophyte (GPT) and/or the sporophyte (SPT). For examples of asexual reproduction see Lab Exercise 1, Part V.

The significance of sexual reproduction is that it is responsible for genetic variation in a population by segregation and recombination of genetic material resulting from, respectively, meiosis and syngamy. In contrast, the significance of asexual reproduction is that it is a "quick and dirty" means for the rapid and immense increase in the number of individuals. Weeds, be it noted, are successful largely due to their great capacity for vegetative reproduction. In addition, in the bryophytes and pteridophytes vegetative reproduction is a means to bypass the rather lengthy and moisture-dependent sexual process. That is, the swimming sperm characteristic of these groups may be a limiting factor in times of moisture stress such as drought. The disadvantage of asexual reproduction is that the new individuals are genetically like the parent ones. In other words, no genetic diversity results from asexual reproduction except in those rare cases where there are mutations.

In summary, asexual reproduction has several significant aspects:

- 1. As in mitosis, the chromosome number is unchanged.
- 2. As in mitosis, the resultant cells are genetically the same.
- 3. As in mitosis, it is common in the life history of certain organisms. It can occur many times and in many places.
- 4. As mitosis, asexual reproduction is *not* a sexual process.
- 5. Asexual reproduction results in the rapid and immense increase in the number of individuals.
- 6. In some plants and fungi, asexual reproduction is a bypass mechanism for the water-dependent sexual processes of the life history.

SUPPLEMENT 5

The Main Types of Life Histories

I. PERSPECTIVE

The *life history* (also called life cycle) of an organism is the genetically programmed sequence of events for a species by which its individuals are produced, grow, develop, and reproduce. In other words, the life history involves the entire sequence of events from zygote formation (via syngamy or fertilization) to gamete formation (via either mitosis or meiosis). Thus, syngamy (fertilization) and meiosis are successive events, so-called sexual events, in the life history:

- syngamy involves $1n$ structures fusing to form $2n$ structures, two items fusing into one;
- meiosis involves $2n$ structures of various types dividing to form $1n$ structures, one item dividing into four items.

Exceptions to this statement occur due to variations in the ploidy level of organisms. Consequently,

- a ***1n phase*** (haplophase) results from meiosis;
- a ***2n phase*** (diplophase) results from syngamy (fertilization).

Generally, syngamy and meiosis are inseparable; if one is present in a life history, so is the other.

Note: Much of the information in this supplement, and specifically some of the diagrams, notably Figs. Sup5–4 and Sup5–5, will only have meaning later in the course!

II. THE MAIN TYPES OF LIFE HISTORIES

There are three main types of life histories, with the haploid-diploid ($1n$-$2n$) type listed having several subdivisions. These types are distributed among the various groups of plants and fungi as follows:

- ***haploid*** ($1n$) ***life history*** (haplontic) (for basic figure see Fig. Sup5–1);
  - ***without a dikaryotic phase;*** most algae (Figs. 3–2, 3–3, Sup4–3B), some fungi (Fig. 12–5);
  - ***with a dikaryotic phase*** (see definition in Supplement 12, Part V): most fungi (Figs. 12–6, 12–7, 13–1);
- ***diploid*** ($2n$) ***life history*** (diplontic) (for basic figure see Fig. Sup5–2): some algae, chiefly diatoms, but also *Fucus* (Figs. 3–1, 3–7, Sup4–3A), and some fungi (Fig. 12–1);
- ***haploid-diploid*** ($1n$-$2n$) ***life history*** (haplodiplontic, diplohaplontic) (for basic figure see Fig. Sup5–3);
  - ***alternation of similar generations*** (isomorphic alternation of generations): some algae (Fig. 3–5), including the red algal variant, and a few fungi (Figs. 12–2, 12–9);
  - ***alternation of dissimilar generations*** (heteromorphic alternation of generations): a few algae, a few fungi, and *all* land plants (= bryophytes + vascular plants);
    - ***gametophyte or sporophyte free-living:*** a few algae (Fig. 3–6) and a few fungi;
    - ***gametophyte dominant, homosporous*** (for basic figure see Fig. Sup5–4): *all* bryophytes (Figs. 4–1, 4–3);

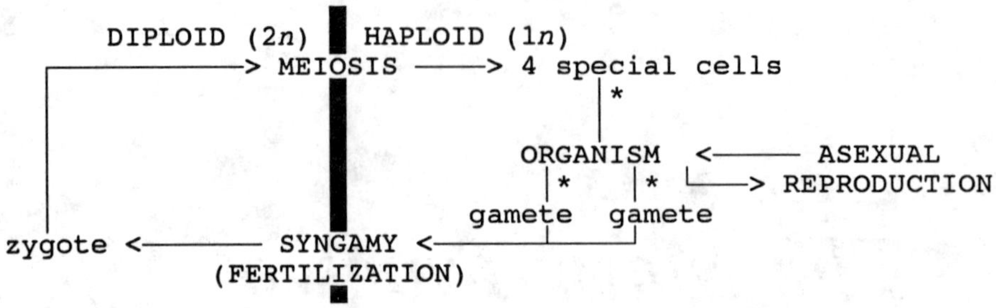

Fig. Sup5-1. Haploid (1*n*) life history (NB: * = mitoses if organism is multicellular). The "special cell" may be a spore or other structure, depending on the organism involved. Oogamy and especially isogamy and anisogamy may occur. The gametes that fuse may come from the same individual or from different individuals. (R. Schmid, original.)

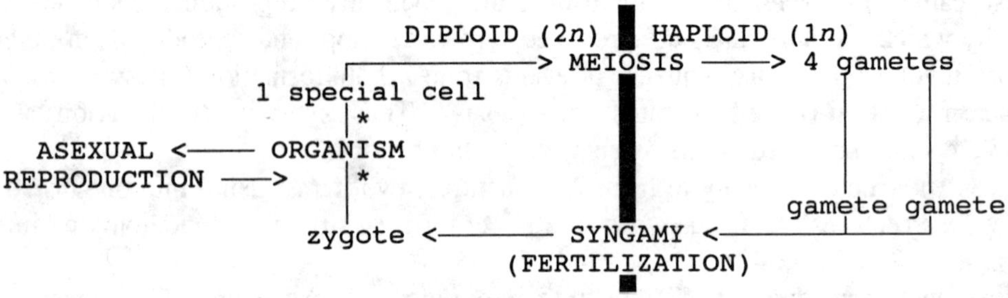

Fig. Sup5-2. Diploid (2*n*) life history (NB: * = mitoses if organism is multicellular). The "special cell" may be a meiocyte (sporocyte) or other structure, depending on the organism involved. Oogamy and especially isogamy and anisogamy may occur. The gametes that fuse may come from the same individual or from different individuals. Humans have this type of life history. (R. Schmid, original.)

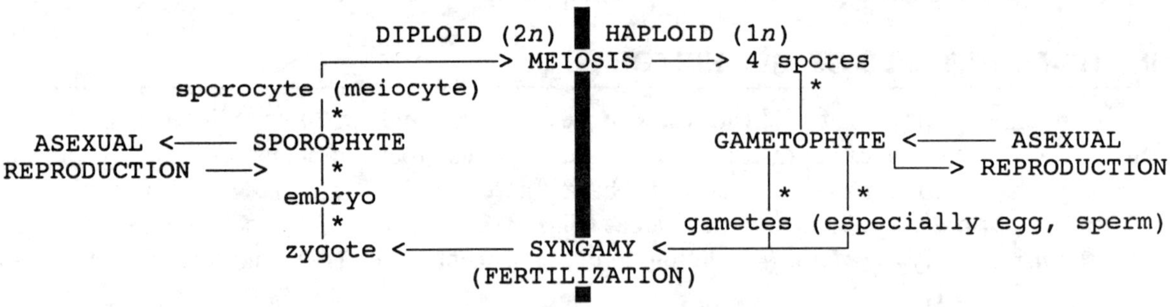

Fig. Sup5-3. Haploid-diploid (1*n*-2*n*) life history (NB: * = mitoses). Note the alternation of two generations (phases), a GPT and SPT, which may be morphologically similar or dissimilar. Isogamy, anisogamy, and especially oogamy may occur. The gametes that fuse may come from the same individual or, more commonly, from different individuals. Asexual reproduction can involve either the SPT or the GPT. (R. Schmid, original.)

- ***sporophyte dominant:*** *all* vascular plants;
    - ***homosporous*** (for basic figure see Fig. Sup5-4): most pteridophytes (Figs. 5-1, 5-3, 5-5);
    - ***heterosporous*** (for basic figure see Fig. Sup5-5): some pteridophytes (Fig. 5-4) and *all* seed plants (Figs. 8-1, 8-2, 9-4).

In the haploid (1*n*) life history the conspicuous parts of the organism are 1*n*, and the only 2*n* structure in the life history is the zygote, which often undergoes a resting or dormant stage. In contrast, in the diploid (2*n*) life history the conspicuous parts of the organism are 2*n*, and the only 1*n* structures

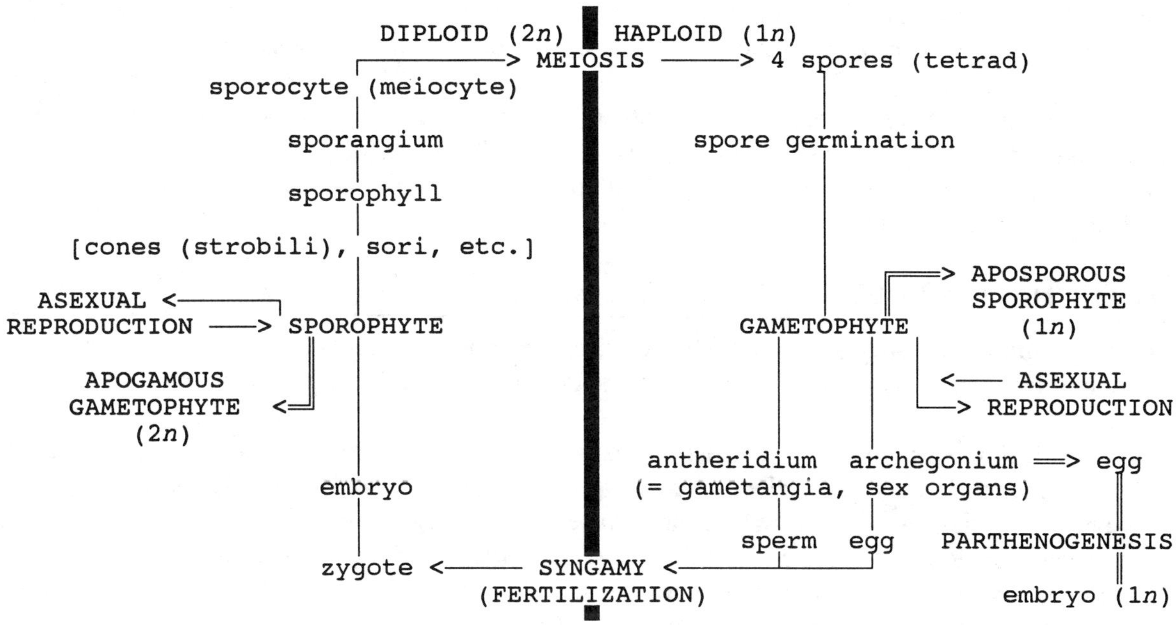

Fig. Sup5–4. Generalized homosporous haploid-diploid ($1n$-$2n$) life history of bryophytes and pteridophytes. A homosporous plant produces morphologically similar spores. Structures in brackets vary depending on the group. Gametophytes of pteridophytes are typically bisexual whereas those of bryophytes are bisexual or especially unisexual. Only oogamy occurs. The gametes that fuse may come from the same individual or, more commonly, from different individuals. Double lines indicate events deviating from the usual life history, i.e., apogamy and apospory, which are not discussed here. Asexual reproduction can involve either the SPT or the GPT. (R. Schmid, original.)

in the life history are the gametes. Part IV details the much more complicated haploid-diploid ($1n$-$2n$) life history. In addition, all three types of life histories have asexual (vegetative) reproduction (see Figs. Sup5–1 to Sup5–3). In the haploid-diploid ($1n$-$2n$) life history asexual reproduction can involve either the SPT or the GPT (Figs. Sup5–3 to Sup5–5). For information on asexual reproduction see Supplement 4, Part V, and Lab Exercise 1, Part V, Lab Exercise 4, Part II-C, and Lab Exercise 12, Parts III-A and B.

III. SPORES

A *spore* is any type of reproductive cell capable of developing into an adult without fusing with another cell. Spores may be sexual or asexual; they may be motile, with flagella, or nonmotile; they may be produced on or within structures called ***sporangia*** (plural, singular ***sporangium***). "Sporangium" means "spore case." There are several types of sporangia:

- ***meiosporangia*** (plural, singular ***meiosporangium***) produce by meiosis sexual spores or ***meiospores***;
- ***mitosporangia*** (plural, singular ***mitosporangium***) produce by mitosis asexual spores or ***mitospores***;
- ***zoosporangia*** (plural, singular ***zoosporangium***) produce motile (flagellate) spores or ***zoospores***;
- ***zygosporangia*** (plural, singular ***zygosporangium***) produce ***zygospores,*** thick-walled, encysted zygotes resistant to adverse environmental conditions;
- ***microsporangia*** (plural, singular ***microsporangium***) are male sporangia that produce by meiosis male sexual spores or ***microspores***;
- ***megasporangia*** (plural, singular ***megasporangium***) are female sporangia that produce by meiosis female sexual spores or ***megaspores.***

For more on micro- and megasporangia and their attendant structures see Part IV. Specialized sporangia such as *asci* and *basidia* and specialized spores such as *conidia, ascospores, basidiospores,* etc., are defined later (see Lab Exercise 12, Parts III-B and III-D). Sexual spores result from meiosis. A ***sporocyte*** or ***meiocyte*** (spore mother cell) produces by meiosis four 1*n* spores (meiospores) that constitute a ***tetrad*** (see Lab Exercise 7, Part IV). Therefore, 10 sporocytes give rise to 40 spores, 100 sporocytes to 400 spores, 1000 sporocytes to 4000 spores, etc., assuming 100% efficiency of meiotic development.

IV. THE HAPLOID-DIPLOID (1*n*-2*n*) LIFE HISTORY

An ***alternation of generations*** (phases) characterizes the haploid-diploid (1*n*-2*n*) life history. Thus:

- a multicellular 2*n* ***sporophyte*** (SPT) or sporophytic (SPTic) or spore-producing phase alternates with
- a multicellular 1*n* ***gametophyte*** (GPT) or gametophytic (GPTic) or gamete-producing phase.

In this life history, the four critical structures involving changes in ploidy level are the sporocyte or meiocyte (spore mother cell), the spore, the gamete, and the zygote (Fig. Sup5–3). In the context of the haploid-diploid (1*n*-2*n*) life histories of the *land plants* (i.e., bryophytes and vascular plants) a "spore" is always a sexual spore (meiospore) because it always results from meiosis in the meiosporangium. Consequently, in the homosporous haploid-diploid (1*n*-2*n*) life history:

- The first stage of the GPTic generation (phase) is always the ***spore,*** which results from *meiosis.*
- The last stage of the GPTic generation (phase) is always the ***gamete,*** two of which fuse in *syngamy* (fertilization).
- The first stage of the SPTic generation (phase) is always the ***zygote,*** the fusion product of the two gametes that develop into the ***embryo,*** the very young SPT.
- The last stage of the SPTic generation (phase) is always the ***meiocyte*** or ***sporocyte*** (spore mother cell), which undergoes meiosis in the sporangium (meiosporangium) to form a tetrad of spores (meiospores).

Naturally, as mentioned in the first paragraph, these sexual events also reflect changes in ploidy level of these "first" and "last" stages.

The four structures shown above in bold italics are produced in definite structures that in turn are associated with other structures. The following chronology for the haploid-diploid (1*n*-2*n*) life history may be helpful:

- A ***sporophyte*** (SPT) ("spore-producing plant" or "spore plant") produces fertile leaves, or sporophylls.
- A ***sporophyll*** ("spore leaf") produces sporangia.
- A ***sporangium*** ("spore case") produces sporocytes (meiocytes).
- A ***sporocyte*** or ***meiocyte*** ("spore site") undergoes *meiosis* and produces four 1*n* spores or meiospores.
- A ***spore*** or ***meiospore*** germinates and grows into the GPT.
- A ***gametophyte*** (GPT) ("gamete-producing plant" or "gamete plant") produces gametangia.
- A ***gametangium*** (sex organ or gamete-producing organ, "gamete case") produces gametes.
- Two ***gametes*** ("sex cells"), egg and sperm, fuse (*syngamy, fertilization*) to form a 2*n* zygote.
- A ***zygote,*** the fusion product of the two gametes, develops into an embryo.
- An ***embryo,*** the very young SPT before the onset of rapid growth (or before germination of the seed in seed plants), develops into the new SPT.

The haploid-diploid (1*n*-2*n*) life history is morphologically the most complex type depending on (1) the similarity or dissimilarity of the GPT and SPT, (2) the relationship of these to each other, and (3) the nature of the spores. The phases of a haploid-diploid (1*n*-2*n*) life history may be morphologically alike or different:

- ***Alternation of similar generations*** (isomorphic alternation of generations) involves a GPT and SPT that are morphologically identical, except obviously for the reproductive structures.
- ***Alternation of dissimilar generations*** (heteromorphic alternation of generations) involves a GPT and SPT that are morphologically different.

As outlined in Part II, there are three possible relationships between the GPT and SPT:

- The GPT and SPT may be totally independent of each other at maturity, each being free-living.
- The GPT may be dominant over the SPT, the latter always dependent on the GPT.
- The SPT may be dominant over the GPT, the former ultimately independent of the GPT.

In the haploid-diploid ($1n$-$2n$) life history, there are two main spore conditions:

- ***homosporous*** (***homospory***), with only one morphological type of spore;
- ***heterosporous*** (***heterospory***), with two morphological types of spores:
  - ***microspores*** (male spores);
  - ***megaspores*** (female spores).

Homospory occurs in the bryophytes and in most pteridophytes, whereas heterospory occurs in some pteridophytes and in *all* seed plants. Figures Sup5–4 and Sup5–5 contrast, respectively, the homosporous and heterosporous types of life histories.

It is important to emphasize that the differences between the homo- and heterosporous life histories involve those stages from the sporophyll through the GPT. In contrast, those stages of the life history from the sex organs or gametangia (if present) or gametes through the mature SPT are comparable in both the homo- and heterosporous type of life history (compare Figs. Sup5–4 and Sup5–5). Because homospory involves only one type of spore, a homosporous life history has only one morphological type of sporangium, sporocyte (meiocyte), and GPT, as shown in Fig. Sup5–4 and as detailed above. In contrast, because heterospory involves two types of spores, a heterosporous life history has both male "micro-" structures and female "mega-" structures (Fig. Sup5–5). These are:

- ***micro-*** and ***megasporophylls,*** which bear
- ***micro-*** and ***megasporangia,*** which produce
- ***micro-*** and ***megasporocytes,*** which produce by meiosis tetrads of
- ***micro-*** and ***megaspores*** (macrospores), which develop into
- ***micro-*** and ***megaGPTs.***

All extant gymnosperms have unisexual cones, that is:

- ***male cones*** (microstrobili, microsporangiate strobili, pollen cones);
- ***female cones*** (megastrobili, megasporangiate strobili, ovule cones, ovulate cones, seed cones).

The male cones bear microsporophylls and their related male structures, whereas the female cones bear megasporophylls and their related female structures. The male and female cones of gymnosperms or the unisexual flowers of angiosperms may be borne on separate, unisexual plants or on the same, bisexual plant, that is:

- the ***dioecious*** condition, with unisexual, male and female plants, ***micro-*** and ***megaSPTs;***
- the ***monoecious*** condition, with bisexual plants, the more common situation.

Most of this male "micro-" and female "mega-" terminology is standard usage in discussions of the seed plants, except that "male cone" and "female cone" seem simpler to use. There are, in fact, many alternative terms, as suggested above for cones. For instance, microsporocytes are also called micromeiocytes, microspore mother cells, pollen mother cells, whereas megasporocytes are also called megameiocytes, megaspore mother cells (but none of these are called "spore father cells"). Indeed, some of this terminology is blatantly sexist (see Schmid 1977).

A consequence of the above is that in the heterosporous haploid-diploid (1*n*-2*n*) life history:

- The first stage of the GPTic generation (phase) is always the ***micro-*** and ***megaspore,*** which develop into, respectively, the ***microGPT*** (male GPT) and the ***megaGPT*** (female GPT).
- The last stage of the GPTic generation (phase) is always the ***gamete,*** either egg or sperm.
- The first stage of the SPTic generation (phase) is always the ***zygote,*** the fusion product of the two gametes that develops into the ***embryo,*** the very young SPT.
- The last stage of the SPTic generation (phase) is always the ***meiocyte*** or ***sporocyte*** (spore mother cell), which undergoes meiosis in the sporangium (meiosporangium) to form a tetrad of spores (meiospores).

This heterosporous life history occurs in *all* seed plants. which differ from the pteridophytes in having ovules/seeds and their concomitant structures pollen grains/pollen tubes and seedlings. These structures are defined in Supplement 8.

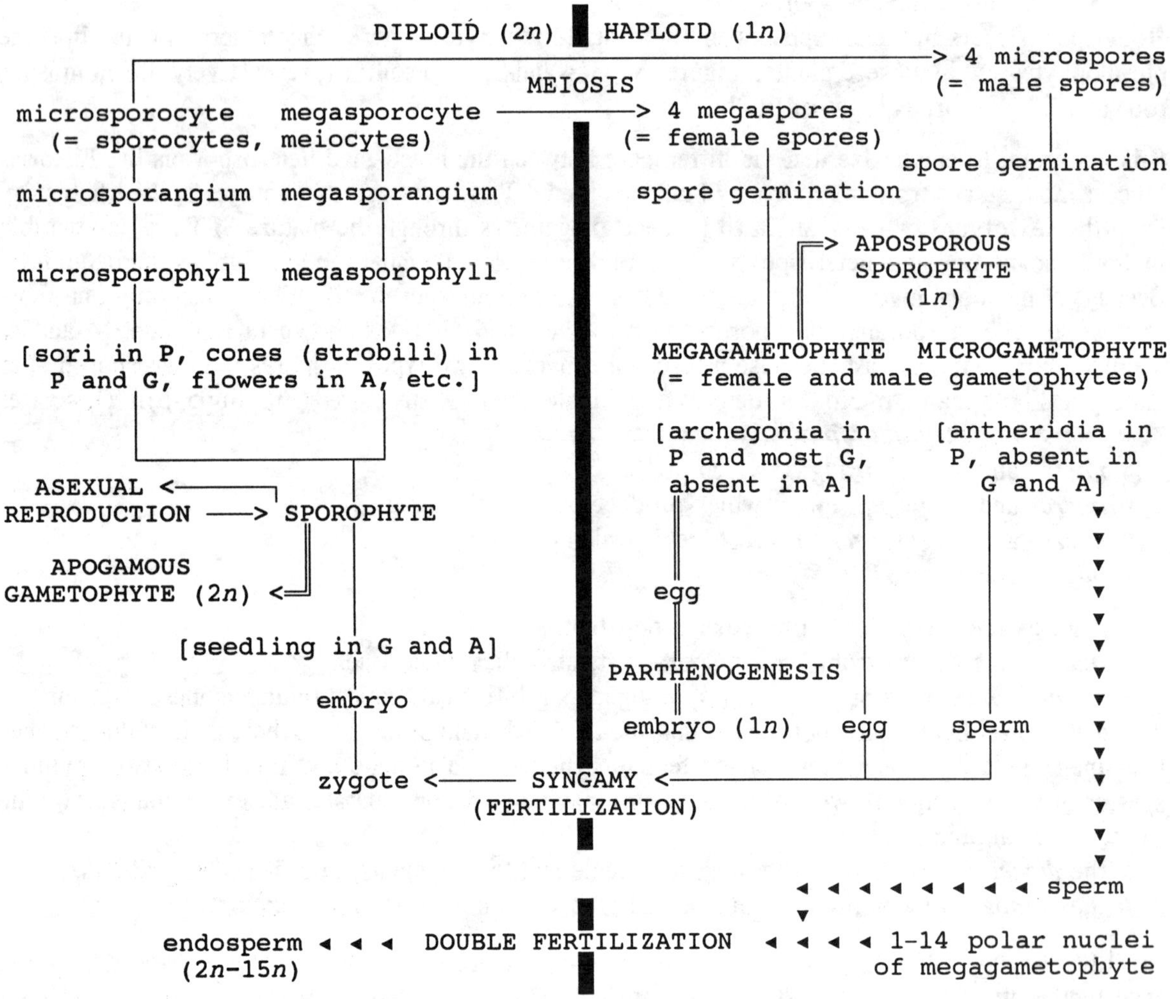

Fig. Sup5-5. Generalized heterosporous haploid-diploid (1*n*-2*n*) life history of pteridophytes and seed plants. A heterosporous plant produces morphologically dissimilar spores. Structures in brackets vary depending on the group (P = pteridophytes, G = gymnosperms, A = angiosperms). Only oogamy occurs. The gametes that fuse may come from the same individual or, more commonly, from different individuals. Double fertilization, represented by the thick arrowed lines, occurs only in angiosperms. Double lines indicate events deviating from the usual life history, i.e., apogamy, apospory, and parthenogenesis, which are not discussed here. (R. Schmid, original.)

LAB EXERCISE **1**

Your First Lab

(Use of the Microscope; Basic Cell Structure; Mitosis; Asexual Reproduction)

OBJECTIVE

To introduce the study of diversity of plants and fungi by examining selected topics: use of the microscope; basic cell structure; mitosis; asexual reproduction.

Note: Read this material *before* coming to the lab; there is insufficient time in the lab to read the material from scratch. Due to the large amount of background information presented here, a ●●● is used to mark those paragraphs calling for actual observations or diagrams in the lab. Most unmarked paragraphs reiterate information presented in the lecture. A bullet (●) signifies an important term or concept.

Suggested Lab Diagrams (see part noted for suggested labels):
A compound microscope (Part I)
Grid (or similar item) at 40X, 100X, and 400X magnification (Part I)
Tradescantia (spiderwort) hair from stamen of flower (Part III)
Allium (onion) root tip l.s. (Part IV)

PERSPECTIVE

This lab exercise introduces you to the subject of plant and fungal diversity. It is now generally accepted that the most fundamental division of living organisms is the distinction between prokaryotes and eukaryotes, which represent taxonomic superkingdoms:

- Prokaryotes *lack* nuclei and other membrane-bounded organelles (mitochondria, chloroplasts, etc.) and also lack microtubules (small tubular structures not membrane-bounded) and 9+2 flagella.
- Eukaryotes, in contrast, *have* all of these features.

The diversity of life can be broadly grouped as shown in the table on the next page. Subsequent lab exercises will examine samples of the diversity of prokaryotic, plant, and fungal life.

Note: Before beginning this lab exercise, you should review:

- Lab Exercise 1, Supplement A, on the use of the compound microscope;
- Supplement 1 on the classification of organisms;
- Supplement 4 on mitosis, sexual reproduction (meiosis and syngamy), and asexual reproduction.

#Prokaryotes: *nuclei and other organelles absent*
*Monerans—bacteria, blue-green bacteria, etc. (3 divisions)
#Eukaryotes: *nuclei and other organelles present*
*Protists—protozoans (many divisions): *cell wall absent; chloroplasts absent*
*Animals—animals (many divisions): *ditto*
*Fungi (7 divisions, 1 form-division): *cell wall present; chloroplasts absent*
*Plants (24 divisions): *cell wall present; chloroplasts present*
Algae (8 divisions)
Bryophytes—mosses, liverworts, hornworts (1 division)

NON-VASCULAR PLANTS
VASCULAR PLANTS

Nonseed plants or pteridophytes (vascular cryptogams)—ferns, club mosses, horsetails, etc. (8 divisions, 4 strictly fossil)
Seed plants (phanerogams)—(7 divisions, 2 strictly fossil)
Gymnosperms—conifers, cycads, *Ginkgo,* etc. (6 divisions)
Angiosperms—flowering plants (1 division)
Dicotyledons
Monocotyledons

Key
= superkingdom; * = kingdom.

Note
"Land plants" = bryophytes + vascular plants.

I. USE OF THE COMPOUND MICROSCOPE

Lab Exercise 1, Supplement A, covers the use of the compound microscope. Your TA or instructor will walk you through this supplementary exercise.

••• As instructed, examine a permanent microscope slide, for example, of wood of *Pinus* (pine), with a compound microscope. Use mainly the two lowest power objectives. What happens to the image when you move the slide to the left? What happens to the image when you move the slide toward you? Lab Exercise 1, Supplement B, describes how permanent microscope slide preparations are prepared.

••• You may be given a microscope slide with a grid or letter on it. Alternately, you may be given a piece of fine graph paper to place over a clean microscope slide. In either case, observe the object with the 4X, 10X, and 40X objectives (*ignore* the oil immersion 100X objective) and note how with progressively higher magnifications you see less and less of the object. This is a useful reminder that when first examining a microscope slide, always use the low power 4X objective. This permits coverage of a wider field and hence is best suited for scanning.

••• *Suggested diagram and labels:* A compound microscope: eyepiece (ocular), objective, observation tube, stage, substage condenser, iris diaphragm, fine and coarse adjustment knobs.

••• *Suggested diagram and labels:* Grid (or similar item) at 40X, 100X, and 400X magnification.

II. USE OF THE DISSECTING MICROSCOPE

Dissecting microscopes bridge the gap between the naked eye and the magnification range of compound microscopes. Using a dissecting microscope is much simpler than using a compound microscope.

●●● Examine a flower of *Tradescantia* (spiderwort) under a dissecting microscope and note especially the hairs on the central yellow-tipped stamens, the male part of a flower. What happens to the image when you move the flower to the left? What happens to the image when you move the flower toward you?

III. BASIC CELL STRUCTURE

See Fig. 1–1. The basic unit or building block of life is the cell. Most bacterial, plant, and fungal cells consist of the following main parts:

- the outer ***cell wall*** ("non-living component") consisting of a variety of chemicals and imparting rigidity to the cell;
- the ***protoplast*** (the living component) consisting of the ***cytoplasm*** and the ***nucleus*** (singular, plural ***nuclei***).

The chief components of the cytoplasm and the nucleus and the chief purpose of these components can be summarized as follows:

COMPONENTS OF THE CYTOPLASM:

- the ***vacuole,*** a large centrally located fluid-filled, membrane-bounded sac containing various storage materials (pigments, etc.);
- ***chloroplasts,*** organelles containing chlorophyll (a green pigment; a ***pigment*** is any substance that absorbs visible light—see "Perspective" of Lab Exercise 3) and that thus are the sites of ***photosynthesis*** (the process of carbohydrate or "food" manufacture from both water and carbon dioxide in the presence of chlorophyll by using light energy—see also Supplement 11);
- ***mitochondria*** (plural, singular ***mitochondrion***), organelles that are the sites of respiration;
- other organelles, for example, ***Golgi bodies*** (dictyosomes), the sites of lipid synthesis;
- ***cytoplasmic membrane*** (plasma membrane, plasmalemma), the outer single membrane enclosing the cytoplasm;

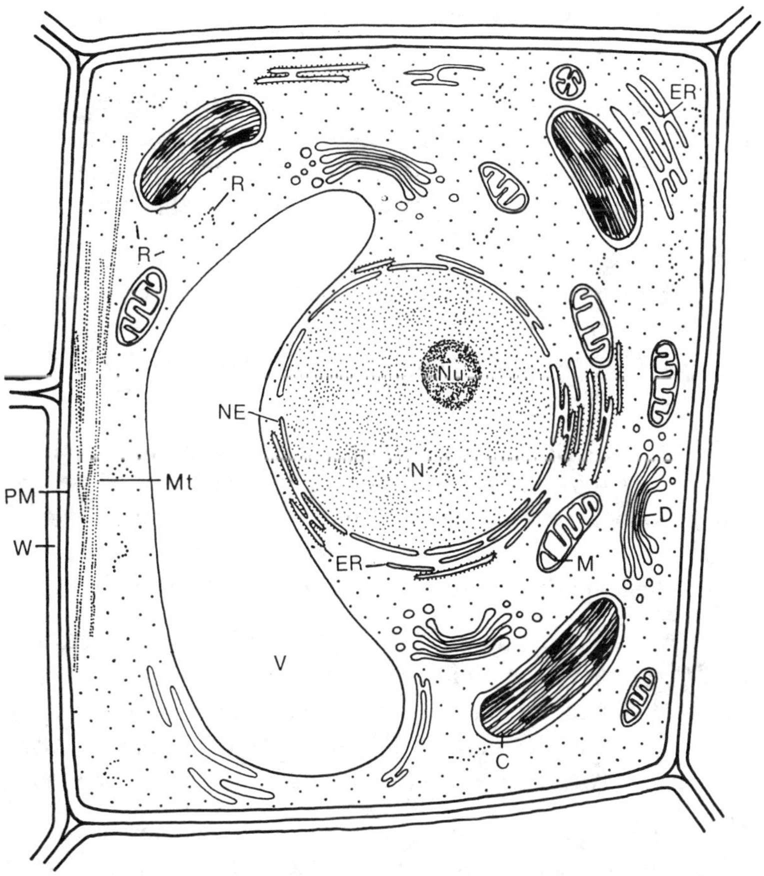

Fig. 1–1. Section of a "typical" plant cell showing the more commonly recognized structures of the protoplasm: C, chloroplast; D, Golgi body (dictyosome); ER, endoplasmic reticulum; M, mitochondrion; Mt, microtubules; N, nucleus; NE, nuclear membrane (nuclear envelope); Nu, nucleolus; PM, cytoplasmic membrane (plasma membrane, plasmalemma); R, ribosomes; V, vacuole; W, cell wall. (From Norstog & Long 1976:93; redrawn and modified from *Botany,* 5th ed., by C. L. Wilson, W. E. Loomis & T. A. Steeves, © 1971 by Holt, Rinehart and Winston, Inc., New York. Reprinted by permission of the publisher.)

COMPONENTS OF THE NUCLEUS:

- ***chromosomes,*** the structures bearing the genetic material (DNA);
- ***nucleolus*** (singular, plural ***nucleoli***), an organelle that is the site of RNA synthesis;
- ***nuclear membrane*** (nuclear envelope), the outer double membrane enclosing the nucleus.

The parts of the cell most relevant to various considerations in this course manual are the nucleus, chloroplasts, vacuole, and cell wall. Indeed, as shown in the "Perspective" and in Supplement 1, Part II, it is the presence or absence of these structures that, among other things, characterize the five kingdoms of organisms: monerans, plants, fungi, animals, and protists.

••• Examine the DEMO models of a plant and an animal cell and compare these with the DEMO diagrams and photographs, plus the explanatory table (from Starr 1991:45–48, 56), of cells of a bacterium, a plant, and an animal. These illustrations contrast the basic structure of cells of prokaryotes and eukaryotes (see "Perspective"). Your TA or instructor will probably discuss the main parts of a cell.

Scientists prefer the metric system over the English system of measurement. Note the following equivalents:

- 1 mm = 0.03937 inch; 1 inch = 25.4 mm or 2.54 cm;
- 1 mm = 1/1000 m; 1 μm ("micrometer" or "micron") = 1/1,000,000 meter or 1/1000 mm.

Cells of a few organisms can be seen with the naked eye (e.g., some cells such as fern tracheids may be up to 50 mm, or 2 in.), but most cells are microscopic. Typical mature plant cells are 25–50 μm in diameter and often 100–200 μm; animal cells are usually much smaller ($< 10\mu$m), and bacterial cells the smallest, about 1 μm (viruses are not cells, but rather replicating entities within cells).

••• Reexamine the permanent slide of *Pinus* (pine) from Part I. Most of the cells you see are dead. Hence mainly cell walls are visible, along with a few protoplasts (the dark-staining cells with contents are living). At 400X, using the 40X objective, how many cells do you see extending across the field of a transection (cross section)? You should see about 13 cells, each about 35 μm (0.035 mm) in diameter.

••• In 1664 the English scientist Robert Hooke discovered cells when he examined bark from a tree of *Quercus suber* (cork oak), the source of bottle corks. Examine the DEMO plate from Hooke's book, *Micrographia* (1664). Hooke estimated the diameter of cork cells to be 0.04 mm (0.001 in.). A cubic inch of cork thus would have about one billion cells (Ray et al. 1983:59). A typical wine bottle cork also has about one billion cork cells.

••• Also examine the DEMO diagram of Hooke's microscope. He used a similar one to prepare his *Micrographia.* Two lenses were placed at the top and bottom of the 15 cm (5.9 in.) drawtube, mounted on a movable ring. The object being viewed was stuck on a pin in the base. Magnification was changed by lengthening the drawtube of the microscope. The light source for viewing the specimen was either the sun (via a mirror) or a flame and lens system.

••• Make a water mount slide preparation of the staminal hairs from a flower of *Tradescantia* (spiderwort). Do this as follows: Put a very small drop of water on a clean slide; with forceps remove some hairs from the flower; gently add a cover glass; remove any excess water with tissue paper. Examine your preparation and locate the following main parts of a cell: cell wall, nucleus, cytoplasm, vacuole.

••• *Suggested diagram and labels: Tradescantia* (spiderwort) hair from stamen of flower: cell showing cell wall, nucleus, cytoplasm, vacuole.

••• Make and examine a water mount slide preparation of a leaf of a shoot of *Elodea* (waterweed). Locate the following main parts of a cell: cell wall, nucleus, cytoplasm, vacuole.

IV. MITOSIS

As shown in Fig. Sup4–1 and elaborated in Supplement 4, Part II, ***mitosis*** is nuclear division involving chromosomes that are replicated and distributed equally between the nuclear progeny; that is, the chromosome number is unchanged. There are four stages in mitosis, namely, prophase, metaphase, anaphase, telophase. ***Interphase*** is the phase between successive mitotic (and meiotic) divisions when the cell grows and when the genetic material (DNA) is actually replicated.

●●● Examine the DEMO light and electron microscope photographs (from W. A. Jensen, unpublished, or from another available source) of interphase and the four stages of mitosis, that is, prophase, metaphase, anaphase, telophase. Then examine a permanent slide of a longisection (longitudinal section) of the root tip of *Allium* (onion) and locate interphase and the various stages of mitosis (metaphase and anaphase are the most conspicuous stages here).

●●● *Suggested diagram and labels: Allium* root tip l.s.: interphase and the four stages of mitosis: prophase, metaphase, anaphase, telophase.

V. ASEXUAL REPRODUCTION

Asexual reproduction (vegetative reproduction), that is, any reproductive process *not* involving meiosis or the union of nuclei, sex cells, or sex organs (i.e., syngamy or fertilization), occurs in both plants and fungi (and also in animals and protists) and, depending on the organism, may involve the gametophyte (GPT) and/or the sporophyte (SPT). It is very common in the life histories of numerous organisms. The new individuals are genetically like the parent ones; in other words, no genetic diversity results from asexual reproduction. See Supplement 4, Part V, for the significance of asexual reproduction in the life history.

There are various types of asexual reproduction:

- ***Fragmentation:*** Here the organism simply fragments or breaks up. For example, with weeds, including lawn weeds, chopping up of underground stems results in fragments, each usually capable of forming a new plant. Potato can be propagated by planting parts of the tuber (underground stem) bearing the "eyes" (shoot buds). In plant propagation, this is commonly called "division."
- ***Special dispersal structures:*** Here there may be special dispersal structures that are quite unlike the parent organism. For instance, in bryophytes and pteridophytes the GPT can disperse special structures called gemmae. Each gemma can form a new plant. Details will be given in Lab Exercise 4, Part II-C.
- ***Mini-adults:*** Here, the asexual reproductive unit is structurally like the parent. Thus a plant may produce ***plantlets*** (little plants) on its leaves or stems.

●●● Examine the DEMO live material of cuttings of *Begonia* (begonia) grown for several weeks in water and note the many root hairs close to the root apices. Cuttings, essentially artificial fragmentation, are a common means of plant propagation. Some persons advise that it is best to plant the fresh cutting in moist dirt because later planting of cuttings grown in water is likely to damage the fragile root hairs.

●●● Examine the DEMO live material of mini-adults. Note that the mechanism of asexual reproduction in the four DEMOs is comparable. However, *Chlorophytum* forms plantlets on stems whereas *Asplenium, Tolmiea,* and *Kalanchoe* form plantlets on leaves.

A. ●●● *Asplenium bulbiferum* (mother fern), native to New Zealand and the southwestern Pacific [*A. daucifolium* (= *viviparium*) of Mauritius is similar]: Many plantlets form on the margin of each heavily divided leaf. These plantlets fall off the leaf, and each can form a new plant that will be genetically like the parent plant.

B. ●●● *Kalanchoe daigremontiana* (maternity plant, devil's backbone) and/or *K. pinnata* (air plant, sprouting leaf), both native to Madagascar (these have, respectively, leaves undivided, gray green, spotted with red, *versus* leaves first undivided, later divided, uniformly gray green): The mechanism is very similar to that of mother fern. Many plantlets genetically like the parent plant are produced in the notches of a leaf and can root even on the parent plant. See the DEMO.

C. ●●● *Chlorophytum comosum* (spider plant), native to southern Africa: Plantlets, complete with roots, form on special stems (like in strawberry) and can form new plants.

D. ●●● *Tolmiea menziesii* (pickaback or piggyback plant), native to coastal western North America north from northern California, is yet another example. In this commonly cultivated ornamental, the plantlets develop at the junction of the leaf stalk (petiole) and blade. These plantlets fall off the leaf, and each can form a new plant.

●●● If there is sufficient material, you may, if you wish, remove a few of the plantlets (wrap them in a moist paper towel) and take them home to plant. *Asplenium* and *Tolmiea* are shade plants and require a fair amount of water; *Kalanchoe* is a sun or light shade plant and requires very little water. *Chlorophytum* requires bright to medium light and is nicely grown in hanging baskets; water thoroughly and then allow the soil to dry out.

NAME ______________________________ LAB SECTION # ______________ DATE ____

TITLE ______________________________

__________________ TOTAL MAG ____

TITLE ______________________________

__________________ TOTAL MAG ______

TITLE ______________________________

__________________ TOTAL MAG ____

TITLE ______________________________

__________________ TOTAL MAG ______

LAB EXERCISE SUPPLEMENT **1A**

Use of the Compound Microscope

OBJECTIVE

To learn the use of a compound microscope designed for student use.

PERSPECTIVE

Microscopes are perhaps the most important tool of biology. This exercise gives notes on the proper use, alignment, and care of the microscope. Following the procedures outlined will result in maximum resolution, image sharpness, and viewer comfort. The following excellent references give detailed methodology and explanations:

Barer, R. 1968. *Lecture notes on the use of the microscope.* 3rd ed. Oxford, England: Blackwell Scientific Publications. vii, 84 pp.

Möllring, F. K. 1966. *Microscopy from the very beginning.* Oberkochen, Germany: Carl Zeiss. 64 pp.

Note: The information below is mainly for reference purposes and thus should not be memorized.

I. PARTS OF A MICROSCOPE

Figure LabSup1–1 shows some of the main parts of a typical student compound microscope. These parts are:

A. The stage

This platform supports the material (i.e., "object") being examined. The object is usually mounted on a glass slide, with a thin cover glass (cover slip) covering the material. The slide is placed over an opening in the stage through which light enters from beneath. A mechanical stage or stage clips are used to hold the slide in place; the slide is moved, respectively, either with the mechanical stage adjustment or with one's fingers.

B. Revolving nosepiece with objectives

The nosepiece located at the lower end of the body tube contains objectives with lenses of various magnifications, typically 4X, 10X, 40X ("high dry"), and the oil immersion 100X (for use see Part V-B). Rotate the nosepiece to change objectives. A faint click denotes when an objective is in proper position.

C. Oculars or eyepieces

The oculars or eyepieces at the top of the body tube each contain a lens of fixed magnification, typically 10X. Total magnification is calculated by multiplying the power of the eyepiece by that of the objective being used; that is, 10X eyepieces and a 4X objective give 40 power, 10X eyepieces

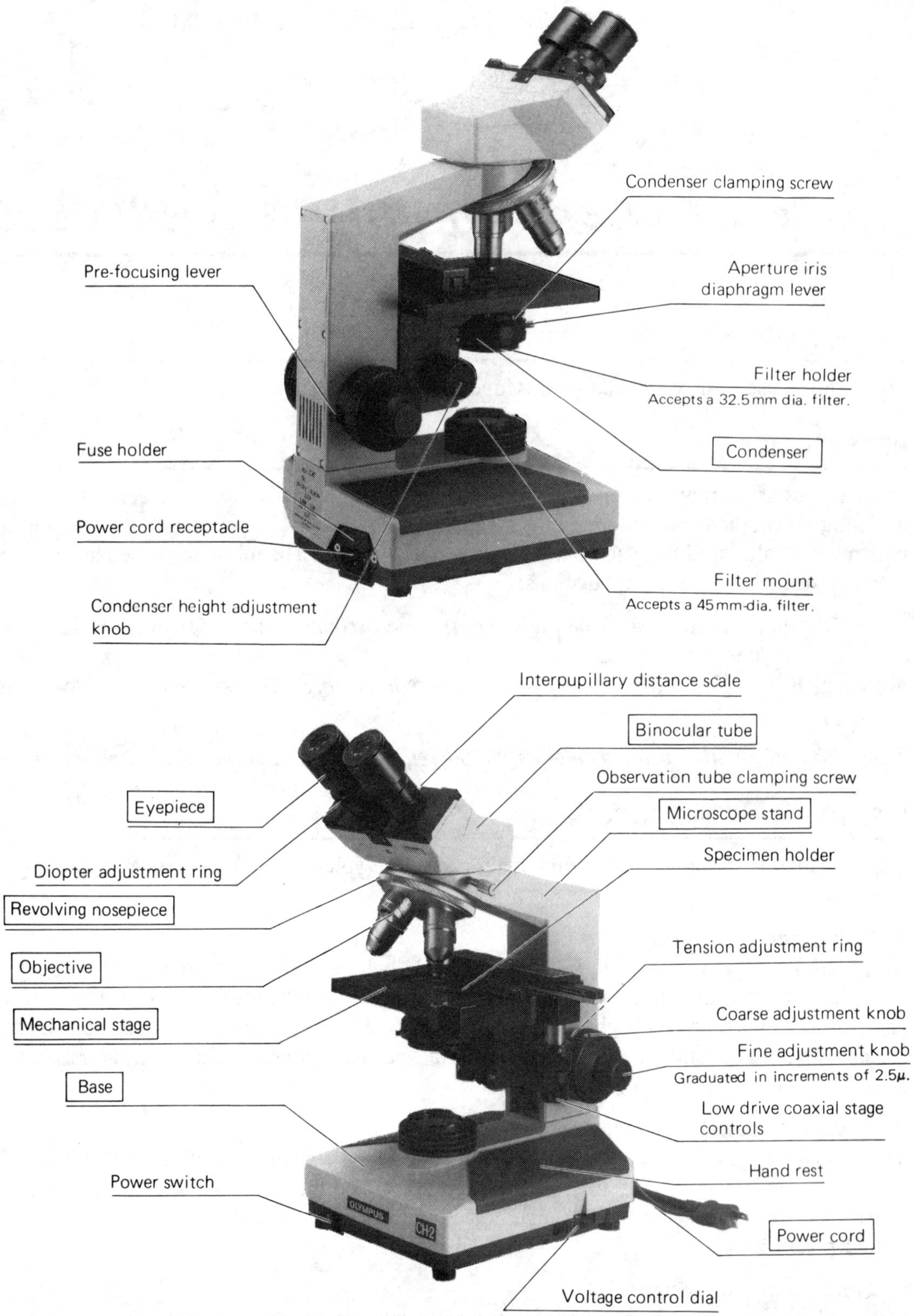

Fig. LabSup1-1. Two views of a modern student compound microscope, the Olympus Model CH-2 Biological Microscope. (From instruction manual, by Olympus Optical Co., © 1989 by Olympus Optical Co. Used by permission of the publisher.)

and a 40X objective give 400 power. One eyepiece of the pair may contain an ocular micrometer for measurement (see Part V-C) and a black hairline for use as a pointer. *Precaution:* The oculars will fall out if the microscope is titled excessively.

D. Binocular observation tube

The binocular tube allows adjustment of the two oculars for the spacing peculiar to your eyes. Set the left ocular to zero unless your eyes need correction. The binocular tube is rotatable so that either the stage or the focusing knobs can be directly in front of you. *Precaution:* If you rotate the binocular tube, be sure to lock it in place, and never carry a microscope by its binocular tube.

E. Light source

The light source (lamp) built into the base of the microscope has at its lower side an on/off switch and a sliding control (rheostat) to regulate light intensity. *Precautions:* When done with your observations, always return the sliding control to zero to prevent premature lamp burnout. Avoid high lamp intensities both to ease eye strain and to prolong lamp life.

F. Substage condenser

The substage condenser beneath the stage has another set of lenses that focus light coming from the light source onto the material being studied. An adjustment knob raises or lowers the condenser. Other adjustment knobs may be present to center the condenser. *Precaution:* The optimal position for the condenser is just below its uppermost position, next to the stage.

G. Iris diaphragm

The iris diaphragm built into the substage condenser consists of adjustable metal plates that move like the iris of a camera lens or the iris of the eye. A lever at the side of the condenser regulates the size of the opening of the iris diaphragm and hence the amount of light entering the microscope. *Precaution:* As explained below, the iris diaphragm must be properly adjusted for maximum resolution.

H. Focusing adjustments

The stage is moved up or down by the focusing adjustments. The large coarse focusing wheel gives rapid movement to locate quickly the object, whereas the small fine focusing wheel gives very slow movement to focus sharply the object. The coarse adjustment knob has an automatic prefocusing lever to prevent the specimen from running into the objective. *Precaution:* Never use the coarse focusing mechanism with the 40X and 100X high power objectives.

II. CARE OF THE MICROSCOPE

The microscope is a delicate and expensive instrument and requires the same care as a camera or a piece of electronic equipment. For emphasis, some of the following statements are repeated from Part I. Other special precautions are noted in Parts III to V.

A. Always carry the microscope with two hands, one hand on the arm of the microscope, the other hand supporting its base.
B. Never lift the microscope by the binocular observation tube, and check that this is locked into place.
C. Do *not* excessively tilt the microscope or else the oculars will fall out of the binocular tube.
D. Before commencing observations, check that the oculars, eyepieces, and other components of the light path are clean (for cleaning see Parts V-A and V-B3).
E. If water or reagents are spilled onto the stage or other part of the microscope, clean up the mess immediately. If immersion oil is used, also clean up this (see Part V-B3).
F. When putting away the microscope, return the light intensity sliding control to zero and turn off the lamp. Set the revolving nosepiece so that the 4X objective is in place. Remove any slide

from the microscope stage and properly file it away. Coil up the electrical cord and cover the microscope with its plastic cover. Properly carry the microscope to its cabinet as outlined above.

III. ALIGNMENT OF MICROSCOPE FOR BRIGHTFIELD ILLUMINATION

The procedures outlined here explain how to align properly the standard type of compound microscope used in introductory courses in biology. Your TA or instructor will probably walk you through some of these procedures and, periodically throughout subsequent lab exercises, offer guidelines on properly using a microscope. *These procedures should be carefully reviewed so that you can obtain maximum effective use of your instrument.* Following them will result in maximum resolution, image sharpness, and viewer comfort.

A. Be sure all glass surfaces in the light path are clean. For the correct procedure for cleaning lenses see Parts V-A and V-B3.

B. Turn on the microscope light, using the switch on the lower side of the microscope.

C. Place a microscope slide on the mechanical stage and center the object on the slide in the middle of the light path.

D. Using a low power (e.g., 4X or 10X objective), coarse focus on the specimen on the microscope slide. *Note:* Always attempt to locate objects on slides by initially using low power and advancing progressively to higher powers as needed.

E. The oculars (eyepieces) on a binocular microscope should always be adjusted to the particular width between your eyes, that is, your interpupillary distance. *Note:* This will always be at the same setting regardless of the microscope used.
   1. Hold the knurled dovetail slides of the right and left eyepiece tubes with both hands and push the tubes together, or pull them apart, whichever is required to achieve perfect binocular vision when you look through the eyepieces with both eyes. Use the scale provided and memorize your interpupillary distance setting for future use.
   2. Rotate the tube length adjustment ring on the *right* eyepiece tube to match your interpupillary distance obtained from the scale.
   3. Look at the image through the *right* eyepiece with your right eye and focus on the specimen with the coarse and fine adjustment knobs.
   4. Looking at the image through the *left* eyepiece with your left eye, rotate the tube length adjustment ring on the left eyepiece to focus on the specimen, *without* using the coarse and fine adjustment knobs.

F. Raise the substage condenser to its maximum height. Remove any granular appearance by lowering the condenser slightly. When using the 4X and 10X objectives, it may be necessary to lower the condenser to eliminate uneven field illumination. If adjustment knobs are available, center the condenser.

G. Swing in the desired objective.

H. Adjust the light intensity with the sliding control lever (see Part IV-B).

I. Adjust the aperture iris diaphragm using the lever at the top of the condenser. Remove an eyepiece from the eyepiece tube. While looking into the eyepiece tube, slowly open (or close) the iris diaphragm of the condenser until the illumination fills about ¾ of the field (back lens) of the objective. *Note:* Properly for maximum resolution, this adjustment should be made for *each* objective, although many researchers do it only for the lowest power objectives, and most students do not do it at all.

J. Replace the eyepiece and fine focus.

IV. GENERAL POINTS

This section relates a miscellany of general points and special precautions.

A. Basic theory

Provided it is smaller than the aperture of the objective, the iris diaphragm of the condenser determines resolution and contrast of the microscope image. The iris diaphragm must *not* be used for control of image brightness (see Part IV-B). The iris diaphragm should be adjusted such that the field (back lens) of the objective being used is ⅔ to ¾ illuminated (Part III-I). Too small an aperture produces a contrasty, double, so-called rotten image; too large an aperture produces a glary, washed-out image. Note that closing the iris diaphragm increases the contrast, but beyond a certain point (less than ⅔ transmission of the objective aperture), the resolving power of the objective is rapidly reduced.

B. Controlling light intensity

Light intensity should always be varied with the rheostat lever located on the lower right side of the microscope. Light intensity should *never* be controlled (1) by varying the substage condenser iris diaphragm or (2) by lowering the condenser. These are common student errors; on student microscopes the condenser is frequently at the opposite end of its range near the base of the microscope.

C. Movement of the image relevant to movement of the slide

Note that the compound microscope inverts and reverses the image. Thus the image will move away from you as the slide moves toward you, and vice versa, or the image will move to the left as the slide moves to the right, and vice versa.

D. Choice of magnification and precautions in focusing

When first examining a microscope slide, always use the low power 4X objective. This permits coverage of a wider field and hence is best suited for scanning. It is also easier to find the plane of best focus. Then, gradually work through the higher power objectives, finally using the oil immersion objective if this is really needed. Always use the fine adjustment focus with the high power objectives (40X and 100X) to avoid running the objective into the slide and damaging either or both. To prevent damage to the slide when using the high power objective, it is always best to focus *away* from the specimen rather than towards it.

Better microscopes are parfocal. That is, the specimen should be in approximate focus despite the objective used, such that with the specimen in focus with one objective only slight adjustment with the fine focus knob is necessary to bring the specimen into focus with another objective. *Never force* an objective into position because the slide and/or specimen on the slide preparation may be too thick.

E. Tension adjustment of coarse adjustment knobs

A tension adjustment ring may be provided next to the coarse adjustment knob. With this device the tension of the coarse adjustment is freely adjustable for either heavy or light movement depending on operator preference. An arrow may indicate increase in adjustment tension. Do *not* loosen the tension adjustment knob too much because this may cause the stage to drop or the fine adjustment knob to slip. *Never* rotate the right and left coarse adjustment knobs simultaneously in the opposite direction.

F. Special precautions

Oil immersion and high dry objectives, because of their extremely small working distances, should never be used with thick slides, either thick whole mount preparations or glass slides that are unusually thick.

G. Use of the mechanical stage

Once you have become accustomed to using a microscope, it may be expedient and simpler to forego completely the use of the mechanical stage except for examinations with oil immersion. It is very quick and easy simply to put a slide on the stage and move it about by gently touching the edges of the slide. With practice you can do this even with 60X objectives and you will automatically compensate for the reversed image effect of the microscope (see point C above). Students often spend much time during a lab uselessly cranking the mechanical stage back and forth.

V. SPECIAL PROCEDURES

These special procedures relate to (A) cleaning lenses, (B) using oil immersion objectives, and (C) measuring objects.

A. Cleaning lenses

Dirty objectives are especially undesirable. Clean all the glass surfaces (eyepiece lenses, objective lenses, condense lenses, glass over light source) of the microscope as follows. Do *not* attempt to remove the objectives from the revolving eyepiece!

1. Remove excess dirt with an air blower (you can effect the same by blowing on the lens, but swallow several times before blowing, and don't do this after eating a peanut butter sandwich!).
2. If necessary, use a brush (e.g., a camel hair brush, or, much better, a temporary brush made from lens tissue folded several times and torn in half) and wipe off any large dirt particles still adhering to the lenses after blowing air onto the lenses.
3. If necessary, use lens tissue (unfolded or once folded) and gently wipe the glass surface with delicate circular motions. Exhaling gently on the glass surface to be cleaned, before wiping, may help. Do *not* use solvents unless absolutely essential (see Part V-B3).
4. *Note:* Some grunge apparent in the light path may *not* be due to dirt on the lenses but rather to so-called mouches volantes. These are particularly evident on specimens with a wide, clear background, which may show strange moving irregular patches shaped rather like worms. These are the shadows of inhomogeneities in the aqueous humor of the eye, which become visible whenever the exit pupil of the microscope becomes very small (i.e., if the iris diaphragm of the condenser is closed down too far or if a high power objective is used). [Incidentally, you can get the same effect if you look at a bright sky or lie on your back in the sun with your eyes closed since in both cases the pupils of your eyes close down greatly.]

B. Use of the oil immersion objective *(for reference information only)*

Use the oil immersion 100X objective only when necessary (e.g., to see minute cell wall details). Many anatomical and morphological details of organisms can be seen adequately with the 40X high dry objective (definitely if a 60X objective is available).

1. *Procedure:* Raise the oil immersion objective as much as possible. Place a *small* drop of immersion oil onto the slide at the small bright spot of light right below the objective. While looking at the microscope from the side, carefully lower the objective until it just contacts the oil. Focus carefully, with the *fine adjustment only*. Be very careful *not* to overfocus and damage the objective and/or slide. The objective should never touch the slide. Be sure the iris diaphragm is properly adjusted (see Part III-I).
2. *Caveats:* Placing oil between the slide and condenser lens is justifiable only in critical work, *never* in classroom work. If you switch to the high dry objective, first clean the immersion oil from the slide (see below).
3. *Cleaning:* When finished with the oil immersion objective, raise it as much as possible. With lens tissue remove excess oil from the objective. If necessary, put a *very small* drop of xylene (beware, xylene is carcinogenic; alcohol or spirit must *not* be used) on new lens tissue and lightly

wipe the surface of the lens. Excess xylene (or other solvent) will dissolve the resin holding the lens elements in place. Also thoroughly *clean the slide* by wiping it with lens tissue and, if necessary, cleaning with alcohol or xylene (excess xylene will dissolve the mounting medium holding on the cover glass). *Do not use the high dry objective until you have cleaned the slide.*

C. Measuring objects *(for reference information only)*
Objects under the microscope are measured by the metric system rather than by the English system. Lab Exercise 1, Part III, gives some equivalents. An ocular micrometer disc, with markings of unknown dimensions, may be permanently inserted into the eyepiece of the microscope and calibrated with a graduated slide, a so-called stage micrometer. Once the divisions of the ocular micrometer have been calibrated in microns (micrometers) for each objective, any object can be quickly measured.

1. Focus, with the light decreased, on the stage micrometer slide with a 10X objective. This slide will have a maximum scale 2 mm long, subdivided into units of 0.1 mm (100 microns) and 0.01 mm (10 microns).
2. In binocular microscopes place the eyepiece with the ocular micrometer on the *non*-focusing eyepiece tube, or on one set for your interpupillary distance. Rotate the eyepiece so that the markings on the ocular micrometer parallel or superimpose those of the slide. The ocular micrometer will have an arbitrary scale of 50 or 100 units or divisions.
3. Adjust the two scales so that the first line of each scale *coincides.*
4. As far as possible from the coincidence of the scales in step 3 find another point at which the lines of the two scales coincide.
5. For each scale count the number of spaces between these points. For the stage micrometer scale, multiply the number of spaces by the length of each space (note that the large spaces each equal 100 microns, the small ones 10 microns).
6. Determine the value (in microns) of each ocular micrometer units as follows:
   Value (in microns) of stage micrometer divisions ÷ Number of ocular micrometer divisions
7. An example to determine the value (in microns) of each ocular micrometer:
   49 ocular micrometer divisions = 68 stage micrometer divisions;
   1 ocular micrometer division = 1.3878 stage micrometer division, each being 0.01 mm;
   1 ocular micrometer division = 0.013878 mm or 13.878 microns (micrometers).
8. There is no further use for the stage micrometer. *The ocular micrometer scale must be calibrated for each objective.* This can be done by steps 1 through 6 above or, more easily but less accurately because of variability among lenses (e.g., a lens marked 40X may really be 39X or 41X), it can be computed mathematically as follows:
   Magnification A x Value ocular micrometer A = Magnification B x Value ocular micrometer B.
9. Sample calibrations: Here are the calibrations for my research microscope:
   4X objective = 24.00 microns (micrometers);
   10X objective = 9.75 microns (micrometers);
   20X objective = 4.82 microns (micrometers);
   40X objective = 2.43 microns (micrometers);
   60X objective = 1.57 microns (micrometers).

   Note that these values are not mathematically equivalent for the various objectives. Also note that I have not calibrated the 100X objective because I rarely use it in my work.

LAB EXERCISE SUPPLEMENT **1B**

Permanent Microscope Slide Preparations

OBJECTIVE

To describe briefly how permanent microscope slide preparations are made for light microscopy so that this description will help in the interpretation of objects seen on microscope slides.

PERSPECTIVE

Slide preparations of biological material are prepared by many different methods but can be broadly characterized as temporary, semi-permanent, or permanent. Temporary slide preparations are commonly prepared of whole objects or objects sectioned by hand, which are then placed in water (the most common solution) beneath a cover glass; stains are also sometimes used. Semi-permanent slide preparations are temporary slide preparations with the cover glass ringed or surrounded by Vaseline, resin, or other material to prevent the water from drying out. Such slides have a life of a day to several weeks or even months. In contrast, permanent slide preparations will last for a hundred years or more, if stored properly. Permanent slide preparations are often desirable for ease of use and for showing material that is difficult to obtain or to prepare.

Note: Most of the terms mentioned below are defined in Lab Exercises 6 and 7. *The information below is mainly for reference purposes and thus should not be memorized.*

I. PREPARING PERMANENT MICROSCOPE SLIDE PREPARATIONS

Many of the slide preparations examined in this course are permanent preparations, either sections or clearings (whole mounts). The following steps are generally involved in making permanent slide preparations (for detailed techniques see the two classical works, Johansen 1940, and Sass 1958):

A. Killing and fixing (pickling) biological material in a fluid such as FAA (formalin, acetic acid, alcohol) or CRAF (chromic acid, acetic acid, formalin) to preserve cytological detail (FAA is a general purpose fixative whereas CRAF yields more precise cytological detail);
B. Dehydrating the material through an alcohol series;
C. Infiltrating the material with wax in a wax-solvent such as xylene until the material is in pure liquid wax (this is done in a 60° C, 140° F, oven);
D. Embedding the material in a wax block;
E. Microtoming, that is, sectioning the wax block with a rotary or sliding microtome;
F. Mounting (with an adhesive) on a microscope slide the resultant wax ribbon with the material;
G. Decerating the slides in xylene to remove the wax;

H. Staining, that is, treatment of the slides in alcoholic or other solutions;
I. Attaching a cover glass with a permanent mounting medium.

II. STAINING TECHNIQUES

A few comments on the results of staining are pertinent as this knowledge will be useful in interpreting the slides examined throughout this course. One to several stains may be used to stain an object. If two or more stains are used, these additional stains are used as a counterstain to provide contrast. Baker (1966) gives an extended discussion on staining. Briefly,

- ***basophilic compounds*** are acidic (negatively charged) and thus attract basic (cationic) dyes;
- ***acidophilic compounds*** are basic (positively charged) and attract acidic (anionic) dyes.

Some dyes are amphoteric, that is, cationic at one pH, anionic at another.

The following are some staining combinations commonly encountered in botanical preparations:

A. **Safranin–fast green:** Safranin is a red basic dye that intensely stains basophilic compounds such as tannins, lignin (especially of fibers, sclereids, tracheary elements), cutin (of epidermal cells), suberin (of cork cells and of Casparian strips of endodermal cells), sporopollenin (of pollen grain walls), and chromatin (of nuclei and nucleoli, i.e., chromosome material). Staining in safranin is several hours to several days. Fast green is used as a counterstain for several seconds to several minutes. Fast green is an acidic dye that stains acidophilic materials such as cytoplasm (except chloroplasts) and especially cellulose of cell walls, notably of meristematic cells, parenchyma cells, and sieve elements.

B. **Quadruple stain:** This staining combination consists of safranin, fast green, orange G, and methyl violet (or crystal violet) and was developed by Johansen in 1939 (Johansen 1940). Johansen's company, the California Botanical Materials Co., generally used this stain, as have Ripon Microslides Laboratory and Triarch, both of Ripon, Wisconsin, among other companies. While this technique produces spectacular results, few researchers use it because of the complexity and extra time involved. The staining effects of Johansen's quadruple stain (with methyl violet) should be as follows (Johansen 1940:89):

> Dividing chromatin red, resting chromatin purplish, nucleoli red (occasionally violet), nucleoplasm colorless or greenish, lignified cell walls bright red, cutinized cell walls reddish-purple, suberized walls red, cellulose cell walls greenish-orange, cytoplasm bright orange, middle lamellae green, starch grains purple with green or orange halos (the color of the halos soon becomes replaced by the purple in some types of material), plastids purplish to greenish, invading fungal mycelium green, the callose portion of the guard cells of stomata bright red and the remainder purple, and Casparian strips red and the remainder of the cell wall of the endodermis yellow. In sections of roots for origin of the lateral roots, the cytoplasm of the latter should be stained green, with purplish nuclei, while the cytoplasm elsewhere should be orange with red nuclei.

Also, sclereids stain a brilliant red and collenchyma bright green or, if immature, yellow green.

C. **Bismarck brown–Prussian blue:** This staining combination is especially favored by Germanophiles. The stain has been used on slides of wood and bark of *Robinia pseudoacacia* (black locust). Here the phloic fibers and wood components stain light blue, the cambial zone dark brown, the other phloic components a brilliant dark brown, and the cork light brown.

D. **Tannic acid–iron chloride:** This stain is used for meristematic tissues. The thin cell walls stain an intense black, the cytoplasm violet or pink, plastids blue, and nucleoli and chromosomes red.

LITERATURE CITED

Baker, J. R. 1966. *Cytological technique: The principles underlying routine methods.* 5th ed. London: Methuen & Co.

Johansen, D. A. 1940. *Plant microtechnique.* New York: McGraw-Hill Book Co.

Sass, J. E. 1958. *Botanical microtechnique.* 3rd ed. Ames, Iowa: The Iowa State College Press.

LAB EXERCISE **2**

Bacteria and Other Prokaryotes

OBJECTIVE

To examine the structure of representative examples of bacteria and other prokaryotes.

Note: Read this material *before* coming to the lab; there is insufficient time in the lab to read the material from scratch. Due to the large amount of background information presented here, a ●●● is used to mark those paragraphs calling for actual observations or diagrams in the lab. Most unmarked paragraphs reiterate information presented in the lecture. A bullet (●) signifies an important term or concept.

Suggested Lab Diagrams (see part noted for suggested labels):
Oscillatoria (a blue-green bacterium) filament (Part II)

PERSPECTIVE

The three divisions of prokaryotes have over 10,100 ***extant*** (living as opposed to fossil or ***extinct***) species (sp. = singular, spp. = plural) according to Raven et al. (1992:728–729):

1. bacteria (Bacteria)—2,600 species, including 100 species archaebacteria;
2. cyanobacteria or blue-green bacteria ("blue-green algae") (Cyanophyta)—7,500 species;
3. Prochlorophyta (proposed 1977)—2 species.

Raven et al. (1992), but not Raven & Johnson (1992), recently divided Kingdom Monera into two kingdoms, Archaebacteria and Eubacteria, the latter with the bacteria and cyanobacteria. The differences between all these groups are mainly chemical. Possessing chlorophylls *a* and *b* and carotenoids like those of the green algae and land plants, the recently described Prochlorophyta are somewhat transitional between the eukaryotes and the other prokaryotes and may represent a chloroplast progenitor of the green algae and land plants (Lewin & Cheng 1989).

I. BACTERIA

Bacteria (Kingdom Bacteria) are exceedingly important organisms from many standpoints. Many bacteria are decomposers of organic material that is in or on soil, or in the intestinal tracts of animals. Other bacteria are parasitic and cause diseases of animals, plants, fungi, and protists. Some bacteria in a process called ***nitrogen fixation*** incorporate nitrogen from the atmosphere into nitrogen compounds usable by plants. The most important nitrogen fixing genus is *Rhizobium,* which forms nodules on the roots of legumes and other plants.

Bacterial cells are the smallest of cells, about 1 μm in diameter. Yogurt is a bacterial culture that ferments the milk and makes it tart. Ordinary yogurts with live bacteria have hundreds of millions, even billions of bacteria per gram (i.e., 1/28.35 of an ounce); for example, one gram of plain, nonfat Dannon yogurt has 2.6 billion bacteria (*The New York times,* 26 June 1991).

Bacteria are characterized on the basis of shape:

- ***rods*** (bacilli), straight, rod-shaped forms;
- ***spheres*** (cocci), spherical forms;
- ***spirals*** (spirilli), long, curved forms.

Bacteria are unicellular, colonial, or filamentous, that is (see detailed definitions in Lab Exercise 3, Parts I-A to C):

- ***unicells,*** completely separate, individual cells;
- ***colonies,*** groups of loosely associated cells;
- ***filaments*** (trichomes), chains of cells.

●●● Examine the DEMO of the washed root system of a legume. Note the tumorlike bacterial root nodules, which are red due to the pigment hemoglobin (this also occurs in blood). Then examine the DEMO slide of a section of a root nodule of *Pisum* (pea). The dark-staining cells contain the innumerable bacteria. Because bacterial cells are the smallest of all cells, about 1 μm, almost no detail is evident from light microscopy.

II. BLUE-GREEN BACTERIA OR CYANOBACTERIA

Blue-green bacteria or cyanobacteria (Cyanophyta, formerly "blue-green algae") occur on land and especially in water. Some blue-green bacteria also fix nitrogen (see Part I).

Fur of polar bears in captivity is often greenish. Lewin et al. (1981) found that the hairs of the bears are hollow and that they are also often broken off when the bears live in zoos because of the rough stone surfaces of their lairs. This allows blue-green bacteria to live inside the hairs and carry on photosynthesis. This is possible because the transparent walls of the hairs allow light to enter, whereas the broken ends allow oxygen to enter. Green rather than blue green is evident due to the absence of certain pigments related to photosynthesis.

Blue-green bacteria are unicellular, colonial (Fig. 2–1A, C, F), or filamentous (Fig. 2–1B, D, E, G to I). Some cells of a filament are structurally very different due to the transformation of ordinary vegetative cells into:

- ***heterocysts,*** thick-walled, usually transparent cells involved in nitrogen fixation (Fig. 2–1D, G to I);
- ***akinetes,*** thick-walled, often large ***spores*** (i.e., reproductive cells of various sorts that are capable of developing into an adult without fusion with another cell) highly resistant to adverse environmental conditions (Fig. 2–1D, G).

Akinetes germinate and form new filaments; heterocysts may do the same. Akinetes stored dry have germinated after 87 years.

●●● Examine the DEMO of a water mount slide preparation of *Gloeotrichia,* which resembles Fig. 2–1G of *Cylindrospermum.* Each filament ends in a heterocyst (transparent cell); an akinete (large thick-walled cell) often occurs next to the heterocyst. The gelatinous matrix around the cells can be seen better with critical microscope adjustment. Nuclei are absent, the hallmark of a prokaryote; all examples in subsequent lab exercises are eukaryotes and thus have nuclei.

●●● See Fig. 2–1B. Then make and examine a water mount slide preparation of *Oscillatoria.* Note the oscillating motion of the individual filaments. The mechanism of this movement is not completely understood but seems due to a gliding movement; an amoeboid motion may also occur. Observe the structure of the cells along the length of the filament. Other than the slightly different apical cell, is there any evidence of regional cell differentiation? Reproduction is vegetative (asexual) by fragmentation of the filament due to the death of one or more cells.

●●● *Suggested diagram and labels: Oscillatoria* (a blue-green bacterium) filament: live cells, dead cells resulting in breakup of filament.

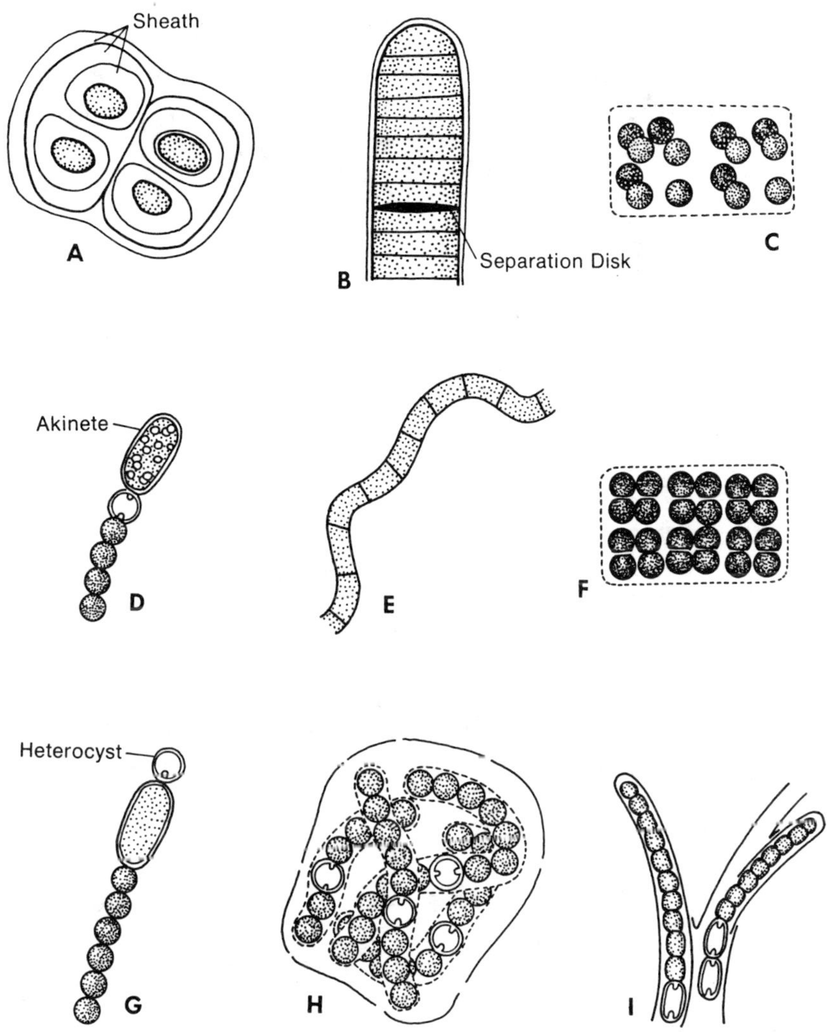

Fig. 2-1. Structure of blue-green bacteria (Cyanophyta) (*A, Chroococcus; B, Oscillatoria; C, Anacystis; D, Anabaena; E, Spirulina; F, Agmenellum; G, Cylindrospermum; H, Nostoc; I, Tolypotrix*). *A, C, F,* colonies; *B, D, E, G-I,* filaments. Note in *D, G-I,* the heterocysts (thick-walled, usually transparent cells involved in nitrogen fixation) and/or akinetes (thick-walled, often large reproductive cells highly resistant to adverse environmental conditions and capable of developing into an adult). Also note the absence of nuclei in these prokaryotes. (From Norstog & Long © 1976:77.)

NAME ______________________________ LAB SECTION # ______________ DATE ____

TITLE ______________________________

__________________ TOTAL MAG ____

TITLE ______________________________

__________________ TOTAL MAG ______

TITLE ______________________________

__________________ TOTAL MAG ____

TITLE ______________________________

__________________ TOTAL MAG ______

LAB EXERCISE 3

Seaweeds and Other Algae

OBJECTIVE

To examine the vegetative structure of representative examples of seaweeds and other algae.

Note: Read this material *before* coming to the lab; there is insufficient time in the lab to read the material from scratch. Due to the large amount of background information presented here, a ●●● is used to mark those paragraphs calling for actual observations or diagrams in the lab. Most unmarked paragraphs reiterate information presented in the lecture. A bullet (●) signifies an important term or concept.

Suggested Lab Diagrams (see part noted for suggested labels):
Diatom (pennate *or* centric) (a yellow-brown alga) (Part I-A)
Spirogyra (a green alga) filament (Part I-C)
Kelp (a brown alga) thallus (SPT) (Part I-I)

PERSPECTIVE

There are at least eight divisions of algae (plural, singular *alga*), involving over 22,650 extant species according to Scagel et al. (1982):

1. diatoms et al. yellow-green/golden-brown algae (Chrysophyta)—6,900 species;
2. dinoflagellates, flame algae (Pyrrhophyta)—over 1,000 species;
3. cryptomonads (Cryptophyta)—100 species;
4. red algae (Rhodophyta)—5,200 species; } These include the seaweeds!
5. brown algae (Phaeophyta)—2,000 species; }
6. green algae (Chlorophyta)—over 6,750 species; }
7. stoneworts (Charophyta)—250 species;
8. euglenoids (Euglenophyta)—450 species.

The ***seaweeds*** are the large macroscopic marine algae, for example, *Ulva* (sea lettuce) and the kelps (Figs. 3-5, 3-6). Algae are important as a nutrient source for many marine animals. According to Bold & Wynne (1985:31), 493 species and 107 genera of marine algae are of economic importance.

The differences among these divisions of algae are morphological and especially chemical:

- nature, number, and insertion of flagella (plural, singular flagellum), or their absence;
- nature of the nutrient reserve, for example, starch or other polysaccharides;
- nature of the cell wall component, for instance, cellulose, silica, calcium, etc.;
- and particularly, nature of photosynthetic pigments (see Lab Exercise 1, Part III).

The last item requires elaboration: All algae have chlorophyll *a* as well as other types of this green pigment, that is, chlorophylls *b* to *e*. These chlorophylls plus other types of ***accessory pigments,*** that is, pigments also relevant to photosynthesis (e.g., xanthophylls, carotenoids, the red and blue biliproteins, etc.) give the distinctive colors to various divisions of algae. The common names of some of these divisions derive from the pigmentation. In many groups the non-green pigments mask the green chlorophyll that is present, so that one winds up with the algae of color noted above.

As elaborated below, the algae include unicellular, colonial, and multicellular organisms. Most species of algae are aquatic, living in fresh water (freshwater algae) or especially salt water (marine algae). However, some algae occur on land (terrestrial algae), and a few live in snow (snow algae) or even in porous rocks in Antarctica (endolithic algae). The green alga *Chlamydomonas nivalis* causes "red snow" due to its red pigment that serves as a light shield to protect the green chlorophyll. Some algae form important associations with fungi called lichens (see Lab Exercise 13, Part II). The aquatic algae can be divided into:

- ***benthic algae*** attached to rocks or other substrates (e.g., coral) at the bottom of bodies of water, in the intertidal zone or in open water;
- ***phytoplankton,*** the free-floating, mainly unicellular, thus mostly microscopic, plant life in water.

However, *Sargassum* is a macroscopic phytoplankton type in the Sargasso Sea, a large area (600 km north-south by 1,200 km east-west; 372.8 by 745.7 mi.) of the Atlantic Ocean between the Azores and the West Indies that has been fabled to strand ocean vessels. Incidentally, some authors (e.g., Raven et al. 1992) use "plankton" instead of "phytoplankton" because they regard the algae as protists and not as plants.

In the ocean, green algae occur closest to the surface, brown and yellow-brown algae occur at greater depths, and red algae extend the deepest, up to depths of about 200 m (656 ft.), much deeper than other algal groups. In October 1984 a corolline rock-encrusting red alga was actually discovered at 268 m (879 ft.), the deepest photosynthesizing organism, where light is only 0.0005% of its value at the ocean surface. Below the 200 m level in oceans there are typically only animals and protists, which get their nutrients from plankton and other debris "raining," that is, falling from the upper waters into the deeper waters.

Light is composed of a spectrum of colors, as in a rainbow, to wit:

ultraviolet——violet——blue——blue-green——green——yellow——orange——red——infrared,

with ultraviolet (UV) light being of short wavelength and infrared light being of long wavelength. The shorter wavelengths, that is, the blue and blue-green, penetrate deeper into water than longer ones. The major groups of algae have different complexes of pigments, which are able to react differently to different wavelengths of light. The net effect, as mentioned above, is that the various divisions of algae tend to occur at different water levels. The red algae occur deepest because they are able to use the blue and blue-green light able to penetrate there. Thus the red algae have accessory photosynthetic pigments (phycoerythrin) that receive this light and transfer it to their chlorophyll *a* component.

Most algae manufacture nutrients via photosynthesis. However, some diatoms, some dinoflagellates, and about two thirds of the species of euglenoids ingest organic matter. The saprobic diatoms and dinoflagellates absorb nutrients, as do the parasitic dinoflagellates and the green, brown, and red algae (Bold & Wynne 1985). Supplement 11 summarizes the major types of biological nutrition.

I. CELL/TISSUE ORGANIZATION IN ALGAE

Rather than focusing on the structural and especially chemical differences of algae, or aspects of their life histories (see Part II), we will concentrate on cell/tissue organization in algae, particularly of the seaweeds, and discuss this from the simplest type (type A, unicells) to the most complex (type H, parenchyma plus phloem). Note that in the unicellular and colonial forms (types A, B) after cell division the parent cell wall is eventually cast off or gelatinized by the sibling cells. In contrast, in multicellular forms (types C-H) after cell division the parent cell wall is retained by the sibling cells.

In true multicellularity, the activities of the individual cells are coordinated and the cells themselves necessarily are in physical contact. True multicellularity occurs mainly in the eukaryotes. Various prokaryotes (bacteria and blue-green bacteria) have filaments that suggest multicellularity, but because the cells are loosely held together by a common sheath these "filaments" in a strict sense are colonies. Some filamentous blue-green bacteria, however, have protoplasmic connections (plasmodesmata) between cells and thus are true multicellular filaments (Bold & Wynne 1985:39).

Any multicellular plant or fungal body *not* differentiated into roots, stems, and leaves is called a ***thallus*** (singular, plural ***thalli***). This refers to all multicellular bodies in the algae, bryophytes, and fungi, but in the vascular plants "thallus" refers only to the GPT.

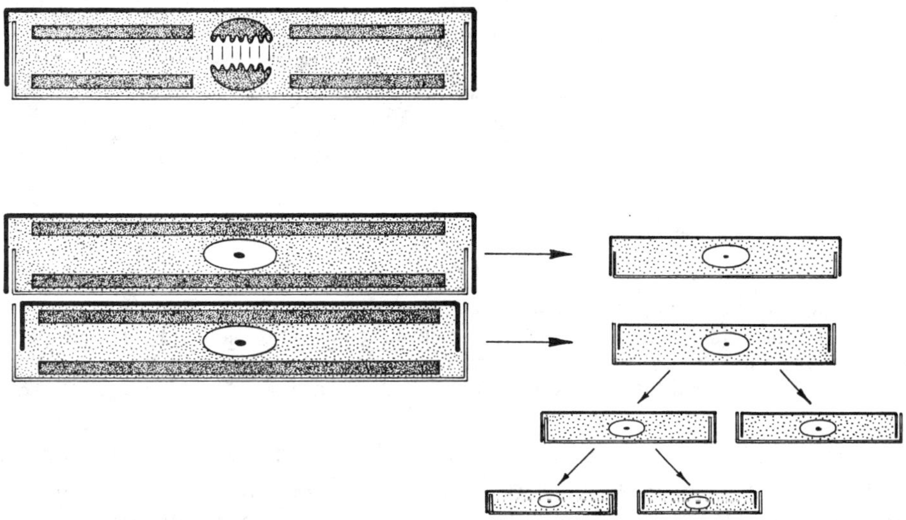

Fig. 3–1. Cell division (mitosis) in a diatom, a unicellular golden-brown alga (Chrysophyta), showing inheritance of valves and subsequent size reduction of certain progeny cells due to the unequal size of the upper and lower valves of the cell wall. Meiosis occurs after an approximately 70% size reduction of the protoplast. (From Norstog & Long 1976:241.)

A. Unicells

See Fig. 3–1. ***Unicells*** are completely separate cells because the parent cell wall is cast off after cell division. Many bacteria (prokaryotes), all dinoflagellates, and most euglenoids and diatoms (yellow-brown algae) are unicellular. Unicells also occur in the green and red algae but not in the brown algae. The dinoflagellates and diatoms are economically the most important unicellular algae.

A dinoflagellate cell has two flagella extending in opposite directions in two grooves, one flagellum encircling the cell like a belt, the other perpendicular to the first flagellum. The beating flagella impart a whirling motion to the cell, which gives the group its name (the "dino" part of name means "to whirl"). Curiously, the stiff plates forming the cell walls of many dinoflagellates are located *inside* the cytoplasmic membrane.

Dinoflagellates cause two common ocean phenomena. Many species emit light (bioluminescence) and can be seen flashing at night due to breaking waves, swimming people, or moving ships. In addition, "red tides" occur when dinoflagellates undergo population explosions, which may reach concentrations of 1–20 million cells per liter (0.95 qt.) of seawater and discolor patches of sea up to several square kilometers (Bold & Wynne 1985:493). The water becomes red due to the reddish pigment of the cells. Dinoflagellates produce potent nerve toxins (neurotoxins) that are poisonous, inhibiting

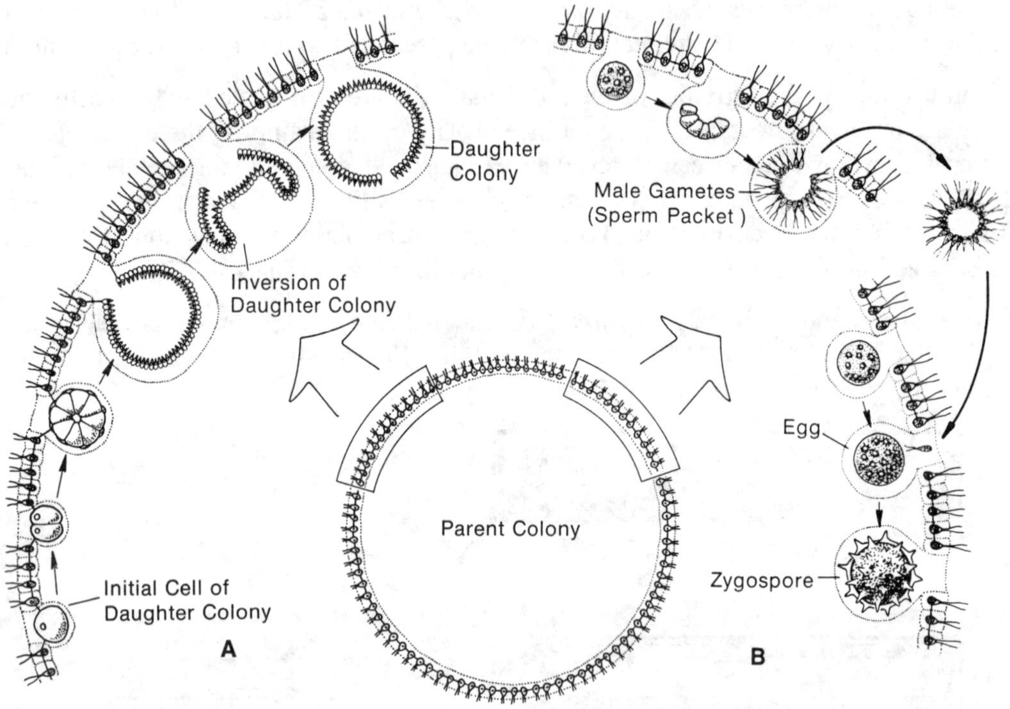

Fig. 3–2. Haploid (1*n*) life history of *Volvox,* a colonial green alga (Chlorophyta). *A,* developmental stages of a daughter colony; *B,* sexual reproduction (oogamy and meiosis). A zygospore is an encysted zygote resistant to adverse environmental conditions. Locate where meiosis and syngamy (fertilization) occur and which structures are 1*n* and 2*n*. (From Norstog & Long 1976:182.)

the diaphragm and thus causing respiratory failure. Fish, birds, and marine mammals may be affected and killed in large numbers; their carcasses will often reduce the tourist trade, as in Florida in August 1987. Shellfish (mollusks) accumulate the toxins but are generally unaffected. However, humans and other animals consuming the shellfish are affected; the toxin can be fatal to humans. In the northern hemisphere shellfish can be safely eaten in the "R" months, September through April, but should be avoided during the other four months.

Diatoms (Fig. 3–1) are mostly strictly unicellular, less often filamentous or colonial. Diatoms are divided into two types based on symmetry:

- the ***pennate diatoms*** with longitudinal (bilateral) symmetry;
- the ***centric diatoms*** with circular (radial) symmetry.

There are some 6,900 extant species but at least 40,000 extinct species (Raven et al. 1992:258).

The diatom cell wall (frustule) consists of two halves (***valves***), a larger ***upper valve*** (epitheca) and a smaller ***lower valve*** (hypotheca) that fit together like the top and bottom of a box (Fig. 3–1). The overlapping sides of the two valves is the ***girdle.*** Due to the unequal valves, mitosis results in progressively smaller cells until meiosis occurs. Both valves have delicate markings consisting of many minute pores or passageways that connect the protoplast with the outside environment. These pores are involved in locomotion. Their markings have traditionally been used to test the resolution quality of microscope lenses. A pennate diatom has a longitudinal groove (raphe) on the upper valve.

Diatoms are useful in forensic pathology to determine if a person found dead in water actually drowned or died before falling into water, as, for instance, in the 1991 death of publishing mogul Robert Maxwell. If a person died before falling into the water there should be no diatoms absorbed into the bone marrow, whereas presence of diatoms in the bone marrow would suggest drowning. Diatoms and other aquatic life in the lungs and stomach would also suggest drowning.

●●● See Fig. 3–1. Then examine the DEMO of a water mount slide preparation of a diatom (*Navicula, Pinnularia,* or other type). Is the diatom pennate or centric? Identify the following structures of a diatom: upper (or lower) valve, longitudinal groove, pores, nucleus.

●●● *Suggested diagram and labels:* Diatom (pennate *or* centric) (a yellow-brown alga): upper (or lower) valve, longitudinal groove, pores, nucleus.

The distinctive wall material of diatoms is silica. Silica is very resistant to degradation. Thus, diatoms have been extensively preserved as fossils, so-called ***diatomaceous earth.*** This is extensively used, as in abrasives, filtering materials (e.g., swimming pools), and insulating materials. The very famous deposits at Lompoc, California, are up to 300 m (985 ft.) thick and have concentrations of over of 6 million diatoms per mm^3 (0.00006 $in.^3$) (Norstog & Long 1976:241); some 270,000 metric tons are mined yearly at Lompoc (Raven et al. 1992:259).

●●● Examine the DEMO piece of diatomaceous earth and the microscope slide prepared of it. Are pennate and/or centric diatoms present? How do you know that this material is derived from diatoms?

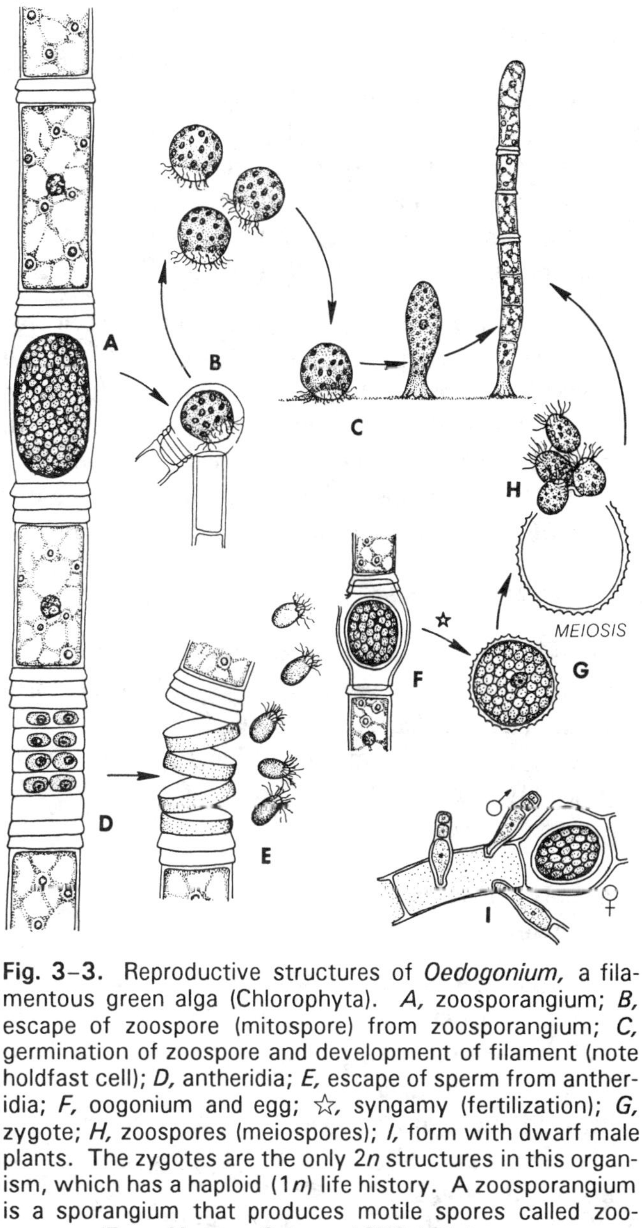

Fig. 3–3. Reproductive structures of *Oedogonium,* a filamentous green alga (Chlorophyta). *A,* zoosporangium; *B,* escape of zoospore (mitospore) from zoosporangium; *C,* germination of zoospore and development of filament (note holdfast cell); *D,* antheridia; *E,* escape of sperm from antheridia; *F,* oogonium and egg; ☆, syngamy (fertilization); *G,* zygote; *H,* zoospores (meiospores); *I,* form with dwarf male plants. The zygotes are the only 2*n* structures in this organism, which has a haploid (1*n*) life history. A zoosporangium is a sporangium that produces motile spores called zoospores. (From Norstog & Long 1976:191.)

B. Colonies

See Fig. 3–2 (also Fig. 2–1A, C, F). ***Colonies*** are groups of loosely associated cells, from few (2, 4, 8, 16) up to about 60,000. The cells may be held together only by a gelatinous sheath. Colonies occur in the blue-green bacteria (see comments above and in Lab Exercise 2, Part II) and in the green, yellow-brown, and red algae, but not in the brown algae. The cells of one colonial diatom, the "broccoli diatom" *Berkeleya,* secrete and live in tubes of mucilage aggregated into tufted thalli that resemble heads of broccoli (Chastain & Stewart 1985).

The green alga *Volvox* is not only the most humongous of all colonial forms, with up to 60,000 cells per colony, but also the most highly differentiated. The colony is a hollow sphere with all the cells in a single surface layer. The cells are held together by a gelatinous matrix, although in some species delicate protoplasmic extensions connect the cells.

●●● See Fig. 3–2. Then examine the DEMO photograph (from Carolina Biological Supply Co., 1987 calendar, or from another available source) of colonies of *Volvox.* Finally examine the DEMO (under the highest power of a dissecting microscope) of a colony on a microscope depression slide. Note the direction of the swimming and the degree of cell differentiation. The large masses of cells in the colony are daughter colonies, which are eventually liberated by enzymatic action.

C. Filaments

See Fig. 3–3 (also Fig. 2–1B, D, E, G to I). ***Filaments*** (trichomes) are chains of cells due to cell division being restricted to mainly one plane. Filaments can be simple, either unbranched or branched, or more complex types (heterotrichs) that are differentiated into an upright part and a prostrate part. Simple filaments occur in the blue-green bacteria (see comments above and in Lab Exercise 2, Part II) and in the green, brown, and red algae. Complexly branched filaments (heterotrichs) occur in the green, brown, and red algae. Sometimes the basal cell is enlarged into an anchoring cell, a ***holdfast cell*** (Figs. 3–3C, Sup4–3B).

●●● *Spirogyra* (watersilk) is a filamentous green alga that often forms bright-green, frothy, and/or slimy masses on small bodies of fresh water. This is the "pond scum" of laypersons. *Spirogyra* inspired the name of Spyro Gyra, an insipid jazz-rock fusion group formed in the late-1970s. Examine the DEMO diagram (from Fuller & Tippo 1954:563) of *Spirogyra.* Depending on the species, a cell has one to several helical, ribbon-shaped chloroplasts. Then make and examine a water mount slide preparation of *Spirogyra* and note its distinctive chloroplast(s) with many conspicuous ***pyrenoids*** (starch accumulating bodies). Some propaganda of Spyro Gyra may also be on DEMO.

●●● *Suggested diagram and labels: Spirogyra* (a green alga) filament: cells, chloroplast, pyrenoids.

D. Coenocytes, tubes, or siphons

See Fig. 3–4. ***Coenocytes, tubes, or siphons*** are basically gigantic cells that are ***multinucleate,*** that is, with many nuclei, 100 or more, and with no cross walls except when reproductive structures are formed. In contrast, types A to C and F to H typically have only one nucleus per cell. Coenocytes are usually branched. The green and yellow-brown algae are the only algae with coenocytes.

●●● Examine under a dissecting microscope the DEMO culture of *Acetabularia* (mermaid's wineglass, mermaid's parasol). This single, gigantic, multinucleate cell has a disclike cap and a long axis that in some species may be several centimeters long.

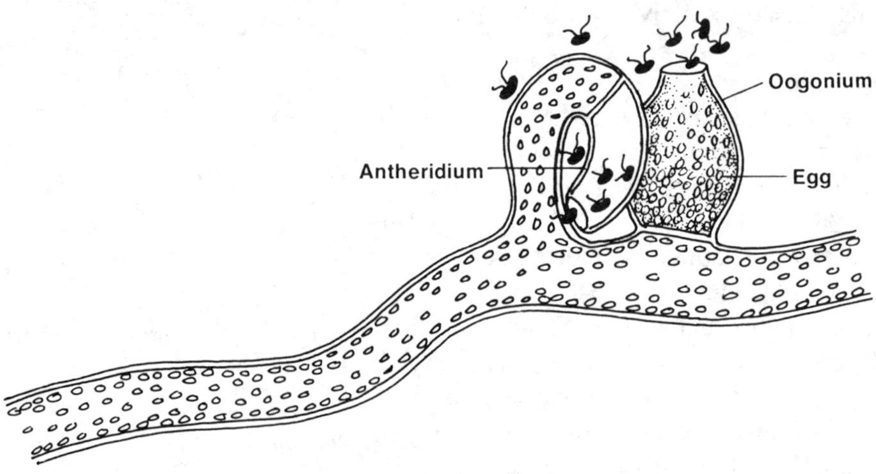

Fig. 3–4. Sexual reproduction in *Vaucheria,* a coenocytic (tubular, siphonous) golden-brown alga (Chrysophyta). (From Norstog & Long 1976:235; redrawn from *Cryptogamic botany,* 2nd ed., vol. 1, by G. M. Smith, © 1955 by McGraw-Hill Book Co., Inc., New York. Reprinted by permission of the publisher.)

E. Membranes

See Fig. 3–5. ***Membranes*** are either one cell layer thick (cell divisions in only two planes) or two cell layers thick (idem, plus one division in a third plane). Both types of membranes occur in the green and red algae, but neither type occurs in the multicellular brown algae.

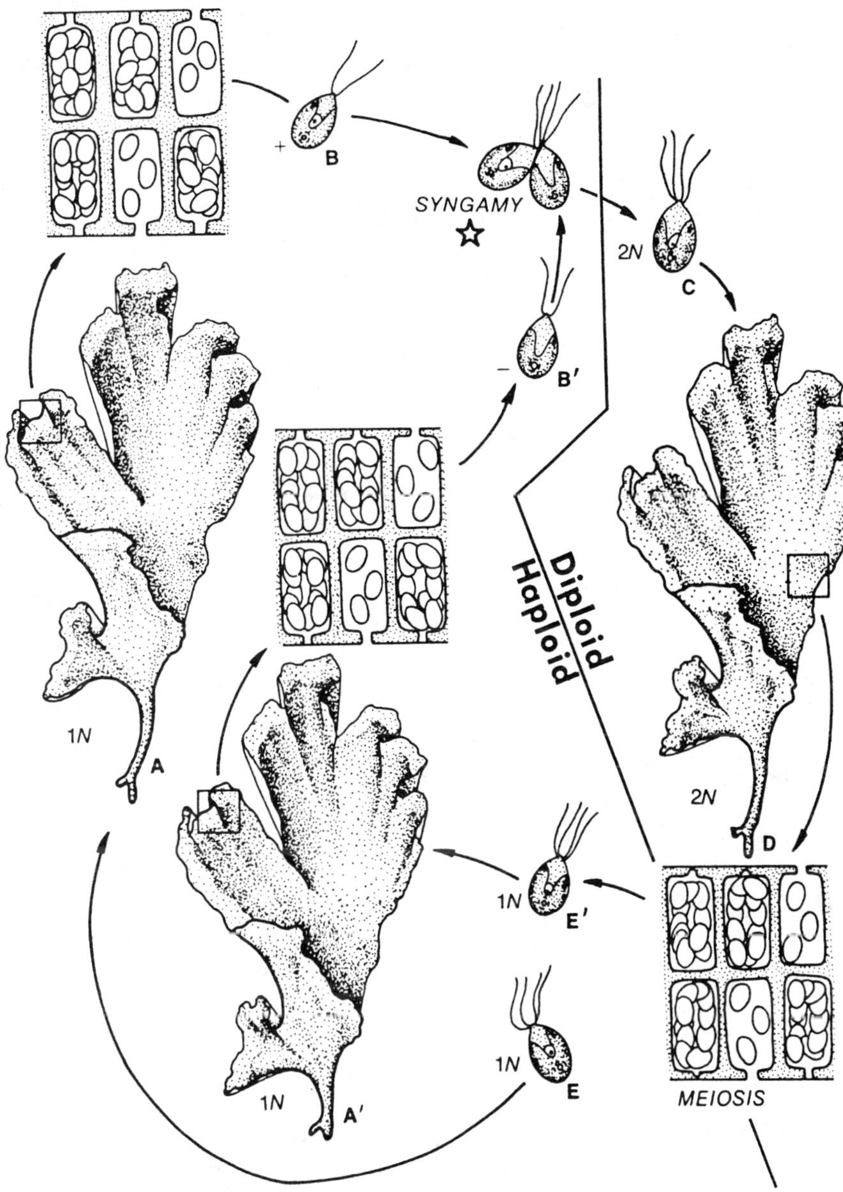

Fig. 3–5. Haploid-diploid (1*n*-2*n*) life history of *Ulva* (sea lettuce), a membranous green alga (Chlorophyta). *A, A′,* GPTs of opposite mating strains; *B, B′,* gametes of opposite mating strains; ☆, syngamy (fertilization); *C,* zygote; *D,* SPT; *E, E′,* zoospores (meiospores) of the two mating strains. Note the alternation of morphologically similar GPTic and SPTic generations (phases). (From Norstog & Long 1976:185; redrawn from *An evolutionary survey of the plant kingdom,* by R. F. Scagel et al., © 1966 by Wadsworth Publishing Co., Inc., Belmont, California. Reprinted by permission of the publisher.)

●●● See Fig. 3–5. Then examine the DEMO live material of the green alga *Ulva* (sea lettuce). The DEMO material under the compound microscope is of a freshly torn blade of *Ulva* and shows along the edge of the tear the two-layered nature of the thallus. Carefully focus (optically focus) the microscope up and down to see the thin, two-layered thallus. Cell division occurs in all three planes, but only once in one plane so that the plant body is only two cells thick.

The red alga *Porphyra* (laver, nori) is composed of one sheet of cells, or two sheets as is the green alga *Ulva.* The Japanese dry *Porphyra* and other algae for human consumption. *Porphyra* is known as "nori" in Japan and is used particularly in sushi. Two common forms of sushi using nori are "tekka maki," which is tuna and rice rolled in nori, and "kappa maki," which is tuna and cucumber rolled in nori. If you have ever eaten these, you know that the surrounding seaweed layer is papery thin, that is, membranous in nature. Incidentally, "nori" appears green because of the photodestruction of the otherwise masking reddish pigments that are characteristic of red algae.

●●● Examine the DEMO material of store-bought nori (*Porphyra*). Some nori sans sushi is available for masticatory sampling. Sorry, but the university is too poor to supply nori with sushi.

F. Pseudoparenchyma

Pseudoparenchyma is a multicellular mass of tissue resulting from the interweaving of filaments to produce a more or less three-dimensional effect. This interweaving may be loose and recognizable as such *or* it may be very tight, with a close resemblance to parenchyma (see next type). Pseudoparenchyma occurs in the red algae (most taxa) and brown algae but not in the green algae. Pseudoparenchyma differs from true parenchyma, which lacks such interweaving and instead results from cells dividing into various planes. We will examine pseudoparenchyma in Lab Exercise 12 on the fungi.

G. Parenchyma

See Figs. 3–6 and 3–7. Parenchyma has extensive cell division in all planes to result in a bulky, very multicellular structure. ***Parenchyma*** usually consists of thin-walled live cells that often have large vacuoles and that usually function in storage or photosynthesis. Parenchyma is common in the red and especially brown algae but does not occur in the green algae. Many of the large brown algae, for example, *Fucus* (rockweed; Fig. 3–7), *Postelsia* (sea palm), and *Laminaria* (oarweed) (Fig. 3–6) and other kelps (see below) are mostly or entirely parenchymatous. Parenchyma

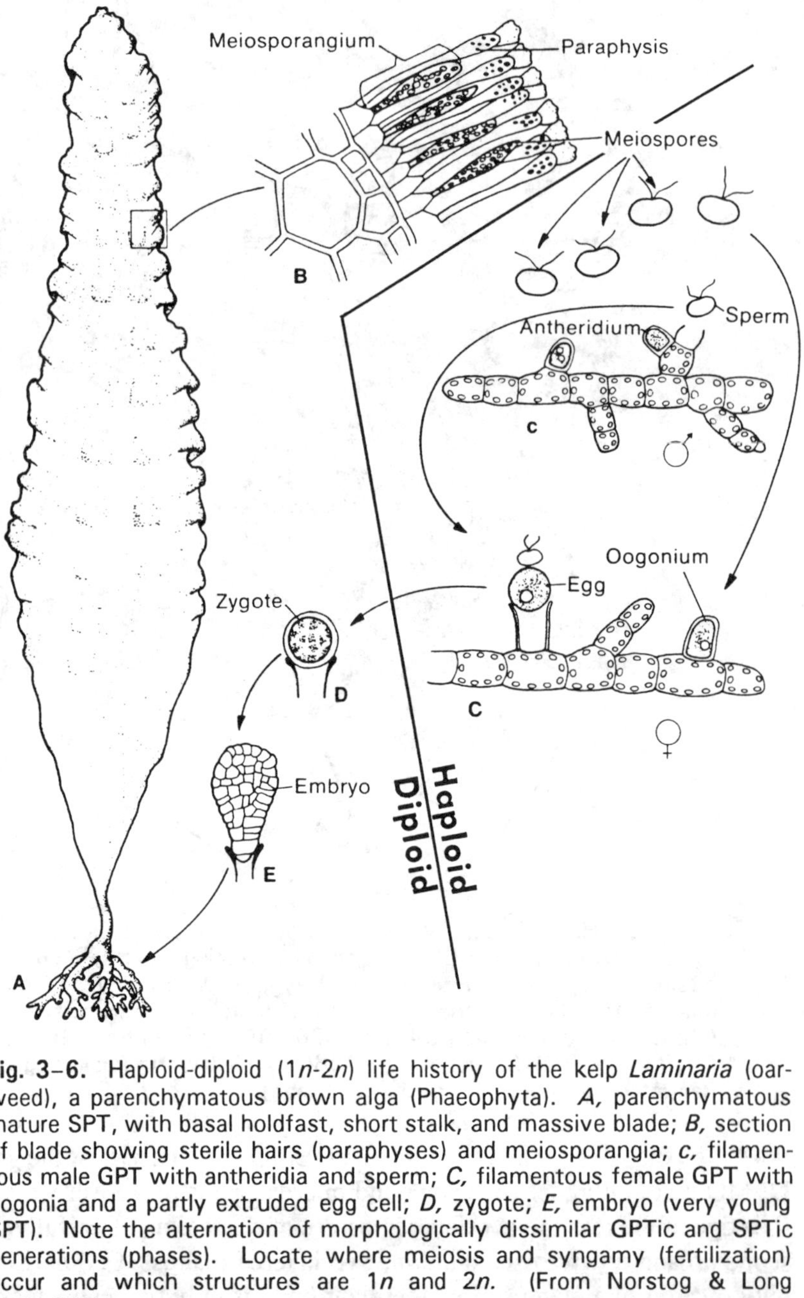

Fig. 3–6. Haploid-diploid ($1n$-$2n$) life history of the kelp *Laminaria* (oarweed), a parenchymatous brown alga (Phaeophyta). *A,* parenchymatous mature SPT, with basal holdfast, short stalk, and massive blade; *B,* section of blade showing sterile hairs (paraphyses) and meiosporangia; *c,* filamentous male GPT with antheridia and sperm; *C,* filamentous female GPT with oogonia and a partly extruded egg cell; *D,* zygote; *E,* embryo (very young SPT). Note the alternation of morphologically dissimilar GPTic and SPTic generations (phases). Locate where meiosis and syngamy (fertilization) occur and which structures are $1n$ and $2n$. (From Norstog & Long 1976:205.)

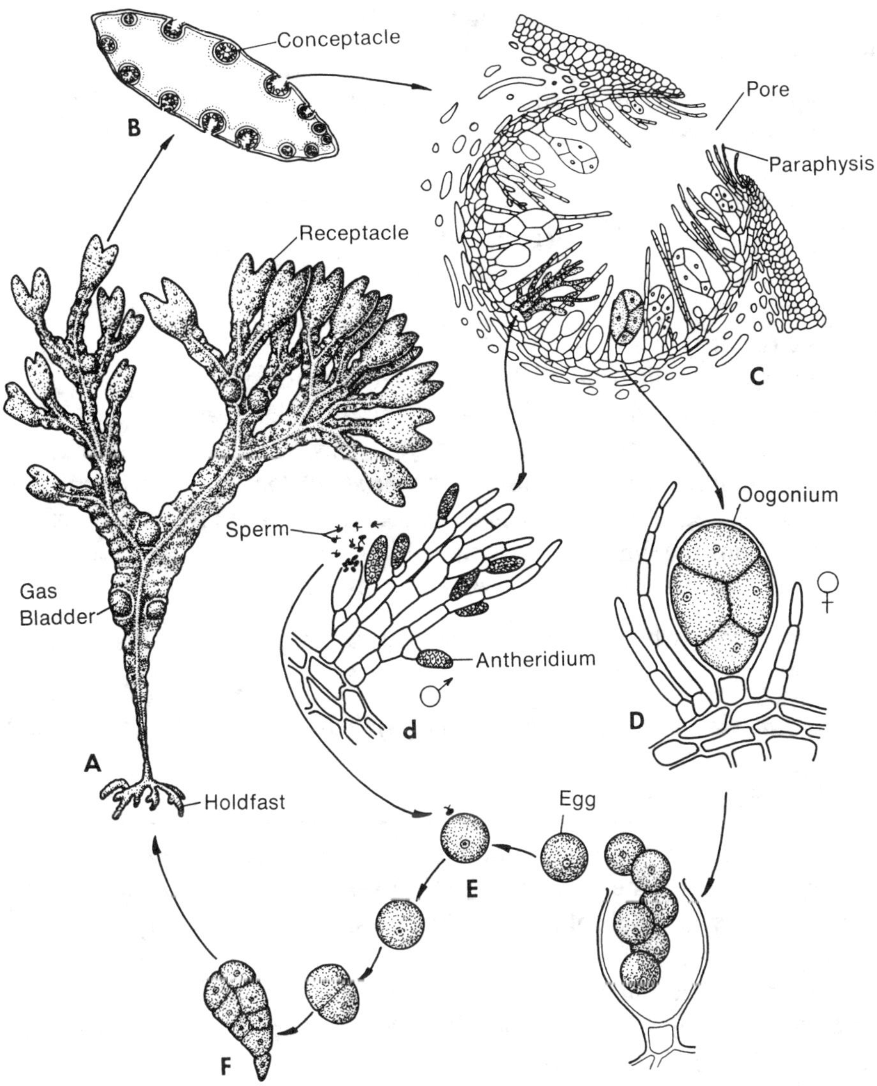

Fig. 3-7. Diploid (2*n*) life history of *Fucus* (rockweed), a parenchymatous brown alga (Phaeophyta). *A,* 2*n* plant with reproductive branch tips (receptacles); *B,* transection of receptacle showing many cavities (conceptacles) with external pores; *C,* longisection of bisexual conceptacle with male and female gametangia (antheridia and oogonia) and sterile filaments (paraphyses); *d, D,* longisections of antheridial branch, each antheridium with 64 sperm (*d*), and oogonium with eight eggs, four shown (*D*); *E,* zygote; *F,* embryo (very young SPT). The gametes are the only 1*n* structures in this organism. Locate where meiosis and syngamy (fertilization) occur and which structures are 1*n* and 2*n*. (From Norstog & Long 1976:208; redrawn from *Cryptogamic botany,* 2nd ed., vol. 1, by G. M. Smith, © 1955 by McGraw-Hill Book Co., Inc., New York. Reprinted by permission of the publisher.)

is also typical of the land plants, that is, bryophytes and vascular plants. We will examine much parenchyma in subsequent labs on the land plants (Lab Exercises 4 to 11).

H. Parenchyma plus phloem

A few brown algal seaweeds have tissue differentiation into parenchyma, representing most of the body, and ***phloem,*** specialized nutrient-conducting tissue. Large brown algal seaweeds such as the kelps *Macrocystis* (giant kelp) and *Nereocystis* (bull kelp) may grow to 60 m (197 ft.) or more in length. The lower reaches of the plant, the stalk (stipe) and holdfast (see below), may be poorly lit compared to the upper parts of the kelp, with concomitant low rates of photosynthesis. The kelps are then able to transport carbohydrates from the blades to the lower parts of the plant by means of specialized phloem cells.

●●● Examine the DEMO herbarium specimens or, possibly, live material of sporophytes (SPTs) of the kelps *Macrocystis* and/or *Nereocystis.*

I. Vegetative morphology

The vegetative body of many algae is differentiated into

- an attaching ***holdfast,***
- a lengthened ***stalk*** (stipe),
- a flattened ***blade*** (lamina), and
- one or more enlarged, air-filled ***floats*** (bladders, air vesicles, pneumatocysts).

These structures are especially common in the larger multicellular seaweeds (types E to H above; Figs. 3–5 to 3–7), although filaments and coenocytes may also have a holdfast (Figs. 3–3C, Sup4–3B). An area of dividing cells (an intercalary ***meristem***) located between the blade and stalk of kelps allows the blade to regenerate if it is eaten or harvested (this is like the situation in grasses). Holdfasts, stalks, and blades are functionally comparable to roots, stems, and leaves of the vascular plants but are not called such because they lack vascular tissue.

●●● Identify the aforenoted parts on the various large algae in the preceding sections as well as on any other seaweeds available for DEMO.

●●● *Suggested diagram and labels:* Kelp (a brown alga) thallus (SPT): holdfast, stalk, blade, float.

II. SAMPLE LIFE HISTORIES

See Supplement 5 for background information on the types of life histories. All of the algal divisions have a sexual life history, except the euglenoids, where sexual reproduction has never been discovered. The algae vary greatly in details of their life histories. All three main types occur in the algae:

- the haploid (1*n*) life history (Figs. 3–2, 3–3, Sup5–1), which is the most common and involves both unicellular and multicellular algae;
- the haploid-diploid (1*n*-2*n*) life history, which is characterized by two multicellular phases, the GPT and SPT (Figs. 3–5, 3–6, Sup5–3);
- the diploid (2*n*) life history (Figs. 3–1, 3–7, Sup5–2), which is the least common, occurring in the diatoms and *Fucus,* which are, respectively, unicellular and multicellular organisms. In diatoms (Fig. 3–1) meiosis occurs after an approximately 70% size reduction of the protoplast.

The life history of some red algae can be obscenely complicated, although it is really a special version of the haploid-diploid (1*n*-2*n*) life history with a secondary 2*n* phase.

A. Haploid life history

Figures 3–2 and 3–3 show a haploid (1*n*) life history of, respectively, colonial and filamentous algae.

B. Diploid life history

Diatoms (Fig. 3–1) and *Fucus* (Fig. 3–7) are two algal groups with a diploid (2*n*) life history.

C. Haploid-diploid life history

The haploid-diploid (1*n*-2*n*) type of life history has an alternation of generations (phases) with either:

- similar SPTs and GPTs that are morphologically identical except for their reproductive structures (Fig. 3–5);

- dissimilar SPTs and GPTs that differ in both vegetative and reproductive morphology, with two variations:
  - the GPT is larger or more conspicuous than the SPT;
  - the SPT is larger or more conspicuous than the GPT (Fig. 3-6).

Ulva, Porphyra, and kelps, respectively, exemplify these three variations.

●●● See Fig. 3-5. Then reexamine the DEMO material of *Ulva* (sea lettuce) from Part I-E. *Ulva* has a haploid-diploid (1*n*-2*n*) life history with similar SPTs and GPTs. Without microscopic examination (do *not* do this!), however, one can not tell whether the material is GPTic or SPTic.

Porphyra also has a haploid-diploid (1*n*-2*n*) life history, but with SPTs and GPTs that are morphologically dissimilar. The GPTic phase of *Porphyra* is the familiar membrane called nori. The SPTic phase is a filament, called *Conchocelis,* that grows on rocks (hence the latter name). For a long time it was not known that *Conchocelis* and *Porphyra* were alternate phases of the same plant. This is why the plant has two names.

●●● See Fig. 3-6. Kelps also have a haploid-diploid (1*n*-2*n*) life history with alternation of dissimilar generations (phases). The SPT is the massive structure previously examined in Part I-H whereas the GPT is a small branched filament.

In *Ulva,* to recapitulate the haploid-diploid (1*n*-2*n*) life history (Fig. 3-5):

- a 2*n* multicellular SPT produces sporangia (Fig. 3-5D), where *meiosis* occurs to produce + and − 1*n* flagellate spores (zoospores, meiospores) (Fig. 3-5E, E′) that are physiologically distinct (of opposite mating strains) but morphologically identical (homosporous);
- each + spore produces by *mitosis* a + 1*n* multicellular GPT (Fig. 3-5A);
- each − spore produces by *mitosis* a − 1*n* multicellular GPT (Fig. 3-5A′);
- the + and − 1*n* GPTs produce gametangia, where *mitosis* occurs to produce + and − flagellate 1*n* gametes (Fig. 3-5B, B′) that are physiologically distinct (of opposite mating strains) but morphologically identical (isogamous);
- a + and a − 1*n* gamete fuse, that is, engage in *syngamy* (fertilization), to form a 2*n* zygote (Fig. 3-5C), which undergoes a resting period;
- the zygote eventually germinates and develops into a new 2*n* SPT (Fig. 3-5D), which begins the life history anew.

Note also these points about the haploid-diploid (1*n*-2*n*) life history of *Ulva* (Fig. 3-5): There are three multicellular plants, a SPT, a + GPT, and a − GPT, which constitute a population of *Ulva.* Most species exhibit anisogamy (Bold & Wynne 1985:197), but Fig. 3-5 shows isogamy. Both the gametes and the spores are motile due to their flagella, two occurring on each gamete (Fig. 3-5B, B′), four on each spore (Fig. 3-5E, E′). Motile spores are called ***zoospores.***

Important note: From the perspective of this course manual, the details of the life histories *of algae* above and in the figures are *not* especially important. What *is* important are the effects of meiosis, syngamy (fertilization), and mitosis on the ploidy level of various structures in the life history, to wit:

- 2*n* structures ——— meiosis ———> 1*n* structures
- 1*n* structures ——— syngamy ———> 2*n* structures
- 2*n* structures ——— mitosis ———> 2*n* structures
- 1*n* structures ——— mitosis ———> 1*n* structures.

For elaboration see Supplements 4 and 5.

NAME ______________________________ LAB SECTION # ______________ DATE ____

TITLE ______________________________

__________________ TOTAL MAG ____

TITLE ______________________________

__________________ TOTAL MAG ______

TITLE ______________________________

__________________ TOTAL MAG ____

TITLE ______________________________

__________________ TOTAL MAG ______

SUPPLEMENT **6**

Characteristics of and Origin of the Land Plants

(Bryophytes versus Vascular Plants)

I. PERSPECTIVE

The bryophytes and the vascular plants comprise the so-called ***land plants.*** Although they share some common characteristics, the two groups are fundamentally different and probably evolved independently.

II. UNIFYING CHARACTERISTICS OF ALGAE AND THE LAND PLANTS

The green algae (Chlorophyta) have all the chemical and structural prerequisites to have been the ancestral group of the land plants, that is, bryophytes and vascular plants. Of the various divisions of algae, the green algae with their approximately 450 genera and 6,750 species exhibit the greatest diversity in form and in life history. The green algae, bryophytes, and vascular plants all have:

- chlorophyll *a* and *b* and similar accessory photosynthetic pigments;
- starch as the nutrient reserve;
- cellulosic, at least polysaccharidic, cell walls.

In addition, at least *some* green algae and *all* land plants have:

- a multicellular body;
- oogamy, gametes differentiated as egg and sperm;
- a haploid-diploid ($1n$-$2n$) life history (see Fig. Sup5–3), with alternation of dissimilar GPTic and SPTic generations (phases).

It is thought that the land plants evolved from some green alga (presumably of aquatic or semiaquatic habitat) of the filamentous or membranous/parenchymatous type. On structural grounds it is highly unlikely that a coenocytic (tubular, siphonous) or colonial type could have been the evolutionary progenitor of the land plants due to various structural limitations imposed by the coenocytic form or the spherical colonial form.

III. UNIFYING CHARACTERISTICS OF BRYOPHYTES AND VASCULAR PLANTS

Bryophytes and vascular plants differ from fungi and the so-called lower plants, the algae, in having:

- 1. multicellular ***gametangia*** (sex organs or gamete-producing organs; male and female gametangia are called, respectively, ***antheridia*** and ***archegonia***) with a sterile jacket surrounding the fertile cells (gametes);

- 2. as noted above, gametes always differentiated as egg and sperm (oogamy); if gametangia occur, there is always one egg but many sperm per gametangium;
- 3. the zygote and very young SPT (i.e., the embryo) retained in the archegonium in the pteridophytes, or in the seed in the seed plants;
- 4. as noted above, exclusively the haploid-diploid (1*n*-2*n*) type of life history (see Figs. Sup5-3 to Sup5-5), and hence always alternation of dissimilar GPTic and SPTic generations (phases).

Features 1 and 3 are clearly protective ones that presumably were significant in the invasion of land by the green algal progenitors of the land plants.

IV. DISTINGUISHING CHARACTERISTICS OF BRYOPHYTES AND VASCULAR PLANTS

Bryophytes differ from the other land plants, the vascular plants, in some important respects:

Extant Bryophytes	*Extant Vascular Plants*
Life histories in Figs. 4-1, 4-3, Sup5-4	Life histories in Figs. 5-1, 5-3, 5-4, 5-5, 8-1, 8-2, 9-4, Sup5-4, Sup5-5
The GPT the conspicuous and dominant generation (phase), with the SPT permanently dependent on the GPT	The SPT the dominant generation (phase), eventually totally independent of the GPT
Vascular tissue absent, although some specialized mosses containing water- and nutrient-conducting cells analogous (functionally equivalent) to xylem and phloem	Vascular tissue present, i.e., water-conducting xylem and nutrient-conducting phloem (see Lab Exercise 6, Part II-D)
True stems, leaves, and roots absent	True stems, leaves, and roots present
Homosporous, producing only one morphological type of spore	Homosporous and mostly ***heterosporous,*** producing two morphological types of spores

V. ORIGIN OF THE LAND PLANTS

In view of the above profound differences, the bryophytes and vascular plants probably evolved independently from their green algal ancestors. The first undisputed fossils of land plants are about 410 million years old. The oldest proven vascular plant, *Cooksonia,* occurs in rocks of this age (Middle Silurian) in Czechoslovakia, Great Britain, and New York. In contrast, the oldest fossil bryophytes appear to be somewhat younger (i.e., more recent) in origin; the oldest specimens date to only about 375 million years ago (Middle and Upper Devonian) and are tentatively assigned to the Marchantiales (Oostendorp 1987), a group of liverworts with mainly extant representatives (see Lab Exercise 4, Part II). Another very ancient plant is *Rhynia* (Fig. Sup6-1), which occurred in the Lower Devonian. *Cooksonia* and *Rhynia* are structurally very simple and not that different from some algae except that they *do* have vascular tissue. Basically, both genera are all stem and sporangia, that is, without roots and leaves. Figure Sup6-2 shows *Rhynia* and other primitive plants in a reconstruction of a Devonian landscape.

Life originated in water. Because the oldest undisputed fossils of land plants date from the Middle Silurian 410 million years ago, the assumption is that land was initially colonized by plants sometime (perhaps even in several attempts) in the Silurian, which was 395-430 million years ago. Precisely how or why plants advanced to the land has been and probably always will be the subject of considerable debate. However, one thing is certain. The first land plants faced environmental stresses totally different from those encountered by their aquatic ancestors. Water, of course, was the critical factor. Water to an aquatic plant effects a number of things:

- a means of support (i.e., size on land may be a limiting factor);
- a source of nutrients (i.e., an aquatic organism is constantly bathed in a nutrient solution and thus nutrients can be easily absorbed by any surface cell);
- a continual supply of moisture and concomitant lack of danger of desiccation (drying out);
- the capability of eliminating waste products by simple diffusion.

[One might also add, protection from forest fires.]

For successful life on land, plants had to evolve a variety of anatomical and morphological adaptations relevant to the aforementioned limitations of a terrestrial environment. These adaptations are detailed in Lab Exercise 6, Part II, and Lab Exercise 11, Part II. In brief, these adaptations involve:

- supportive structures, that is, supportive tissues (xylem, phloem, collenchyma, sclerenchyma) and other, mainly morphological specializations to support the land plant;
- nutrient absorption by roots and other nutrient absorptive mechanisms (note that the rhizoids encountered in bryophytes and extant pteridophytes are GPTic structures, *not* SPTic ones);
- moisture retention, that is, cuticle and other water-proofing layers (periderm, spore walls) to retain water inside the organism;
- waste elimination, that is, stomatal pores in the epidermis, and other means, for diffusion of gases between the inside and the outside of the organism.

Land plants have at least some of these features, and many plants have all of them. The Devonian plant *Rhynia* (Fig. Sup6–1) had most of these features, but it lacked roots and had rhizoids instead. Plants living under moist conditions, as in swamps, usually have these features less accentuated, whereas plants living under dry conditions, as in deserts, generally have them more accentuated.

The evolution of land plants was not a linear sequence. The various modifications probably originated separately and at different rates. In addition, different evolutionary strategies or lines of evolution involved the GPT as opposed to the SPT. However, because we know essentially nothing of the GPTs of fossil plants (the GPTs of the extinct bryophytes resemble the GPTs of the extant bryophytes), most discussion pertains specifically to the SPT. The GPT of especially the bryophytes and pteridophytes lives under moist conditions, and here the sperm must still swim to the eggs. Consistently, the GPTs of the pteridophytes are smaller and structurally simpler than their SPTs.

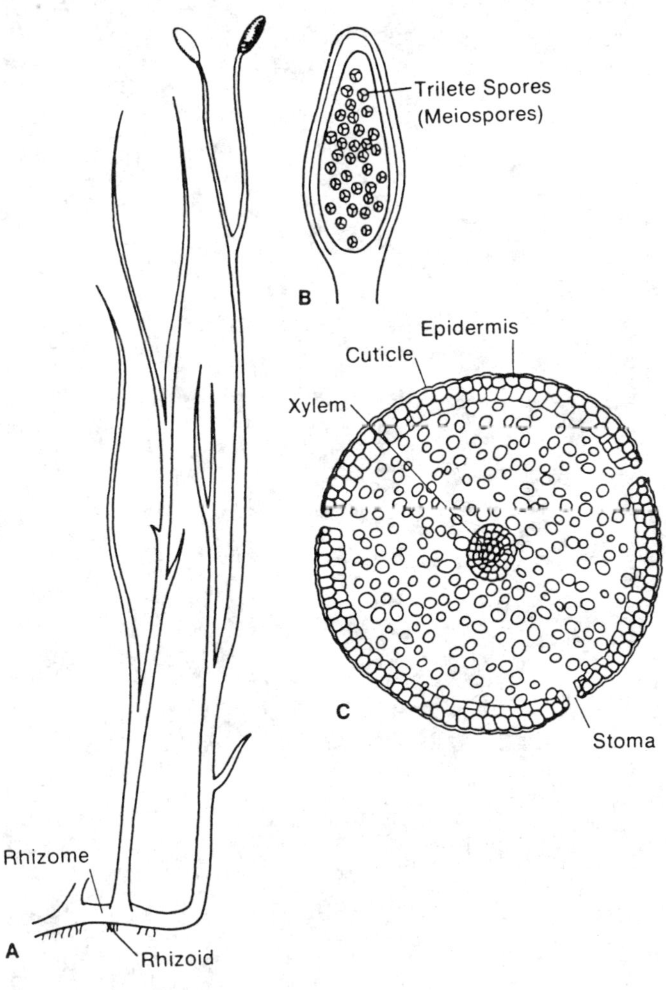

Fig. Sup6–1. Reconstruction of *Rhynia,* a Lower Devonian plant. *A,* habit; *B,* longisection of sporangium with spores; *C,* transection of stem. (From Norstog & Long 1976:319; *A* redrawn from *Morphology of vascular plants,* by D. W. Bierhorst, © 1971 by The Macmillan Co., New York. *B, C* redrawn from *Cryptogamic botany,* 2nd ed., vol. 2, by G. M. Smith, © 1955 by McGraw-Hill Book Co., Inc., New York. Reprinted by permission of the publishers.)

Fig. Sup6–2. Reconstruction (by Priscilla Fawcett) of a Devonian landscape. The plants in the foreground by the lobefin fish (sarcopterygians) are *Rhynia*. Primitive lycopods, horsetails, and ferns occur in the background. (From Norstog & Long 1976:312.)

LAB EXERCISE **4**

Mosses and Other Bryophytes

OBJECTIVE

To examine the vegetative and reproductive structure of representative examples of mosses and other bryophytes.

Note: Read this material *before* coming to the lab; there is insufficient time in the lab to read the material from scratch. Due to the large amount of background information presented here, a ●●● is used to mark those paragraphs calling for actual observations or diagrams in the lab. Most unmarked paragraphs reiterate information presented in the lecture. A bullet (●) signifies an important term or concept.

Suggested Lab Diagrams (see part noted for suggested labels):
Marchantia polymorpha (a liverwort) thallus (GPT) (Part II)
Polytrichum (a moss) (or *Funaria* or similar moss) GPT and SPT (Part III)

PERSPECTIVE

The bryophytes and the vascular plants comprise the so-called land plants. Although they share some common characteristics, the two groups are fundamentally different and probably evolved independently. As discussed in Supplement 6, Part II, the land plants most likely evolved from the green algae.

The bryophytes comprise a single division (Bryophyta), with three main groups (taxonomic classes) containing over 22,400 extant species according to Scagel et al. (1982:443):

1. liverworts (Hepaticopsida)—8,000 species;
2. hornworts (Anthocerotopsida)—100 species;
3. mosses (Muscopsida)—over 14,300 species.

We will not examine the hornworts, although they are structurally a most interesting group.

These three groups of bryophytes have been treated as taxonomic classes within the Bryophyta or as separate divisions, namely, Hepatophyta, Anthocerotophyta, and Bryophyta. Bold et al. (1987), for instance, find the sum total of the 14 features (there are others) they tabulate (p. 311) so impressive that they treat each class as a separate division. In their opinion (Bold et al. 1980:275, a less strong statement in 1987:312) "there is only a single unique major attribute uniquely characteristic of both liverworts [including hornworts] and mosses: All these plants are similar in their life cycle, which involves the regular alternation of a free-living gametophyte and a sporophyte that is permanently attached to the gametophyte." The latest editions of the basic textbooks by Raven & Johnson (1992) and Raven et al. (1992) also accept three divisions of bryophytes, whereas their earlier editions accepted only one division.

Unfortunately, almost all of the distinguishing characters of these three groups overlap. Schofield (1985:1–2), for example, lists ten features shared by all bryophytes. Because of these overlapping features, this course manual and Schofield's (1985) recent book on bryophytes follow the more conservative and traditional approach of a single division Bryophyta for all three classes.

Bryophytes and pteridophytes have a strictly haploid-diploid ($1n$-$2n$) type of life history (see Figs. Sup5–3 to Sup5–5). The alternation of dissimilar GPTic and SPTic generations (phases) in the bryophytes involves a GPT dominant over the SPT, the SPT always dependent on the GPT. In contrast, the pteridophytes have the SPT dominant over the GPT and ultimately independent of it.

Compared to the vascular plants, the bryophytes lack true stems, leaves, and roots. However, the GPTs of most bryophytes have leaflike and stemlike structures. Because these are GPTic rather than SPTic in nature, and because they lack vascular tissue, the bryophytes technically do not have true leaves or true stems. The leaflike and stemlike structures are nevertheless comparable or analogous to leaves and stems, and thus these terms are commonly used in discussions of bryophytes.

Note: Before beginning this lab exercise, you should review:
- Supplement 5 on the main types of life histories;
- Supplement 6 on the characteristics of and origin of the land plants.

I. GENERAL ON BRYOPHYTES

Mosses and liverworts have attracted much attention because of their beauty and intricacy. The mosses, liverworts, and the moss *Sphagnum* (peat moss) differ in various technical respects.

●●● Examine the exceptionally fine pictures in the DEMO book by Johnson (1983). Although this book is titled "mosses," it also deals with liverworts. Incidentally, this book is actually a children's book intended for ages ten to 12!

Moisture is a critical factor for two vital processes in the life histories of bryophytes *and also* pteridophytes:
- For syngamy (fertilization) moist or humid conditions are *good,* but dry conditions are *bad.*
- For spore dispersal, however, dry conditions are *good,* but moist or humid conditions are *bad.*

Thus moist or humid conditions are needed for syngamy because the sperm have to swim to the eggs. However, for spore dispersal the converse conditions apply because ideally the spores should be spread as far away from the parent plant as possible, and this is much more likely under dry rather than moist or humid conditions.

Many spore discharge mechanisms thus are ***hygroscopic*** or moisture sensitive, that is, being affected by moisture uptake or moisture loss (note that general dictionaries define "hygroscopic" as only moisture uptake). Such mechanisms typically involve dead cells and hence are reversible. Other non-hygroscopic mechanisms of spore discharge exist. A water-pressure or so-called ***turgor*** mechanism typically involves live cells actively taking in water and thus is *not* reversible.

Perhaps the main difference between liverworts, mosses, and *Sphagnum,* at least for the purpose of this course manual, is their mechanism of spore dispersal, as summarized here and described below:

	LIVERWORTS	"TRUE MOSSES," *Andreaea*	PEAT MOSSES
Discharge mechanism	Hygroscopic, cells dead	Hygroscopic, cells dead	Turgor, cells alive
Cover (operculum)	Absent	Mostly present*	Present
Peristome teeth	Absent	Mostly present*	Absent
Elaters	Mostly present	Absent	Absent

**Andreaea* (Fig. 4–4D, E) and its relatives lack a cover and a peristome.

●●● See Fig. 4–4. Then examine the DEMO diagrams (from Raven et al. 1981:301, 1986:287; the figure in Raven et al 1992:306 is not as good) of spore dispersal in bryophytes as you examine the live material in the following parts of this lab exercise.

II. LIVERWORTS

The liverworts include both leafy and leafless (thallose, non-leafy) types. *Marchantia polymorpha,* which is structurally not typical of liverworts, is a leafless liverwort of worldwide distribution. This species grows on moist to wet soils and is characteristic of recently burnt and cleared sites.

●●● Consult Fig. 4–1 of the haploid-diploid ($1n$-$2n$) life history (see also Figs. Sup5–3 and Sup5–4) of the liverwort *Marchantia* as you examine the material in the following sections.

●●● *Suggested diagram and labels: Marchantia polymorpha* (a liverwort) thallus (GPT): thallus, rhizoids, gemma cups, gemmae, antheridiophores, archegoniophores. Note where the SPTs form!

A. Gametophytes (GPTs)

As with all bryophytes, in the liverworts the GPT is the conspicuous and dominant generation (phase), with the SPT permanently dependent on it. In contrast, in the vascular plants the SPT is the dominant generation and is eventually totally independent of the GPT.

●●● See Fig. 4–2. Then examine the live material of *Marchantia,* including a very small piece under a dissecting microscope. The GPT or thallus branches evenly (dichotomously) and has a flattened structure. Note the conspicuously outlined, polygonal areas visible on the upper (dorsal) surface of the thallus. These areas represent the air chambers located just beneath the surface of the thallus. The center of each air chamber has a very prominent barrel-shaped air pore (analogous to a stoma—see Lab Exercise 6, Part II-B1) that connects the chamber with the external environment.

●●● Now observe the lower (ventral) surface of the thallus. This bears ***rhizoids,*** delicate anchoring, absorbing structures. There are two types of rhizoids (Fig. 4–2), but these may not be clearly distinguishable in the material available:

1. One type is unicellular and smooth-walled, grows vertically directly into the soil, and thus functions in anchorage. Possibly some water and nutrient absorption occurs via these rhizoids. However, most or all absorption occurs directly and rapidly through the thallus.
2. The other type consists of unicellular but rough-walled rhizoids that run horizontally in association with scales and apparently function like a wick for horizontal water conduction.

The GPTs of *Marchantia polymorpha* are strictly unisexual. Unlike many liverworts, *Marchantia* produces its gametangia or sex organs on special elevated structures:

- ***antheridiophores*** (male gametophores, antheridial branches), male stalked structures resembling umbrellas and bearing many erect antheridia sunken internally in cavities connected by a narrow canal to a surface pore (Fig. 4–1a, b);
- ***archegoniophores*** (female gametophores, archegonial branches), female stalked structures resembling palm trees and bearing many archegonia externally (Fig. 4–1A, B).

The archegoniophores bear archegonia exposed on their apparent bottom surfaces [Overgrowth on the upper (dorsal) surface of the archegoniophore displaces the archegonia to its lower (ventral) surface]. The antheridia and archegonia of liverworts and mosses are structurally similar (compare Figs. 4–1b, B, and 4–3a, A). Like all land plants, bryophytes have oogamy and, like the pteridophytes, require moisture for syngamy (fertilization) because sperm have to swim to the eggs.

●●● See Fig. 4–1a, b, A, B. Then examine the live material of *Marchantia* for the male and female reproductive structures described above. Note that if reproductive material is sparse, it will be supplied as a DEMO under a dissecting microscope.

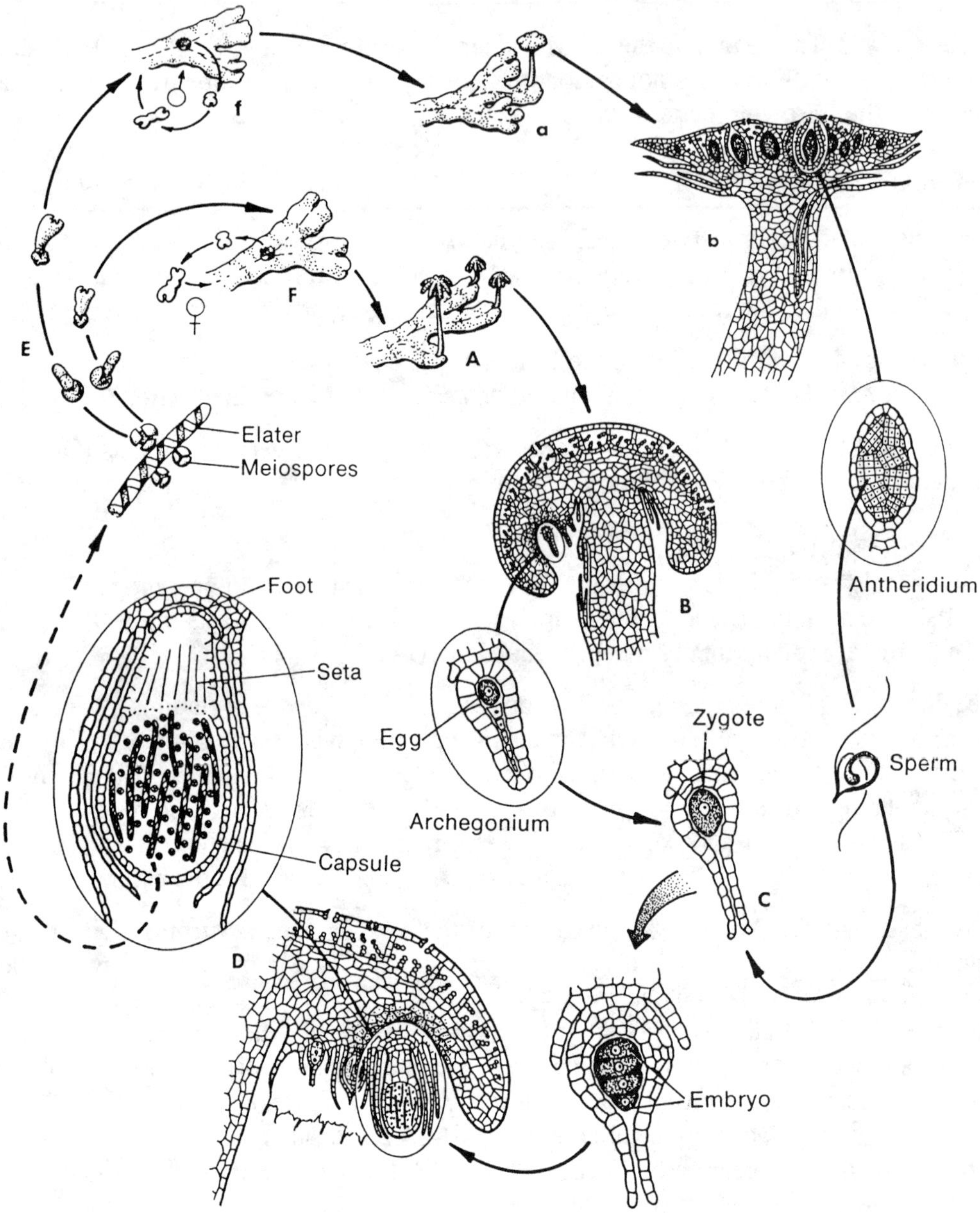

Fig. 4–1. Haploid-diploid (1*n*-2*n*) life history of the liverwort *Marchantia*. *a,* male GPT with antheridiophore (antheridial branch); *A,* female GPT with archegoniophores (archegonial branches); *b, B,* longisections of antheridiophore (*b*) and archegoniophore (*B*); *C,* longisection of archegonium with zygote; *D,* longisection of archegoniophore with young and mature SPTs (note in the sporangia the intermingled elaters and spores); *E,* germinating spores; *f,* male and, *F,* female GPTs reproducing asexually (vegetatively) by gemmae prior to initiation of sex organs. Note the alternation of morphologically dissimilar GPTic and SPTic generations (phases), with the GPT dominant over the SPT, the SPT always dependent on the GPT. Locate where meiosis and syngamy (fertilization) occur and which structures are 1*n* and 2*n*. (From Norstog & Long 1976:305.)

B. Sporophytes (SPTs)

See Fig. 4–1C, D. After syngamy (fertilization), the zygote in the archegonium develops into a SPT that remains permanently attached to the GPT. The SPT is smaller than the GPT and consists of several parts:

- an anchoring, absorbing ***foot*** (this permanently anchors the SPT to the GPT!);
- a rather short ***stalk*** (seta);
- a small ***capsule*** (spore capsule, spore case) or ***sporangium*** with several parts:

- a thin enclosing ***wall*** (2*n*) that is only one layer thick;
- many 1*n* ***spores*** that had resulted from meiosis;
- ***elaters,*** 2*n* elongate cells with helical wall thickenings (as shown in Fig. 4–1D, the elaters are intermingled with the spores).

The SPT is enveloped in GPTic tissue, that is:

- a ***hood*** (calyptra, cap), which represents the enlarged remnants of the archegonium and which thus is actually 1*n;*
- proliferations of other parts of the GPT.

The hood and associated proliferations of the GPT initially prevent the developing sporangium from drying out but eventually these GPTic parts rupture due to limited elongation of the stalk of the SPT and are shed prior to spore release. The sporangium ***dehisces*** (opens up) irregularly into several segments bearing clusters of elaters and spores. The elaters are hygroscopic and aid in spore dispersal by twisting and turning with changes in moisture, that is:

- moist or humid conditions = elaters straight, spores attached;
- dry conditions = elaters twisted, spores released, wafted off.

The pressure of the moving elaters effects both gradual dehiscence of the sporangium and a loosening of the spore mass (Bold et al. 1987:209).

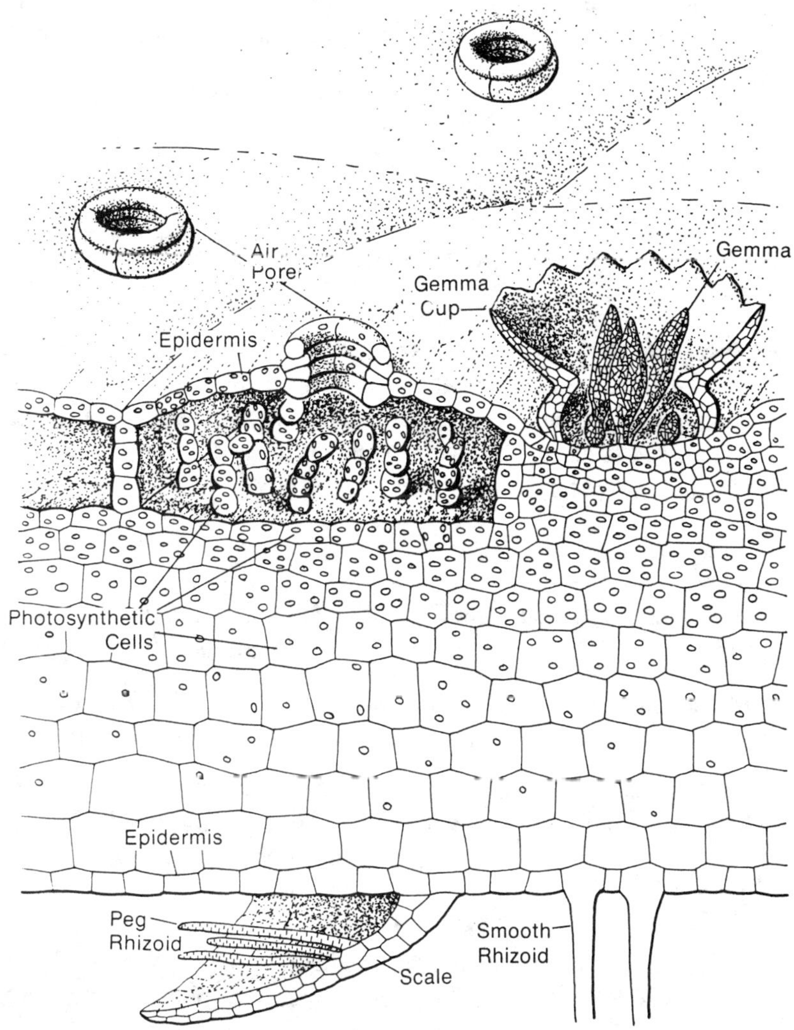

Fig. 4–2. Transection of the GPT of the liverwort *Marchantia,* with cell types illustrated. Note the gemmae responsible for asexual (vegetative) reproduction. (From Norstog & Long 1976:304.)

●●● Examine the live GPTic material of *Marchantia;* this should have SPTic capsules (sporangia). Again, if reproductive material is sparse, it will be supplied as a DEMO under a dissecting microscope. Mature SPTs are yellowish and the sporangia may have dehisced and shed their spores. Unopened sporangia examined under a dissecting microscope may well dehisce. This may take a while. Heat helps. Why?

C. Asexual reproduction

Many of the land plants reproduce asexually (vegetatively) by one or more means (see Lab Exercise 1, Part V). Asexual reproduction can involve either the SPT or the GPT (Figs. Sup5-3 to Sup5-5). The advantages of asexual (vegetative) reproduction (see also Supplement 4, Part V) over sexual reproduction include:

- rapid and immense increase in the numbers of individuals;
- the capability (in the bryophytes and pteridophytes) of bypassing the somewhat lengthy and moisture-dependent sexual process (motile sperm are a limiting factor in times of moisture stress).

Asexual reproduction in *Marchantia* occurs by fragmentation of the thallus and by discrete units of vegetative tissue, the ***gemmae*** (Figs. 4–1f, F, 4–2). Gemmae are produced in ***gemma cups*** and on germination develop into entirely new plants. The gemmae are splashed out of the gemma cups by falling rain drops and may be ejected for distances up to 160 cm (63 in.) (Schmid, unpublished).

••• See Figs. 4–1f, F, and 4–2. Find gemma cups and gemmae in the live material of *Marchantia.* You might test dispersal of gemmae by dropping water onto the gemma cups using a water dropper. What is the ploidy level of the gemmae and their genetic nature compared to the rest of the plant?

D. Other liverworts

The roughly 8,000 species of liverworts have very divergent morphology. *Marchantia* is not really typical of the liverworts.

••• Some other liverworts may be available on DEMO, for example: *Lunularia,* which is rare in nature but common in greenhouses (Bold et al. 1987), has moon-shaped gemma cups (hence the generic name). *Conocephalum* lacks gemma cups but does produce gemmae at the tips of the thallus. *Ricciocarpus* is an aquatic form; some species float on water whereas other species are submerged just below the water surface.

III. "TRUE MOSSES"

As noted in the "Perspective," "leaf" and "stem" are commonly used for analogous structures in the bryophytes, even though technically they do not have true leaves or true stems (they also lack roots).

••• Consult Fig. 4–3 of the haploid-diploid (1*n*-2*n*) life history (see also Figs. Sup5–3 and Sup5–4) of a typical moss as you examine the material in the following sections.

••• *Suggested diagram and labels: Polytrichum* (a moss) (or *Funaria* or similar moss) GPT and SPT: leafy GPT, location where archegonia would be, SPT attached to GPT, and for SPT: stalk (seta), capsule (spore case, sporangium), hood (calyptra).

A. Gametophytes (GPTs)

The GPT of a moss is the part most evident to the layperson (Fig. 4–3F, h, H, and leafy parts in Fig. 4–4). The GPT forms the dominant generation (phase), with the SPT permanently dependent on it.

Moss GPTs are either ***unisexual,*** with separate male and female plants (Fig. 4–3a, A, h, H), or ***bisexual,*** with male and female gametangia produced on the same plant either on separate branches or intermixed within an apex. Antheridia and archegonia occur at the tips or sides of the GPTs (Fig. 4–3a, A, h, H). Apices of particularly male individuals or branches often appear distinctively cuplike because of the closely packed leaves (Fig. 4–3h), which may be red or purple. Multicellular sterile hairs (paraphyses) separate the gametangia (Fig. 4–3a, A) and probably have a protective function. The sperm swim in heavy dew from the male GPTs to the female GPTs or are distributed by a splash-drop mechanism like that for the gemmae of *Marchantia.* This mechanism, however, is not as efficient as the gemma-mechanism in *Marchantia* because almost all of the sperm not splashed directly into the female cups are lost and hence wasted.

••• Examine a live GPT of *Polytrichum* (hairy cap moss), *Funaria* (this seems to lack a common name), or a similar moss. Note its multicellular rhizoids and the delicate leaves that are one cell lay-

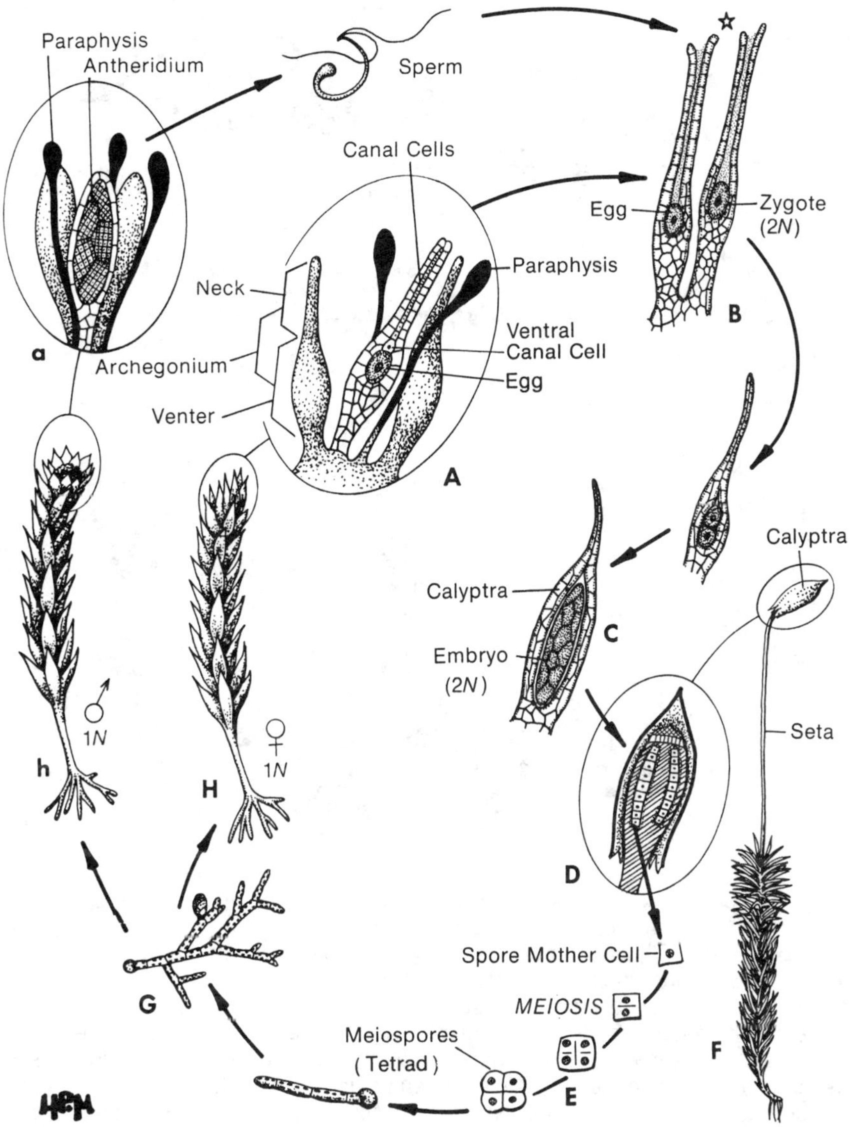

Fig. 4–3. Haploid-diploid (1*n*-2*n*) life history of a moss. *a,* apex of male GPT showing antheridia and sterile hairs (paraphyses); *A,* apex of female GPT showing archegonia and sterile hairs (paraphyses) (note in the longisection of archegonium its neck canal cells, ventral canal cell, egg, and neck and venter regions); ☆, syngamy (fertilization); *B,* longisections of archegonia showing egg and zygote (the neck canal cells disintegrate before syngamy); *C,* longisections of embryos (very young SPTs), each covered by a hood (calyptra), which is the proliferated archegonium; *D,* longisection of maturing capsule (sporangium) of SPT (the GPTic hood is still present) with sporocytes (meiocytes, spore mother cells); *E,* spore tetrads after meiosis of sporocytes; *F,* GPT with mature SPT consisting of foot (not shown) embedded in GPT, stalk (seta), and capsule (sporangium) covered by hood (calyptra); *G,* filamentous GPT resulting from spore germination; *h,* mature male GPT; *H,* mature female GPT. Note the alternation of morphologically dissimilar GPTic and SPTic generations (phases), with the GPT dominant over the SPT, the SPT always dependent on the GPT. (From Norstog & Long 1976:296.)

er thick except at the middle. The GPTs of *Polytrichum* are unisexual, whereas those of *Funaria* are bisexual. Are any gametangia evident? If these are not readily evident, it is OK to skip them. In *Polytrichum* what sex is the plant bearing the SPT? Why?

●●● Some other mosses may be available on DEMO.

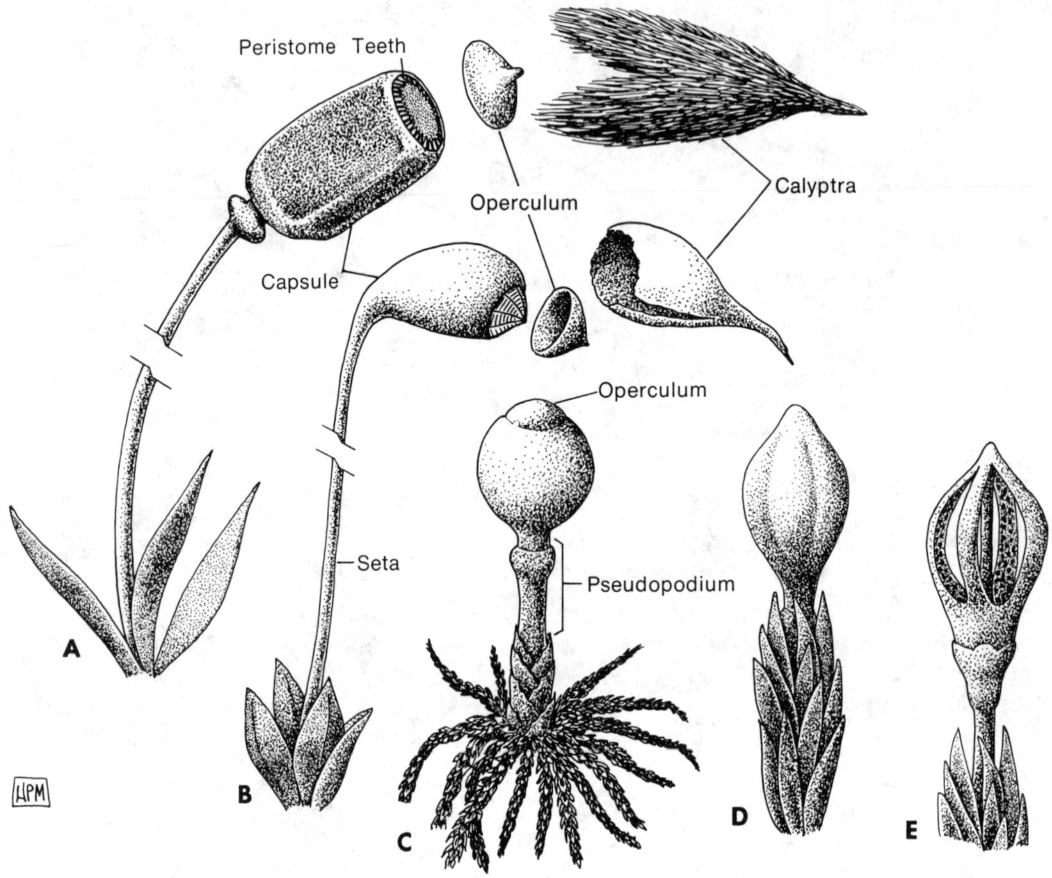

Fig. 4–4. SPTic structure of mosses. *A, B, Polytrichum* (hairy cap moss) and *Funaria,* respectively, each SPT with a foot (not shown), stalk (seta), and capsule (sporangium, spore case), the last consisting of the cover (operculum, lid), peristome, and wall. The stalks are much longer; the breaks represent deleted parts. The GPTic hood (calyptra, cap) and SPTic cover have detached to reveal the peristome, which effects spore dehiscence by a hygroscopic mechanism; *C, Sphagnum* (peat moss, bog moss, sphagnum); *D, E, Andreaea* (granite moss), respectively, unopened (*D*) and opened (*E*) capsules. The SPTs of *Sphagnum* and *Andreaea* differ from those of most mosses (the "true mosses") in lacking a stalk (the pseudopodium is a leafless GPTic stalk) and a peristome. *Sphagnum* has a cover and releases spores by an air-gun mechanism). *Andreaea* lacks a cover; its capsule wall splits hygroscopically into four longitudinal parts. (From Norstog & Long 1976:299.)

B. Sporophytes (SPTs)

After syngamy (fertilization), the zygote in the moss archegonium develops into an ***embryo,*** the very young SPT before the onset of rapid growth (or before germination of the seed in seed plants). The embryo then develops into a SPT that remains permanently attached to the GPT (Figs. 4–3B, C, D, F, 4–4A, B). The SPT is usually smaller than the GPT and in many mosses consists of:

- an anchoring, absorbing ***foot*** (this permanently anchors the SPT to the GPT!);
- a ***stalk*** (seta) that is usually long and delicate;
- a ***capsule*** (spore capsule, spore case) or ***sporangium*** with several parts:
  - a thick enclosing ***wall*** (2*n*);
  - many 1*n* ***spores*** that had resulted from meiosis, up to 50 million spores per sporangium (Raven et al. 1992:311) (no elaters are present);
  - a ***peristome,*** a 2*n* structure in the tip region involved in spore discharge;
  - a ***cover*** (operculum, lid), the detachable 2*n* tip region of the sporangium;
  - enclosing the capsule, a ***hood*** (calyptra, cap), which represents the enlarged remnants of the archegonium and which thus is actually 1*n* GPTic tissue.

The SPTs of *Funaria* (Fig. 4–4B) may exceed 5.1 cm (2 in.), whereas those of *Polytrichum* (Fig. 4–4A) may be up to 15.2 cm (6 in.) (Bold et al. 1987:287); the stalk makes up most of this length. The hood initially protects the developing sporangium. As the SPT enlarges due to the lengthening stalk, the hood ruptures at its base. In *Funaria* this rupture occurs when the SPT is about 6 mm (0.24 in.) long (Bold et al. 1987:284). The hood then is carried aloft by the enlarging SPT; that is, the hood overlies the sporangium at the end of the lengthened stalk (Figs. 4–3F, 4–4A, B).

Sporocytes (meiocytes) in the capsule (sporangium) undergo meiosis to produce 1*n* spores (meiospores); each sporocyte forms four spores, that is, a ***tetrad*** (Fig. 4–3D, E; Lab Exercise 7, Part IV). The ***peristome*** consists of one or more rows of a variable number of long, triangular peristome teeth that form a ring around the opening (mouth) of the sporangium. Spore release from a mature SPT involves the following activities (Fig. 4–4A, B):

- shedding of the protective hood;
- shedding of the cover;
- folding back of the hygroscopic (moisture sensitive) peristome teeth; as they lose moisture, they fold back to release the spores, which are wafted out, that is:
  - moist or humid conditions = peristome teeth straight, the sporangium closed;
  - dry conditions = peristome teeth folded back, the sporangium open, spores released, wafted off.

Wind usually disperses moss spores, but insect dispersal occurs in *Splachnum* and other members of the Splachnaceae. These grow on unusual objects: dung of herbivorous mammals, skeletal remains, bone, anthers, stomach pellets of predatory birds, corpses, and enriched gravel (Koponen 1990).

●●● See Fig. 4–3F. Then examine the DEMO live GPTs of *Polytrichum, Funaria,* or a similar moss bearing young and mature SPTs. The DEMO is mainly for orientation; dissect your own material if sufficient is available! Identify on a young (greenish) SPT the following structures: the stalk and the hood covering the capsule (sporangium).

●●● See Fig. 4–4A, B. Then place a mature (brownish) SPT under your dissecting microscope and shine a microscope lamp on the tip of the sporangium. Flip off any remaining hood and cover. Watch the hygroscopic movement of the peristome teeth. With a water dropper wet the peristome teeth after they have dried out. What happens?

Some questions: How are the spores normally dispersed from the capsule of a moss? That is, what structures must be shed or displaced before spores are released? What is the adaptational advantage of this type of spore dispersal mechanism? What is the fate of the SPT after it has released its spores? What is the fate of most of the spores produced by the moss?

●●● Some other mosses may be available on DEMO.

IV. THE MOSS *SPHAGNUM* (PEAT MOSS)

The mosses described in Part III are the "true mosses" because they have the SPTic structure shown in Figs. 4–3 and 4–4A, B. However, some mosses have SPTic structure differing from that of such true mosses. *Andreaea* (granite moss) (Fig. 4–4D, E) is a queer moss that is blackish rather than greenish; its ten species in North America occupy acidic rock surfaces and seeps at high altitudes. *Sphagnum* (Fig. 4–4C) is a very distinctive moss with some 150 species distributed worldwide, particularly in the temperate northern hemisphere. *Sphagnum* grows in wet areas such as bogs, and the remains of the plant become compacted to form dense deposits called ***peat.*** The common names of *Sphagnum* are "sphagnum," "bog moss," and especially "peat moss."

Peat is very acidic and extremely imbibing, when moistened absorbing 16 to 26 times its dry weight. Because of this, *Sphagnum* is ecologically and commercially the most valuable of all bryophytes.

Peat is used in North America chiefly in gardening as an acidic and moisture soil conditioner for plants. In less developed areas, peat has been and still is used for fuel and packing materials, and even as lamp wicks. Under wartime conditions, peat was used to dress wounds. Its absorbent quality soaked up blood, whereas its acidic nature prevented infection. Peat is also a good preserving agent of other organisms. More than 2,000 human bodies have been dug out of European peat bogs.

●●● The most famous peat body, the Lindow Man, is excellent testimony to the acidic and preservative properties of peat bogs. Examine the DEMO text and figure (from Wernick 1988:146) of the Lindow Man. In 1984 a naked dead man was discovered in a peat bog at Lindow in Cheshire, England. At first he was thought to have been recently murdered, but later radioactive carbon dating showed that the man was about 2,000 years old. Apparently, he was ritually murdered, perhaps as part of Druid rites. The man's facial features and body organs, including his stomach contents, were nearly perfectly preserved by the acidic peat. The preservation was much better than that of Egyptian mummies. Incidentally, the lower part of the Lindow Man was lost to science because it was chopped up by a peat-cutting machine. [*News flash:* A 4,600-year-old mummified man discovered 19 Sep. 1991 in a glacier in the Austrian alps is the most complete Bronze Age find in Europe and offers great research potential for understanding past climates and cultures. However, this guy has nothing to do with botany.]

●●● See Fig. 4–4C. Then examine the DEMO diagrams (from Fuller & Tippo 1954:720) of *Sphagnum* and its leaf anatomy, the DEMO live material of *Sphagnum* (most likely sterile), and the DEMO store-bought material of "peat moss." Note that some commercially available, so-called peat moss may *not* be *Sphagnum* but rather another moss. Squeeze the water from some plants and note their absorbent nature. Finally examine the DEMO of a water mount slide preparation of a live leaf of *Sphagnum.* The leaves are differentiated into

- small live cells that contain chloroplasts (the cells appear green), and
- large dead (hence colorless) cells that have helical wall thickenings and many pores to allow for direct entry of water.

Because the dead cells of *Sphagnum* relate to its great absorptive ability, there is no real absorptive difference between the live material and the dead, store-bought peat moss (if it is really *Sphagnum*).

The mechanism of spore discharge of *Sphagnum* (Fig. 4–4C) is quite different from that of *Marchantia* (Fig. 4–1D, E) or a typical moss (Fig. 4–4A, B). *Sphagnum* lacks both elaters and peristome teeth. The capsule (sporangium) of *Sphagnum* dries out differentially and contracts in mid region, placing its spores and other contents under great pressure, up to five atmospheres (Schofield 1985:40). Eventually the internal pressure is too much, the sporangium audibly blows its cover (hence it is not a good candidate for the CIA), and the spores are shot out as a gaseous cloud for distances up to 15 cm (5.9 in.) (Bold et al. 1987:265). Thus there is explosive dehiscence of the sporangium via an air-gun mechanism, actually a turgor mechanism (see Part I).

V. "FALSE" OR "UNTRUE MOSSES"

The term "moss" has been applied to diverse organisms that superficially resemble moss GPTs but that are not mosses: algae, lichens (*Cladonia rangifera* or reindeer moss, *Ramalina menziesii* or California Spanish moss, *Usnea, Alectoria, Bryoria,* etc.), and particularly vascular plants such as the lycopods *Lycopodium* (club moss) and *Selaginella* (spike moss) or the flowering plants *Tillandsia usneoides* (Spanish moss), *Polemonium* (moss pink), and *Silene acaulis* (moss campion).

A rolling stone gathers no moss.—Publilius Syrus, *Maxim 524,* 1st century B.C.

NAME ______________________________ LAB SECTION # ______________ DATE ____

TITLE ______________________________

__________________ TOTAL MAG ____

TITLE ______________________________

__________________ TOTAL MAG ______

TITLE ______________________________

__________________ TOTAL MAG ____

TITLE ______________________________

__________________ TOTAL MAG ______

LAB EXERCISE SUPPLEMENT **4**

The Peel Technique in Paleobotany

(Preparation of a Peel from a Coal Ball)

OBJECTIVE

To make one or more peels from a fossil coal ball by means of the paleobotanical peel technique.

PERSPECTIVE

Of the various techniques available to paleobotanists, the so-called peel technique for coal balls is among the simplest and most important. ***Coal balls*** are rocks containing fossil materials embedded in calcium carbonate and are restricted in occurrence to Carboniferous rocks, where they are found associated with seams of bituminous coal. The resultant peel of the fossil in the coal ball may show exquisite cellular detail and hence often reveals considerable information about internal structure. The modern technique dates to Joy et al. (1956), who introduced the use of preformed cellulose acetate in preparing sections.

Figure LabSup4–1 and the explanation below give steps in the preparation of a fossil peel. The acid etches the cut surface of the rock, dissolving away the supporting rock matrix, but leaving the carbonized cellular material standing out in relief above the surface. The acetone flooded on the surface of the rock partly dissolves the overlying sheet of cellulose acetate, which adheres to the exposed cellular material. After drying, the film is peeled away from the rock, resulting in a thin layer of carbonized fossil material adhering to the film. About 500 peels can be prepared from a fossil of 25 mm (1 in.) thickness.

MATERIALS NEEDED

Fossil coal balls
Sheet of plate glass (or grinding lap)
#400 grit (or comparable grade) Carborundum
Water
Dilute (3–5%; I prefer 5%) hydrochloric acid (HCl) or hydrofluoric acid (HFl) [HCl is used for carbonates, HFl for quartz (silicates). However, HFl is too dangerous for class use and also must *not* be used with or stored in glass containers.]
Large enamel or glass tray to hold the HCl (or plastic container for HFl)
Acetone in dropper or squirt bottle

Sheet of cellulose acetate 0.0118 mm (0.003 in.) thick (or comparable thickness)
Scissors
Tray with sand to support fossil
Heat source for drying peel (goose neck lamp, electric hair-dryer, etc.).

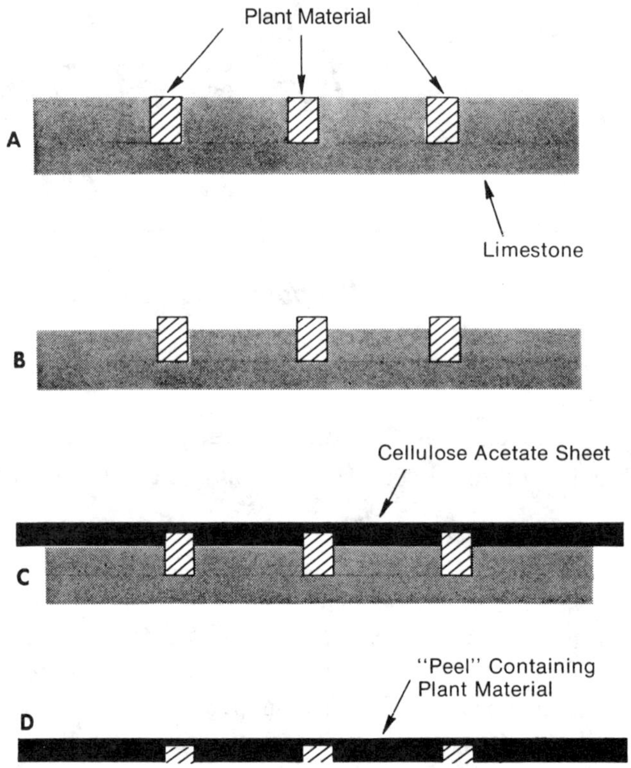

Fig. LabSup4–1. Sequences in preparing a peel from a fossil coal ball. *A,* limestone rock with fossil plant material is cut and ground flat; *B,* surface is etched with hydrochloric acid, some plant material remaining untouched and standing in relief; *C,* surface is flooded with acetone, which dissolves enough plastic film (cellulose acetate) so that it adheres to all surfaces of the matrix; *D,* when dry, plastic film is peeled off, removing a thin layer of the fossil plant material. (From Norstog & Long 1976:320; redrawn from *Evolution and plants of the past,* by H. P. Banks, © 1970 by Wadsworth Publishing Co., Inc., Belmont, California. Reprinted by permission of the publisher.)

PROCEDURE *Important: Read this completely before starting!*

1. Smoothen the surface of the fossil by grinding it on the sheet of plate glass. Put a pinch of the Carborundum on the glass and moisten it slightly with water to make a slurry. Grind the surface of the fossil until smooth; 25 to 30 circular motions should be sufficient. It is assumed that the coal ball has already been cut flat with a diamond saw and that its surface has been roughly ground with coarser Carborundum.
2. Gently wash off the grinding paste with water. Drain off any excess water.
3. Etch the ground surface of the fossil with the acid. The etching time will vary with the material and the concentration of the acid but should be somewhere between 15 to 30 seconds [from 6 seconds for Belgian coal balls to about 3 minutes for dolomitic (magnesium carbonate) coal balls from Germany, with 15–17 seconds for most American coal balls (Phillips 1971)]. Do not let the acid etch too long or too deeply into the rock. Etching of the acid dissolves away the mineral

matter and leaves the cell walls of the fossil standing in relief. Unlike HFl, dilute HCl will not harm your hands; however, avoid contact with cuts, jewelry, or clothing.

4. After etching, wash the specimen with a gentle stream of water and allow it to dry. Artificial heat can be used to hasten drying. Do not touch the etched surface, which is very fragile. Cut a piece of the cellulose acetate film so that it is slightly larger than the fossil surface.
5. Support the fossil so that it is *level* and with the etched surface upward (a flat tray filled with sand is excellent for this). Flood the surface of the fossil with a thin layer of acetone and, before the acetone evaporates, quickly and gently roll the sheet of cellulose acetate onto the fossil surface. Bowing the film slightly and starting at one edge of the fossil will eliminate most or all air bubbles, which are undesirable. The acetone will dissolve enough cellulose acetate so that it adheres to all surfaces of the fossil. Do not press on the film at any time!
6. Let the fossil dry for about 20 to 30 minutes to evaporate the acetone and harden the cellulose acetate. When dry, carefully peel the film from the fossil. Some plant material will be lifted (peeled) off. Be especially careful on peeling the film from the edges of the rock.
7. Look at your peel with a dissecting or compound microscope. Small pieces of the film can be mounted on microscope slides, either as temporary mounts (without water) or as permanent mounts using an appropriate resinous mounting medium and a cover glass. Mount the peel smooth side up and put resin both under and over the peel. *Note:* This quick method of mounting peels may not be appropriate for research and photography, for which any remaining carbonates on the peel must be removed (see Phillips 1971).
8. If you wish to prepare another peel from the same rock, it must be reground (step 1).
9. ***Achtung Baby! When you are finished making fossil peels, clean up your mess!***

LITERATURE CITED

Joy, K. W., A. J. Willis & W. S. Lacey. 1956. A rapid cellulose peel technique in palaeobotany. *Ann. Bot.* 20:635–637.

Phillips, T. L. 1971. *A laboratory handbook for paleobotany.* Unpublished course manual.

SUPPLEMENT **7**

Characteristics of Vascular Plants

(Pteridophytes versus Seed Plants)

I. PERSPECTIVE

Supplement 6, Part IV, lists the unifying characters of *all* extant vascular plants and thus their differences from the extant bryophytes (see Lab Exercise 4). The vascular plants comprise two main groups that are fundamentally separated on the basis of the absence or presence of ovules and seeds:

- ***pteridophytes*** (nonseed plants, cryptogams), without ovules/seeds and their associated structures—about 10,300 species, 230 genera (Kubitzki 1990);
- ***seed plants*** (spermatophytes, phanerogams), with ovules/seeds and their associated structures—about 240,750 species, 13,808 genera.

In turn, the seed plants comprise two main groups that are fundamentally separated on the basis of the enclosure of the ovule or seed:

- ***gymnosperms,*** with exposed or naked ovules/seeds—750 species, 83 genera (Kubitzki 1990);
- ***angiosperms*** (flowering plants), with enclosed ovules/seeds—about 240,000 species, 13,725 genera (Mabberley 1987).

There are, however, numerous other differences between these groups.

The divisions of extant pteridophytes are:

1. whisk ferns (fork ferns, psilophytes) (Psilophyta)—12 species, 2 genera;
2. lycopods (lycophytes, club mosses) (Lycophyta)—about 1,260 species, 4–6 genera;
3. horsetails (sphenophytes) (Sphenophyta)—29 species, 1 genus;
4. ferns (Pterophyta)—about 9,000 species, 214 genera.

The divisions of extant seed plants are:

1. cycads (Cycadophyta)—137 species, 11 genera;
2. *Ginkgo biloba* (maidenhair tree) (Ginkgophyta)—1 species;
3. conifers (Coniferophyta)—536 species, 68 genera;
4. gnetophytes (Gnetophyta)—76 species, 3 genera;
5. flowering plants, angiosperms (Anthophyta)—about 240,000 species, 13,725 genera (Mabberley 1987).

In addition, there are four divisions of extinct pteridophytes and two divisions of extinct seed plants.

II. DISTINGUISHING CHARACTERISTICS OF PTERIDOPHYTES AND SEED PLANTS

The unifying characters of *all* extant pteridophytes (see Lab Exercise 5) and hence their differences from the extant seed plants (see Lab Exercises 8 to 11) are:

Extant Pteridophytes	*Extant Seed Plants*
Life histories in Figs. 5–1, 5–3, 5–4, 5–5, Sup5–4, Sup5–5	Life histories in Figs. 8–1, 8–2, 9–4, Sup5–5
Herbaceous, lacking secondary growth, very rarely slightly woody, i.e., with secondary growth (e.g., *Botrychium, Isoetes*)	Herbaceous or especially woody, i.e., with secondary growth
Mostly homosporous, with spores that are structurally the same, but occasionally heterosporous, with large female and small male spores	Entirely heterosporous, with spores that are structurally different, i.e., with female spores and male spores
Pteridophytic (cryptogamic) life history, i.e., sporangia and spores present, but seed plant features such as ovules/seeds, pollen grains/pollen tubes, and seedlings absent	Seed plant life history, i.e., sporangia and spores present, but also ovules/seeds, pollen grains/pollen tubes, and seedlings present
Sperm always flagellate (bi- or multiflagellate), swimming to eggs	Sperm mostly *not* flagellate (if so, multiflagellate), but rather usually conveyed to eggs directly by the pollen tube

In the above respects, the pteridophytes are more similar to the bryophytes than to the seed plants. As noted in Supplement 4, Part V, swimming sperm are a weak point in the life history due to the necessity of moisture for syngamy (fertilization). To reiterate from Lab Exercise 4, Part I, moisture is a critical factor for two vital processes in the life histories of pteridophytes and also bryophytes:

- For syngamy (fertilization) moist or humid conditions are *good,* but dry conditions are *bad.*
- For spore dispersal, however, dry conditions are *good,* but moist or humid conditions are *bad.*

["Four legs good, two legs bad."—George Orwell, *Animal farm,* 1945.]

The following table contrasts the extant pteridophytes with the two main groups of seed plants, the gymnosperms and the angiosperms:

Extant Pteridophytes	*Extant Gymnosperms*	*Extant Angiosperms*
Life histories in Figs. 5–1, 5–3 to 5–5, Sup5–4, Sup5–5	Life histories in Figs. 8–1, 8–2, Sup5–5	Life history in Figs. 9–4, Sup5–5
Mostly herbs	*All woody (mostly trees)	*Woody (trees, shrubs), herbs
Vessels mostly absent	*Vessels mostly absent	*Vessels mostly present
Cones often present	*Cones usually present	*Inflorescences, flowers, and fruits present (cones absent)
Hetero-, mainly homosporous	#All heterosporous	#All heterosporous
Pteridophytic life history, i.e., no seed plant features present	#Seed plant life history and attendant structures, i.e., ovules/seeds present, and, concomitantly, pollen grains/pollen tubes and seedlings	#Same as gymnosperms
Ovules/seeds absent	*Ovules/seeds present, naked	*Ovules/seeds present, enclosed
GPTs large to small	#GPTs greatly reduced	#GPTs very greatly reduced
Antheridia present	#Antheridia absent	#Antheridia absent
Sperm all flagellate	#Sperm mostly not flagellate (multiflagellate in cycads and *Ginkgo*)	#Sperm never flagellate
Archegonia always present	*Archegonia mostly present (absent in two genera)	*Archegonia always absent

The *-marked items contrast gymno- and angiosperms; #-marked items are shared by them (see elaboration in Supplement 9).

III. EVOLUTIONARY RELATIONSHIPS OF VASCULAR PLANTS

Figure Sup7–1 shows the suggested relationships of various groups of pteridophytes and gymnosperms and their evolution from Devonian pteridophytes. The progymnosperms (see Lab Exercise 5, Part III-B) are a strictly Devonian group and the pteridophytic progenitors or ancestors of the gymnosperms. The angiosperms originated from some group of gymnosperms.

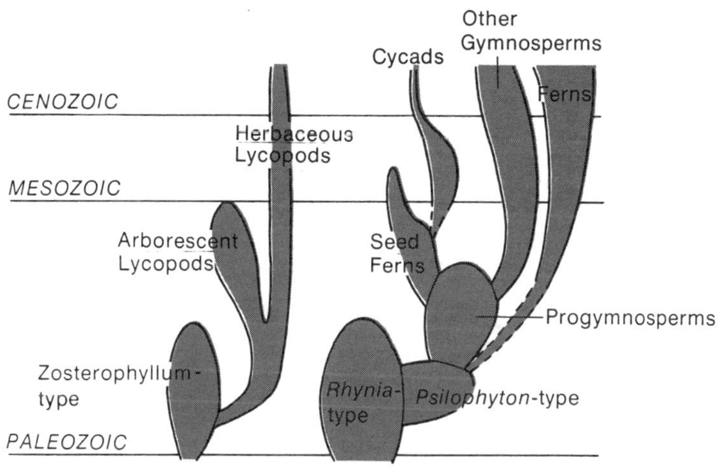

Fig. Sup7–1. Suggested relationships of various groups of pteridophytes and gymnosperms and their evolution from Devonian pteridophytes. (From Norstog & Long 1976:339; redrawn from *Evolution and plants of the past,* by H. P. Banks, © 1970 by Wadsworth Publishing Co., Inc., Belmont, California. Reprinted by permission of the publisher.)

All the [plant] evolution after Devonian time was just 'frosting on the cake.'
—Theodore Delevoryas, *Plant science bulletin,* 1970

LAB EXERCISE **5**

Ferns and Other Pteridophytes

God made ferns to show what He could do with leaves.—Thoreau

OBJECTIVE

To examine the structure of representative examples of ferns and other pteridophytes.

Note: Read this material *before* coming to the lab; there is insufficient time in the lab to read the material from scratch. Due to the large amount of background information presented here, a ●●● is used to mark those paragraphs calling for actual observations or diagrams in the lab. Most unmarked paragraphs reiterate information presented in the lecture. A bullet (•) signifies an important term or concept.

Suggested Lab Diagrams (see part noted for suggested labels):
Polypodium (polypody, a fern) (or similar live fern) shoot (Part I-A)
Cyrtomium falcatum (holly fern) (or other fern) mature sorus l.s. (Part I-A)
Equisetum (horsetail) shoot (Part II-B)

PERSPECTIVE

The pteridophytes have four divisions of extinct plants and four divisions of extant plants, with about 10,300 extant species and about 230 extant genera (Kubitzki 1990). The divisions of extant taxa are:

1. whisk ferns (fork ferns, psilophytes) (Psilophyta)—12 species, 2 genera;
2. lycopods (lycophytes, club mosses) (Lycophyta)—about 1,260 species, 4–6 genera;
3. horsetails (sphenophytes) (Sphenophyta)—29 species, 1 genus;
4. ferns (Pterophyta)—about 9,000 species, 214 genera.

The strictly extinct divisions include various Devonian groups: Rhyniophyta, Zosterophyllophyta, Trimerophytophyta, and the progymnosperms (Progymnospermophyta). The United States and Canada have 406 extant native and naturalized (i.e., introduced taxa that are flourishing as if native) species, subspecies, and varieties of pteridophytes, plus many hybrids (Lellinger 1985).

The ferns and the flowering plants are the only two groups of vascular plants expanding and undergoing rapid evolution today. The other groups of vascular plants had their zenith in earlier geologic times. The renowned expert on the gymnosperms, Charles J. Chamberlain, believed that the cycads, for example, will be extinct in the next geologic age (see Arnold 1964). In contrast, the ferns, according to Arnold (1964:65), one of the foremost American paleobotanists, "will survive the present [geologic] epoch, and probably the next one. Beyond that, [he does] not predict."

Pteridophytes and bryophytes have a strictly haploid-diploid ($1n$-$2n$) type of life history (see Figs. Sup5–3 to Sup5–5). The alternation of dissimilar GPTic and SPTic generations (phases) in the pteridophytes involves a SPT dominant over the GPT and ultimately independent of it. In contrast, the bryophytes have the GPT dominant over the SPT, the SPT always dependent on the GPT.

Note: Before beginning this lab exercise, you should review:

- Supplement 5 on the main types of life histories;
- Supplement 6 on the characteristics of and origin of the land plants;
- Supplement 7 contrasting pteridophytes and seed plants.

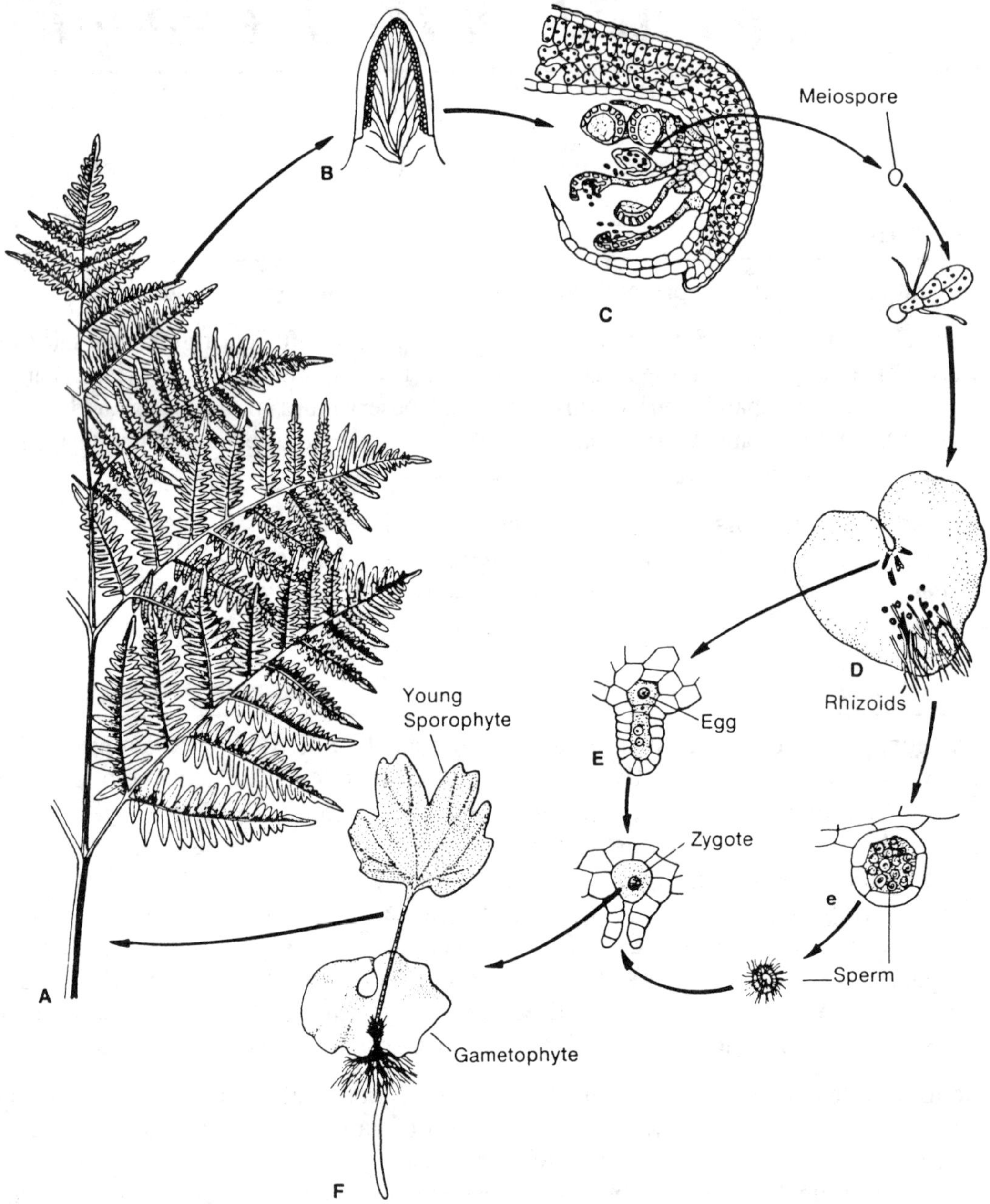

Fig. 5-1. Homosporous haploid-diploid (1*n*-2*n*) life history of *Pteridium aquilinum* (bracken fern) (Pterophyta). *A,* leaf of SPT; *B,* fertile leaf segment, abaxial view; *C,* section of a sorus; *D,* GPT viewed from beneath; *e, E,* longisections of antheridium and sperm (*e*) and archegonia with egg and zygote (*E*); *F,* GPT with attached young SPT. Note the alternation of morphologically dissimilar GPTic and SPTic generations (phases), with the SPT dominant over the GPT and ultimately independent of it. A homosporous plant produces morphologically similar spores (compare with Fig. 5-4). Locate where meiosis and syngamy (fertilization) occur and which structures are 1*n* and 2*n*. (From Norstog & Long 1976:360.)

I. FERNS (PTEROPHYTA)

The ferns (Pterophyta) occur in a variety of habitats, not only in moist tropical and temperate forests, but also in many areas where moisture is limiting, including on mountains and in deserts.

The extant ferns are characterized by the following features:

- typically large leaves, coiled when young (circinate vernation) into fiddleheads (croziers);
- sporangia borne *not* in cones (strobili), but in sori, usually abaxially directly on leaves;
- mostly homosporous, rarely heterosporous life histories.

Most ferns are homosporous, that is, producing spores that are morphologically the same. However, 93 species (Kubitzki 1990) of aquatic ferns are heterosporous, producing spores that morphologically dissimilar.

●●● Consult Fig. 5–1 of the homosporous haploid-diploid (1*n*-2*n*) life history (see also Figs. Sup5–3 and Sup5–4) of a typical fern as you examine the material in the following sections.

A. Sporophytes (SPTs)

The pteridophytes are the first of the various groups of organisms studied up to this point to have a typically underground ***root*** system plus a ***shoot*** system consisting of stems and leaves. These are all SPTic structures. In contrast, the GPT (prothallus) lacks such organs.

The typical fern SPT has the following general ***habit,*** or external, morphological appearance:

- a ***rhizome,*** a stem that is underground, at least partly so;
- ***adventitious roots,*** that is, roots borne on the stem (rhizome);
- ***fiddleheads*** (croziers), young, coiled unfolding leaves;
- ***fronds,*** mature leaves that are either undivided or ***simple*** (Fig. 5–2A), or variously divided or ***compound*** (Figs. 5–1A, 5–2B to D);
- ***sori*** (plural, singular ***sorus***), clusters or groups of sporangia (Fig. 5–1B, C).

Fern roots are characteristically associated with a leaf and are usually adventitious roots because the first-formed root is ephemeral. Fern stems are typically rhizomes. If the leaf is compound, its subdivisions are called ***leaflets*** (pinnae); these are attached to the main axis of the leaf (rachis). Compound leaves may be once divided (pinnate) (Fig. 5–2B), twice divided (bipinnate) (Fig. 5–2C), thrice divided (tripinnate) (Fig. 5–1A, 5–2D), etc. Fern sporangia are variously borne: terminal on ultimate axes, marginally on fronds, or, usually, abaxially on fronds (Fig. 5–1B, C).

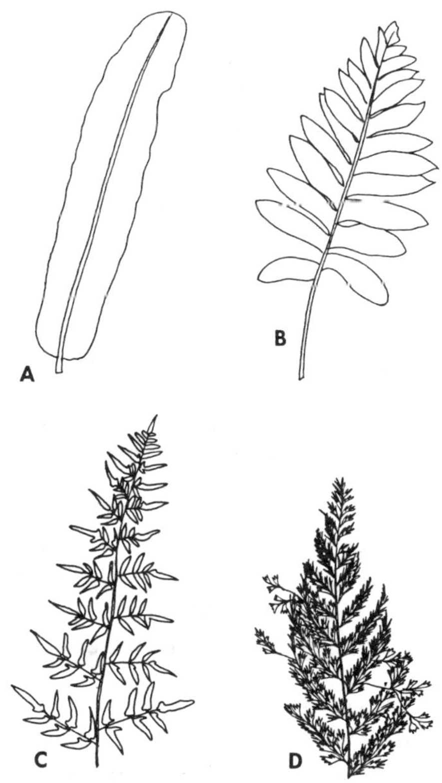

Fig. 5–2. Leaf types of ferns. *A,* simple leaf; *B,* once-divided compound leaf; *C,* twice-divided compound leaf; *D,* thrice-divided compound leaf. The main subdivisions of compound leaves are leaflets (pinnae). (From Norstog & Long 1976: 350.)

Ferns, "a site for sori"

(from: University of California, Davis, T-shirt.)

●●● Examine the DEMO live material (adult SPT) of *Polypodium* (polypody), or a similar fern, and note its general habit. Identify the following structures: rhizome, adventitious roots, any fiddleheads, mature fronds divided into leaflets (pinnae) attached to the main axis of the leaf (rachis), abaxial sori. How are the fronds divided? Are the sori present young (i.e., green) and/or mature (i.e., brown)?

••• *Suggested diagram and labels: Polypodium* (polypody, a fern) (or similar live fern) shoot: rhizome, leaf (frond), leaflet, sori.

••• See Fig. 5-1A. On DEMO are the highly divided fronds of the world's most common fern, the bracken fern *Pteridium aquilinum.* The single species of this genus occurs worldwide, even on remote islands, which testifies to the dispersibility of its spores! Bracken is also one of the few pteridophytes with nectaries, which occur on its fronds.

••• Fiddleheads of some species are cooked and eaten. In eastern North America the young coiled leaves of *Matteuccia struthiopteris* var. *pensylvanica* (ostrich fern) are highly esteemed by connoisseurs of natural foods and are marketed fresh, canned, and frozen; these fiddleheads taste a tad like asparagus but are crisper (Lellinger 1985:39). Examine the DEMO of canned fiddleheads of ostrich fern. For instance, the Belle of Maine brand offers the following nutritional information:

serving size	1 cup	protein	5 g
calories	65	carbohydrate	4 g
ash	2 g	fat	0 g

Although this fern is apparently safe to eat, be forewarned that some ferns that are eaten, notably *Pteridium aquilinum* (bracken fern), are carcinogenic; a high incidence of stomach cancer in the Orient has been associated with extensive consumption of bracken fiddleheads (Lellinger 1985:38).

••• Tree ferns can attain heights of 15 m (50 ft.). Examine the DEMO live frond of a tree fern and the DEMO photograph of tree ferns growing in Java. Some tree ferns can be seen growing on the more temperate college campuses, for example, in Berkeley along Strawberry Creek near the Faculty Club of the University of California.

The typical fern sporangium is delicate and consists of (Fig. 5-1C):

- a long thin ***stalk*** supporting the capsule;
- a small ***capsule*** (spore capsule, spore case, sporangium proper) bearing an annulus and a mouth and containing the spores, as follows:
  - the ***annulus,*** the ridge of unevenly thick-walled cells (thickened mostly on three of their six walls) along one edge of the sporangium;
  - the ***mouth*** (stomium), the region of thin-walled cells (oft four) 90° to the annulus, through which spore discharge occurs;
  - few ***spores*** per sporangium, usually 16, 24, 32, 48, 64, at times 128, 256, 512, rarely 1024.

The annulus (Fig. 5-1C) is responsible for the explosive, forceful ejection of spores by a hygroscopic mechanism (see Lab Exercise 4, Part I). Water loss from cells of the annulus results in pressures of 300 atmospheres or more, rupture of the wall at the weak mouth cells, and catapulting of the spores a centimeter or so out of the sorus (Gifford & Foster 1989:262).

The sporangia of many fern species have ***indusia*** (plural, singular ***indusium***); an indusium is a flap of tissue protecting the sporangium (Fig. 5-1B, C). Sterile protective hairs (paraphyses) are often present. The indusia and sterile hairs, if present, are for protection. As sporangia mature, the indusia shrivel up and pull backwards to expose the sporangia. Usually the sporangia are in different stages of development. With adverse environmental conditions as a freeze or drought, some sporangia probably will survive to mature later and thus release spores.

••• See Fig. 5-1C. Then take from the live plant of *Polypodium* a very *small part* of a leaflet with mature (i.e., brown) sori and make a water mount slide preparation of a sorus. Save the fragment of the leaflet for use below. Next, in conjunction with a slide of a longisection of a mature sorus of *Cyrtomium falcatum* (holly fern), identify on both the temporary and permanent slides the following structures of a sporangium: stalk, spore case, spores, and sporangial wall, including its annulus and mouth. The sorus of *Cyrtomium* differs from that of *Polypodium* in having an umbrellalike indusium. Note that a permanent slide will show sporangia in a variety of views: stalk only, spore case only, and the complete sporangium; the annulus is thick-walled and red-stained.

●●● *Suggested diagram and labels:* *Cyrtomium* (or other fern) mature sorus l.s.: indusium, stalk of indusium (if shown), sporangium and its parts (stalk, spore case, annulus, mouth, spores).

●●● Examine the DEMO diagram (from Gifford & Foster 1989:262) of dehiscence of a fern sporangium. Then observe your fragment of a leaflet of *Polypodium,* with the sori upwards, under a dissecting microscope and shine a microscope lamp onto the sori to dry out the sporangia. What happens? What are the similarities and differences of the mechanism of sporangial discharge and spore dispersal of ferns and bryophytes?

B. Gametophytes (GPTs)

See Fig. 5-1D, e, E, F. All ferns have very small GPTs, a maximum 10 mm (0.4 in.) or so. Most fern GPTs are green and free-living. Fern GPTs are often heart-shaped (cordate) and typically bisexual, bearing both

- antheridia, containing many swimming sperm (Fig. 5-1e) that bear many flagella, and
- archegonia, each with one nonmotile egg (Fig. 5-1E).

Both male and female gametangia may not always be evident on a given GPT. Why? The gametangia are located on the lower (ventral) surface of the GPT rather than on its upper (dorsal) surface. What advantage might this location of gametangia have?

●●● See Fig. 5-1D, e, E. Then take a GPT from the live material of *Pteridium* and mount it *upside down* in water on a slide, using a cover glass. Observe the morphology of the GPT, noting the following: its delicate nature (vascular tissue is absent), the archegonia located towards the apical notch, and the antheridia located among the rhizoids away from the apical notch. You should be able to observe swimming sperm and perhaps even sperm being liberated from an antheridium.

●●● See Fig. 5-1F. A DEMO of an older stage of *Pteridium* shows young SPTs attached to their GPTs. The young leaves, which are initially coiled (fiddleheads), do not resemble the highly divided leaves of adult SPTs. After a short time the GPTs will die and the SPTs will become physically and nutritionally independent.

One Good Fern Deserves Another

A tree may be a prologue when it has a hyper bole.

Prothallia of ferns are always haploid
Producing sperms and eggs that seize the procreative role
When, of a dampness, they unite to form a diploid.

Up springs the frondly sporophyte,
with rhizome, root, and rachis
And a meristem that's apical and tight.
It uncoils; but on a leaf that is preparing for meiosis
Sporangia in clusters make a very sori sight.

(from: J. M. Burns, 1975, *BioGraffiti: A natural selection,* p. 55.)

In summary, compared to the bryophytes, in ferns and all other vascular plants the SPT is the dominant generation (phase) and is eventually totally independent of the GPT. Initially the SPT is attached to the GPT (Fig. 5–1F), but eventually the SPT becomes physically and nutritionally independent.

II. OTHER EXTANT PTERIDOPHYTES

Sporangia of most fern species are aggregated into sori rather than ***cones*** (strobili). In contrast, the sporangia of lycopods and horsetails are generally aggregated into cones consisting of ***sporophylls*** (specialized reproductive leaves) bearing the sporangia. Extant species of these two divisions also have small leaves compared to the usually much larger leaves of ferns.

A. Lycopods (Lycophyta)

The lycopods (Lycophyta) exhibit both the homosporous and the heterosporous types of life histories (for details see Supplement 5). Figures 5–3 and 5–4 show, respectively, the homosporous life history of *Lycopodium* and the heterosporous life history of *Selaginella;* compare also Figs. Sup5–4 and Sup5–5. The extant lycopods are characterized by the following features:

- simple, generally small leaves, each with a single vein (some fossil species had large leaves with two veins);
- sporangia borne on the adaxial surfaces of the sporophylls;
- the sporophylls aggregated into compact cones (strobili) in most species.

The two largest genera are *Lycopodium* (club moss), with about 450 species found in temperate and tropical areas, and *Selaginella* (spike moss), with about 700, chiefly tropical species.

●●● Examine the DEMO live material of *Lycopodium*. Note the size and arrangement of the leaves, and if any cones are present. Most species of *Lycopodium* have their sporophylls aggregated into cones, but a few species (e.g., *L. lucidulum*) have very leaflike sporophylls and lack definite cones. Some species even have branched cones.

●●● See Fig. 5–3A. Then examine the DEMO slide of a longisection of a mature cone of *Lycopodium*. Identify the following structures: sporophylls bearing sporangia, each sporangium with a stalk, spore case, wall (only one cell layer persists in the mature wall), and many spores. Are the sporangia on the ad- or abaxial surfaces of the sporophylls? To reiterate, this position of the sporangia is typical of all lycopods and not of any other pteridophytes.

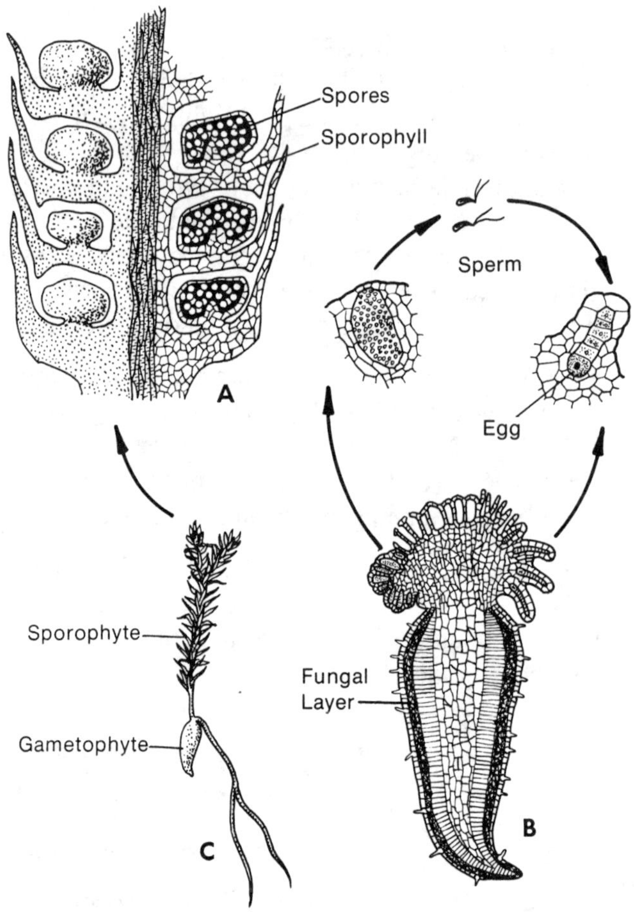

Fig. 5–3. Reproduction in *Lycopodium* (club moss), a homosporous lycopod (Lycophyta). *A,* longisection of cone; *B,* longisection of subterranean GPT (note rhizoids and mycorrhizal fungus) of *L. phlegmaria,* with antheridia (left) and archegonia (right); *C,* young SPT still attached to GPT. Note the alternation of morphologically dissimilar GPTic and SPTic generations (phases), with the SPT dominant over the GPT and ultimately independent of it. A homosporous plant produces morphologically similar spores (compare with Fig. 5–4). Locate where meiosis and syngamy (fertilization) occur and which structures are 1*n* and 2*n*. (From Norstog & Long 1976:331.)

●●● *Lycopodium* powder is actually a mass of spores. The spores contain up to 47% oil and only 4% moisture. The powder has been variously used, especially in former times: (1) flash powder in photography, (2) theatrical effects and in fireworks, and (3) because the spores resist moisture due to their high oil content, as a dusting powder in pill boxes to prevent pills from sticking together, in baby talcum powder, and, recently, to coat condoms. Examine the DEMO of spore powder from *Lycopodium clavatum* and of the article (Balick & Beitel 1989) on *Lycopodium* spores used to dust condoms.

●●● See Fig. 5-4A, which shows *Selaginella kraussiana,* a species with paired leaves of two sizes: pairs of small leaves are on the upper (dorsal) side of the stem, whereas pairs of large leaves are on the lower (ventral) side. Then examine the DEMO live material of *Selaginella,* which may have a habit very different from that shown in Fig. 5-4A. Note the size and arrangement of the leaves, and if any cones are present. Cones of some species of *Selaginella* are rather inconspicuous and may be

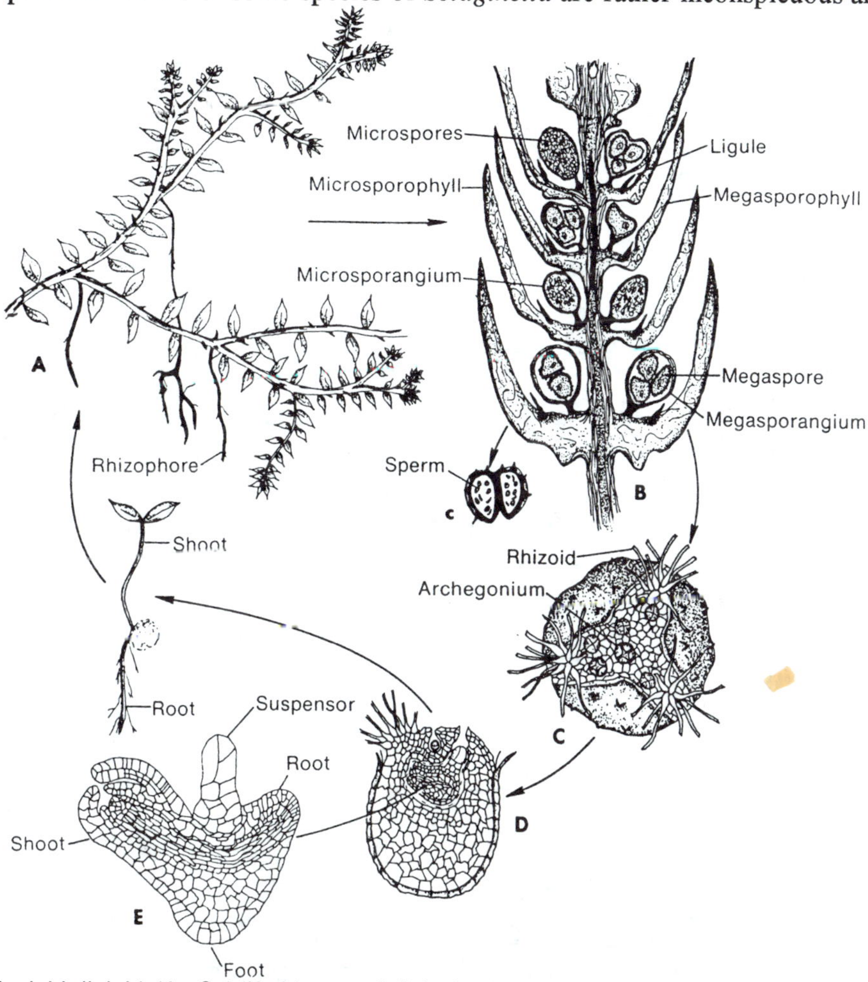

Fig. 5-4. Haploid-diploid (1*n*-2*n*) life history of *Selaginella* (spike moss), a heterosporous lycopod (Lycophyta). *A,* mature SPT of *S. kraussiana,* a species with two sizes of leaves; *B,* longisection of cone with male and female structures, respectively, "micro-" and "mega-" (the ligule is a small leaflike appendage); *c,* section of microGPT (male GPT); *C,* megaGPT (female GPT) protruding through the opening in megaspore wall caused by separation along the trilete mark; *D,* longisection of megaGPT; *E,* longisection of embryo (very young SPT); *F,* young SPT still attached to megaGPT. Note the alternation of morphologically dissimilar GPTic and SPTic generations (phases), with the SPT dominant over the GPT and ultimately independent of it. A heterosporous plant produces morphologically dissimilar spores (compare with Figs. 5-1, 5-3). Locate where meiosis and syngamy (fertilization) occur and which structures are 1*n* and 2*n*. (From Norstog & Long 1976:337; redrawn from *Cryptogamic botany,* 2nd ed., vol. 2, by G. M. Smith, © 1955 by McGraw-Hill Book Co., Inc., New York. Reprinted by permission of the publisher.)

"buried" in the middle of a shoot. In other words, a shoot produces a cone and then the tip of the cone continues to grow into a vegetative shoot.

B. Horsetails (Sphenophyta)

All the extant horsetails (Sphenophyta) have a homosporous life history, as shown in Figs. 5–5 and Sup5–4. The extant horsetails are characterized by the following features:

- jointed, ribbed stems with leaves (these are small in extant taxa) (Figs. 5–5A, 5–6) in whorls, the stems divided into
  - ***nodes,*** the regions where the leaves are attached, and
  - ***internodes,*** the regions between the nodes (see also definitions in Lab Exercise 6, Part I-A);
- sporangia borne on highly modified stalks (sporangiophores) that in turn are aggregated into terminal cones (strobili) (Fig. 5–5B);
- silica in the epidermal cell walls.

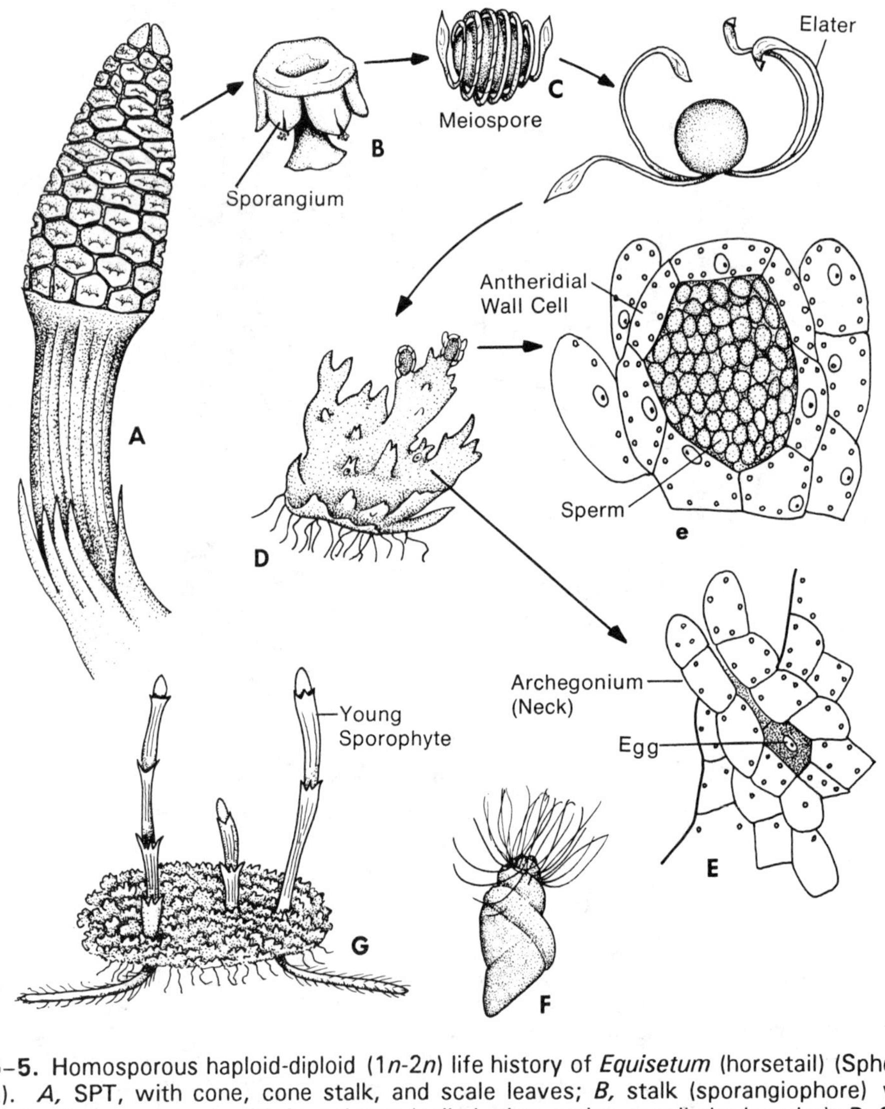

Fig. 5–5. Homosporous haploid-diploid (1*n*-2*n*) life history of *Equisetum* (horsetail) (Sphenophyta). *A,* SPT, with cone, cone stalk, and scale leaves; *B,* stalk (sporangiophore) with sporangia; *C,* spores, each with four elaters (coiled when moist, uncoiled when dry); *D,* GPT; *e, E,* longisections of antheridium (*e*) and archegonium (*E*); *F,* sperm (each cell has about 80 flagella); *G,* GPT with attached young SPTs. Note the alternation of morphologically dissimilar GPTic and SPTic generations (phases), with the SPT dominant over the GPT and ultimately independent of it. A homosporous plant produces morphologically similar spores (compare with Fig. 5–4). Locate where meiosis and syngamy (fertilization) occur and which structures are 1*n* and 2*n*. (From Norstog & Long 1976:326.)

The only extant genus of Sphenophyta is *Equisetum* (horsetail, scouring rush). The genus has about 29 species that are cosmopolitan (except Australasia), occurring chiefly in wet and swampy places. Some South American species of *Equisetum* grow up to 13 m (43 ft.) tall, but only 2 cm (0.8 in.) thick, all this height being due exclusively to primary growth (Mabberley 1987). The silica in the epidermal cells of *Equisetum* imparts an abrasive texture to the shoots. The common name "scouring rush" derives from the colonial use of *Equisetum* to scrub pots and pans ("Colonial Brillo"). Children like to pull apart the internodes to make a popping sound.

●●● See Fig. 5-5A. Then examine the DEMO live material of *Equisetum.* Note the general habit, the size and arrangement of the leaves at the nodes, the long ribbed internodes, and any cones present.

●●● *Suggested diagram and labels:* *Equisetum* (horsetail) shoot: stem, scale leaves, node, internode, and, if present, cone (strobilus).

Each spore of *Equisetum* has four spoonlike ***elaters*** (Fig. 5-5C), which represent part of the outer spore wall (not a cell, no ploidy, i.e., 0*n*) and thus are not complete cells or clusters of cells as are, respectively, the elaters (2*n* cells) of the liverworts and hornworts of the bryophytes (see Lab Exercise 4, Part II-B). However, the elaters of *Equisetum,* like those of the bryophytes, are hygroscopic and aid in spore dispersal by twisting and turning with changes in moisture (Fig. 5-5C), that is:

- moist or humid conditions = elaters coiled;
- dry conditions = elaters uncoiled, spores wafted off.

Apparently, the uncoiling of the elaters aids in dehiscence of the sporangium, and the tangling up of the elaters releases clumps of spores.

C. Whisk ferns, psilophytes (Psilophyta)

The whisk ferns (Psilophyta) are characterized by the following features:

- roots absent (*Psilotum* also lacks true leaves);
- axes (often referred to as stems) repeatedly forking (dichotomously branched) in a regular manner;
- compound sporangia borne on specialized, very short, lateral branches.

The whisk ferns have only two genera and 12 species found in the subtropics and tropics of both the northern and particularly the southern hemispheres. The division has no fossil record, and its relationships are obscure.

●●● Find the aforenoted structural features on the DEMO live plant of *Psilotum nudum.*

III. EXTINCT PTERIDOPHYTES

The ferns (Pterophyta), lycopods (Lycophyta), and horsetails (Sphenophyta) were well represented in the coal swamps of the Carboniferous 280 to 345 million years ago, and especially the latter two groups had many tree (arborescent) members in the past. Compared to the ferns, which are still today an actively evolving group with many representatives, the lycopods and horsetails are remnant groups that were much better developed in the past (Fig. 5-6). All extant pteridophytes are herbs (herbaceous) and thus lack the woody tissue (see Lab Exercises 6 and 7) found in so many seed plants. Although extant pteridophytes almost entirely lack secondary growth, some species become quite large, for example, tree ferns (see Part I-A).

A. Carboniferous pteridophytes

The coal swamps of the Carboniferous Period contained great diversity of plant and fungal life.

●●● See Fig. 5-6. Then examine the DEMO of the famous reconstruction (from the Field Museum of Natural History, Chicago) of a Carboniferous coal swamp and the explanatory DEMO diagram (from Norstog & Long 1976:332). This reconstruction shows various fossil horsetails, seed ferns, and especially lycopods. Finally examine the DEMO diagrams (from Hirmer 1927:302, 413, and Norstog & Long 1976:322, 334) of reconstructions of fossil lycopods and horsetails. Some actual fossil material of Carboniferous pteridophytes may also be available on DEMO.

B. Progymnosperms (Progymnospermophyta)—progenitors of the seed plants

The progymnosperms (Progymnospermophyta) are a strictly fossil (Devonian) group that was proposed in 1960 and that has proven to be of great evolutionary significance. The progymnosperms are characterized by the following contrasting features:

- vegetative anatomy like that of the conifers, that is, with large woody, pinelike trunks, but
- reproductive morphology like that of the pteridophytes, that is, with sporangia and spores but *no* seeds.

Because of this blend of coniferophytic vegetative anatomy and pteridophytic reproductive morphology, the progymnosperms are regarded as the pteridophytic progenitors or ancestors of the seed plants, specifically of the gymnosperms.

IV. CODA

••• Examine the DEMO of the militantly pteridological cartoon (from Canright 1962:161), which appeared during World War I. "The figures (Horse-tail Marines) firing the annulus-operated sporangial catapults are represented by *Equisetum* spores with their extended elater-like appendages. The clever reference to ammunition of two calibres made by the 'Selaginella Heterospory Company' is self-explanatory" (quote from Canright 1962:161).

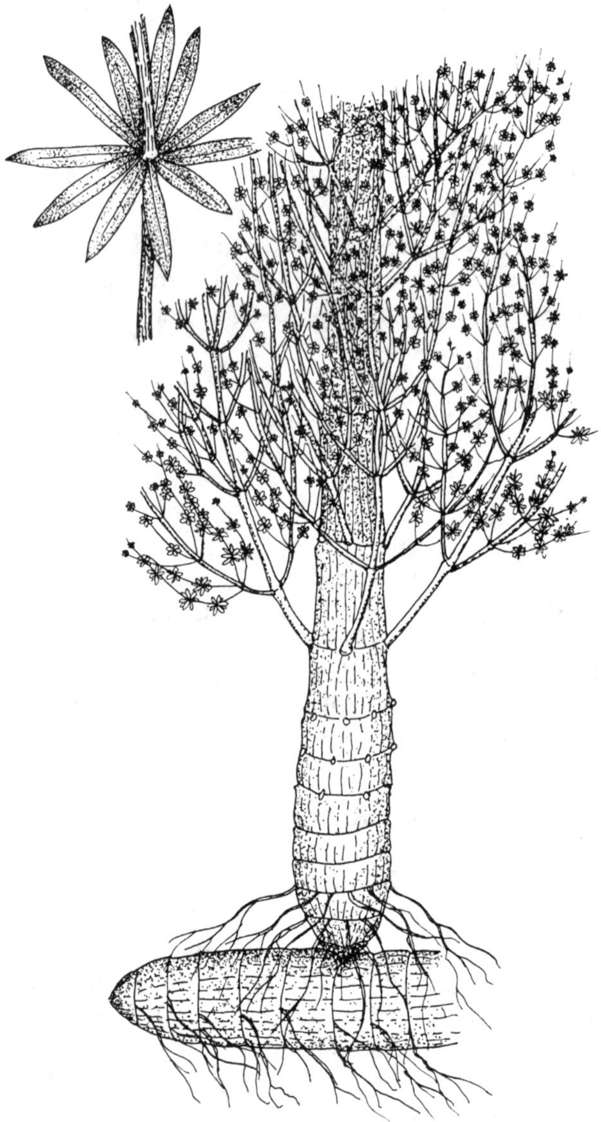

Fig. 5–6. Reconstruction of *Calamites,* an Upper Devonian and Carboniferous horsetail (Sphenophyta) somewhat resembling the modern-day *Equisetum.* (From Norstog & Long 1976:322; redrawn from *Cryptogamic botany,* 2nd ed., vol. 2, by G. M. Smith, © 1955 by McGraw-Hill Book Co., Inc., New York. Reprinted by permission of the publisher.)

NAME ______________________________ LAB SECTION # ______________ DATE ____

TITLE ______________________________

__________________ TOTAL MAG ____

TITLE ______________________________

__________________ TOTAL MAG ______

TITLE ______________________________

__________________ TOTAL MAG ____

TITLE ______________________________

__________________ TOTAL MAG ______

LAB EXERCISE **6**

Basic Morphology and Anatomy of Vascular Plants I

(Basic Organization of the Vascular Plant Body; Structural Adaptations for a Terrestrial Life)

OBJECTIVE

To examine the salient features of vegetative morphology and anatomy of representative examples of gymnosperms and especially angiosperms, particularly in the context of the basic organization of the vascular plant body and structural adaptations for a terrestrial life.

Note: Read this material *before* coming to the lab; there is insufficient time in the lab to read the material from scratch. Due to the large amount of background information presented here, a ●●● is used to mark those paragraphs calling for actual observations or diagrams in the lab. Most unmarked paragraphs reiterate information presented in the lecture. A bullet (●) signifies an important term or concept.

Suggested Lab Diagrams (see part noted for suggested labels):
Allium cepa (onion, an angiosperm and monocotyledon) root apex l.s. (Part I-D2)
Coleus (coleus, flame nettle, painted leaf, an angiosperm and dicotyledon) shoot apex l.s. (Part I-D3)

PERSPECTIVE

The vascular plants are plants with conducting or vascular tissue, that is, water-conducting xylem and nutrient-conducting phloem (see Part II-D). Although they exhibit appreciable structural diversity, the typical vascular plant can be resolved into the basic morphology and anatomy described below and in Lab Exercise 7. Morphological and anatomical specializations may occur so that some of these structures are absent in a given plant. Although this lab exercise and the following one focus on the seed plants, most of this information also applies to the pteridophytes. Most extant pteridophytes, however, lack secondary growth. Note that almost everything mentioned in this and the following lab exercise pertains to the SPT rather than to the GPT. In addition, the "typical" cases described below have many exceptions, as shown by standard books on plant anatomy (Esau 1965; Fahn 1990).

Note: Before beginning this lab exercise, you should review:
- Lab Exercise 1, Supplement B, on permanent microscope slide preparations;
- Supplement 2 on basic descriptive terminology of morphology and anatomy;
- Supplement 6 on the characteristics of and origin of the land plants.

I. BASIC ORGANIZATION OF THE VASCULAR PLANT BODY

The following sections focus on (1) organs of vascular plants (external morphology), (2) tissues and tissue systems, (3) basic tissue regions of stems, roots, and leaves, (4) apical and transitional (primary) meristems, (5) primary versus secondary growth, and (6) simple versus complex tissues. Part II discusses additional aspects of tissue structure, especially in the context of ecological and physiological adaptations, whereas Parts I and II of Lab Exercise 7 focus on the anatomy of typical stems, roots, and leaves and reprise information presented earlier.

A. Organs of vascular plants (external morphology)

The ***habit*** of a plant is its external, morphological appearance. The typical vascular plant consists of:

- a usually below-ground (underground, subterranean) ***root system;***
- a usually above-ground (aerial) ***shoot system*** consisting of:
  - ***leaves,*** the lateral, green photosynthetic appendages;
  - ***stems,*** the axis part to which the leaves are attached (a ***rhizome*** is a stem that is underground, at least partly so);
  - ***axillary buds,*** which are condensed shoots that develop into branches (lateral shoots).

Rhizomes are especially common in pteridophytes and monocotyledons. An axillary bud usually occurs in the ***axil*** (the upper angle between a leaf and its stem) of each leaf. Horticulturalists know that pinching the apical or terminal bud will cause a plant to become bushy by the axillary buds growing into side branches. This eliminates the apical dominance resulting from hormones produced by the terminal bud; therefore, the axillary buds can now expand.

As the shoot grows it forms new increments or units of stem and new leaves in a direction from the base of the plant upward (i.e., acropetally). Thus the oldest part of the shoot is at the basal (proximal) region and the youngest (most embryonic) part at the apical (distal) region. The point of insertion of one or more leaves is termed the ***node,*** and the region of stem between two nodes is called the ***internode.*** In contrast, roots are simpler in structure and are not differentiated into nodes and internodes.

Leaves exhibit much morphological diversity. As in the pteridophytes (see Fig. 5-2 and Lab Exercise 5, Part I-A), leaves of the seed plants may be ***simple*** (undivided) or ***compound*** (divided), in the latter case divided once to several times into subdivisions called ***leaflets.*** Leaves (and reproductive parts) exhibit various arrangements on stems:

- in the ***alternate*** condition one leaf occurs at the same node;
- in the ***opposite*** condition two leaves occur opposite each other at the same node;
- in the ***whorled*** condition three or more leaves occur at the same node.

The typical simple leaf consists of:

- a stemlike ***petiole*** attached to the actual stem (but many leaves lack petioles);
- a flattened photosynthetic ***blade*** (lamina).

The shoot system just described is the ***vegetative shoot system.*** Vascular plants bear reproductive structures of diverse morphology (see Lab Exercises 5 and 8 to 10). Some of these, for example, cones and flowers, have been interpreted as ***reproductive shoot systems.***

••• Examine the DEMO live plants of *Coleus* (coleus, flame nettle, painted leaf). Identify the following structures: shoot system with stems and leaves, nodes, internodes, petioles and blades of leaves, axillary buds. The leaves are arranged on the stem in pairs at right angles to each other; that is, each node has two leaves. At the terminal or distal regions of the shoot note that the leaves are smaller and that the internodes are not as long. Also note the extensive root system of the plant that had soil washed from its roots.

B. Tissues and tissue systems

Tissues are groups of similar cells organized into a structural and functional unit, that is, parenchyma, epidermis, the conducting tissues xylem and phloem, the supportive tissues collenchyma and sclerenchyma, and cork. Part II below treats these tissues in more detail.

A tissue or group of tissues in a plant or plant organ can also be structurally and functionally organized into a unit or ***tissue system.*** There are three main tissue systems, each comprised of the following structures:

- the ***dermal system,*** the bounding layer of plants: epidermis, periderm;
- the ***vascular system*** (fascicular system, conducting system): vascular tissue (xylem and phloem), both primary and secondary;
- the ***ground system*** (fundamental system): pith, cortex, mesophyll, endodermis, that is, all parts of the plant other than the dermal and vascular systems.

These tissue systems are found in both vegetative and reproductive organs. Part II treats the functions of these tissue systems. Actually, a fourth tissue system unique to reproductive organs should also be distinguished, namely:

- the ***reproductive tissue system,*** spore-producing (sporogenous) tissue and associated structures (e.g., GPTic tissue of seed plants), the function of which is to perpetuate the species via sex, that is, meiosis and syngamy (fertilization).

C. Basic tissue regions of stems, roots, and leaves

A ***tissue region*** is simply a geographic region comprised of one or more tissues, that is, epidermis, cortex, endodermis, pericycle, xylem, phloem, pith, mesophyll, and periderm. Part II below and Lab Exercise 7, Parts I and II, will treat these tissue regions in detail. In brief, however, plant organs (stems, roots, leaves, and reproductive parts) have the following parts:

- an ***epidermis,*** the outermost, bounding layer, typically a single layer of cells;
- conducting or ***vascular tissue*** consisting of
  - ***phloem,*** the nutrient-conducting tissue, and
  - ***xylem,*** the water-conducting tissue;
- ***ground tissue*** (fundamental tissue), all the tissue between the vascular tissue and the epidermis.

In transections (cross sections) xylem cells typically appear larger and thicker-walled than phloem cells and stain red with the histological stains commonly used to prepare permanent microscope slides (see Lab Exercise 1, Supplement B). The ground tissue internal to the vascular tissue is the ***pith,*** whereas that external to the vascular tissue is the ***cortex.*** Stems typically have both cortex and pith, but that roots generally have only cortex. In leaves (and reproductive organs) the ground tissue is called ***mesophyll*** rather than cortex or pith.

●●● Examine the DEMO diagram (from Gifford & Foster 1989:34) of the three tissue regions of organs. DEMO models of a stem, root, and leaf are also available. Your TA or instructor will probably discuss their main parts.

D. Apical and transitional (primary) meristems

Plants grow by their shoot and root tips due to the activity of ***meristems,*** undifferentiated regions of usually active cell division (i.e., mitosis). The very distal tip of the shoot is the ***shoot apical meristem.*** It and the parts just below it comprise the ***shoot apex.*** Roots comparably have a ***root apical meristem*** and a ***root apex.***

D1. Transitional (primary) meristems

Apical meristems are permanent and self-perpetuating and produce (immediately beneath themselves) three regions of derivatives, the ***transitional meristems*** (primary meristems). The transi-

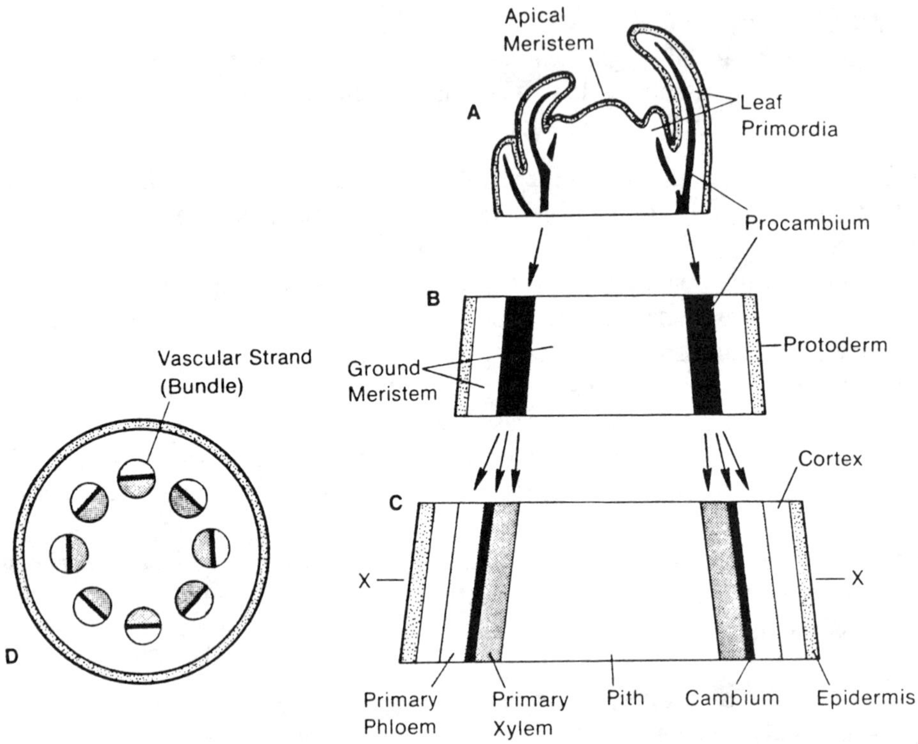

Fig. 6–1. Longisection of shoot apex of a dicotyledon. *A,* apical region composed of meristematic cells; *B,* region of continuing cell division, cell elongation, and differentiation of fundamental tissues; *C,* region of cell specialization and maturation; *D,* transection of *C* at X-X (note the vascular bundles forming a ring) (compare with Fig. 7–4B). (From Norstog & Long 1976:457.)

tional meristems are *non*-self-perpetuating in that they differentiate into the mature primary tissues, as follows (Figs. 6–1, 6–2):

- ***protoderm,*** the outermost transitional meristem that is always a single layer of cells and that will differentiate into the epidermis;
- ***provascular tissue*** (procambium), the internally located transitional meristem (with long narrow cells) that is continuous with the conducting or vascular tissue and that thus will differentiate into the xylem and phloem;
- ***ground meristem,*** all the intervening or "filler" tissue between the protoderm and provascular tissue; this will differentiate into the pith and cortex of stems and roots or into the mesophyll of leaves and reproductive organs.

The cells of the apical meristem are characteristically small and have nearly equal amounts of nuclear and cytoplasmic material; that is, the cells are not appreciably vacuolate and hence stain densely. Further from the apical meristem the cells become more vacuolate and thus stain *less* densely. Cell differentiation is occurring, and this is the region of the transitional meristems. Further yet from the apex are the mature tissues. In other words, as one passes down the axis of the shoot there is a gradient from a completely meristematic state (apical meristem) through a transitional state (transitional meristems) to a mature or nearly mature state (primary tissues).

D2. Root apical structure

See Fig. 6–2. Lacking lateral structures (leaves and branches), root apices are simpler than shoot apices. Roots have a structurally distinctive region, the ***root cap,*** covering and protecting the root apical meristem and the rest of the young, growing root apex. Cells of the root cap are relatively large and vacuolate. The root apical meristem differentiates proximally the three transitional meristems noted above. The root cap secretes mucilage and continually sloughs off cells as the root pushes through the soil.

●●● See Fig. 6–2. Then examine the DEMO model of a longisection of a root apex and compare it with a slide of a longisection of the root apex of *Allium cepa* (onion). On both identify the following structures: root apical meristem, root cap, and transitional meristems (protoderm, provascular tissue, ground meristem). Incidentally, the slide of *Allium* nicely shows interphase and the stages of mitosis (see Lab Exercise 1, Part IV).

●●● *Suggested diagram and labels: Allium cepa* (onion, an angiosperm and monocotyledon) root apex l.s.: apical meristem, transitional (primary) meristems (protoderm, provascular tissue, ground meristem), region of mature primary tissues, root hair region. Note the absence of leaf primordia, young leaves, and axillary buds!

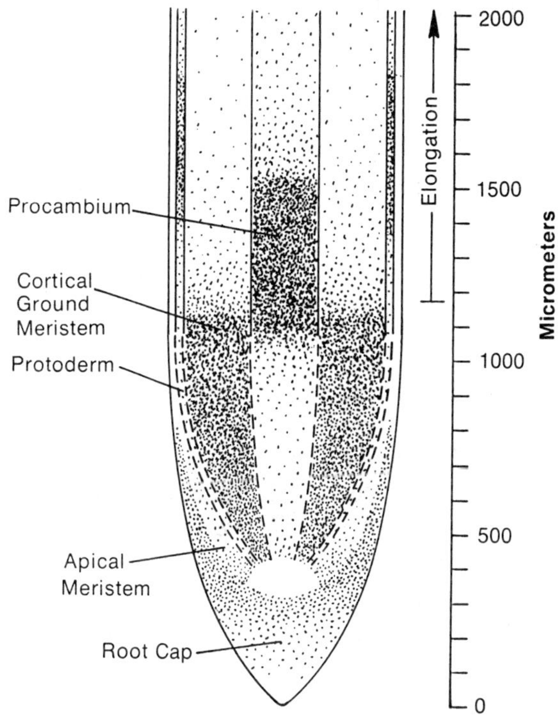

Fig. 6–2. Longisection of root tip of *Allium cepa* (onion) based on the work of Jensen & Kavaljian (1958). The relative number of cell divisions (mitoses) in each part corresponds to the intensity of the stippling. The unshaded area in the apical meristem showed no mitoses at noon, but these occur here at other times. (From Norstog & Long 1976:483; redrawn from *Botany,* 5th ed., by C. L. Wilson, W. E. Loomis & T. A. Steeves, © 1971 by Holt, Rinehart and Winston, Inc., New York. Reprinted by permission of the publisher.)

Root hairs are outgrowths of epidermal cells and function in water uptake. Root hairs occur near the root tip, grow rapidly, and are usually short-lived. They become closely intertwined with soil particles and greatly increase the total surface area of the root tip. Root systems are very extensive, even in small herbs. One plant of *Secale cereale* (rye) was estimated to have over 14 million small roots with over 14 billion root hairs in 0.057 m^3 (2 $ft.^3$) of soil after only four months of growth; the total surface area of the root hairs approximated that of a tennis court (Norstog & Long 1976:486).

●●● Examine the DEMO live material of cuttings of *Begonia* (begonia) grown for several weeks in water. Note the many root hairs near the tips of the adventitious roots (see also comments on *Begonia* in Lab Exercise 1, Part V).

D3. Shoot apical structure

See Fig. 6–1. The shoot apical meristem forms leaves in very close succession without very marked internodal elongation between the times of leaf inception. Hence the young leaves appear closely packed together and form what is called a ***terminal bud.*** The shoot apex bears ***leaf primordia*** (i.e., the earliest stage of leaf differentiation) of various size. Axillary buds develop in the shoot apex (though *not* in the apical meristem) in the axils of the leaves, frequently in the axil of the second or third leaf (or leaf pair) from the apex. Proximally, active cell division will result in growth in length of the stem and differentiation of tissues in it.

●●● See Fig. 6–1. Then examine a slide of a longisection of the shoot apex of *Coleus,* which is also shown in a DEMO photograph (from W. A. Jensen, unpublished, or from another available source). The shoot apical meristem appears as a small, rather densely staining dome between the pairs of the most recently initiated leaf primordia, that is, the youngest or smallest leaf pairs. The slide and DEMO photograph will reflect the leaf arrangement observed in the mature plant (see Part I-A). That is, *Coleus* has leaves oppositely arranged at right angles to the leaves of the next pair; thus the two leaves of a node are evident in one plane of the longisection, but the two leaves of the adjacent node will *not* be (note that Fig. 6–1A shows alternate leaf arrangement). Identify the following structures: shoot apical meristem, transitional meristems (protoderm, provascular tissue, ground meristem), leaf primordia, and axillary buds.

••• *Suggested diagram and labels: Coleus* (coleus, flame nettle, painted leaf, an angiosperm and dicotyledon) shoot apex l.s.: apical meristem, leaf primordia and young leaves, axillary bud, transitional (primary) meristems (protoderm, provascular tissue, ground meristem), region of mature primary tissues.

D4. Shoot versus root tips

Compare Figs. 6–1 and 6–2. Shoot and root apices differ fundamentally as follows:

- The shoot apex is exposed (root cap absent), has leaves and thus is segmented into nodes and internodes, and produces the leaves and branches (the axillary buds) externally.
- The root apex is covered (root cap present), lacks leaves and thus is not segmented into nodes and internodes, and produces its branches (lateral roots) internally some distance from the apical meristem.

Stems may also bear adventitious roots.

E. Primary versus secondary growth

The structures referred to above are *all* derived directly from the shoot and root apical meristems and thus are ***primary*** in nature. Therefore, tissues derived directly from the apical meristems, via the transitional meristems, are called ***primary tissues*** (i.e., epidermis, parenchyma, collenchyma, sclerenchyma, primary xylem, primary phloem), and any growth producing these tissues and lateral appendages (leaves, branches) is called ***primary growth.*** In summary, primary growth results especially in increase in length of organs due mainly to the activity of the apical meristems. In some cases ***intercalary meristems*** located at nodes and away from the apices are also involved in primary growth, for instance, in the growth of grass leaves. Here, leaves still continue to grow due to their basal, intercalary leaf meristems, even though a lawn mower had cut off the leaf tips.

Many stems and roots increase in diameter through the process of ***secondary growth*** due to the addition of ***secondary tissues*** by the activity of ***lateral meristems*** (Figs. 7–4, 7–5). There are two lateral meristems:

- The ***vascular cambium*** (singular, plural ***cambia***) produces ***secondary xylem*** (wood) internally and ***secondary phloem*** externally; these tissues comprise the ***secondary vascular tissue.***
- The ***cork cambium*** (phellogen) produces ***cork parenchyma*** (phelloderm) internally and ***cork*** (phellem) externally; these tissues comprise the ***periderm.***

Both meristems are a single layer of cells. Due to the activity of the vascular or cork cambia, the resultant secondary tissues usually exhibit (as seen in transection) cells nicely aligned in a radial direction from the center of the axis. Secondary growth results mainly in increase in width (girth, thickening) of organs. Woody plants usually have large amounts of ***wood*** or secondary xylem.

The ***primary body*** of a plant is that part of the plant, or the entire plant if no secondary growth occurs, that arises from the apical meristems (both shoot and root), any intercalary meristems present, and their derivative transitional meristems. In contrast, the ***secondary body*** of a plant is that part of the plant (i.e., secondary xylem, secondary phloem, cork, cork parenchyma) that is added to the primary body by the activity of the lateral meristems (vascular and cork cambia).

All plants undergo primary growth, but only some plants undergo secondary growth. Herbs (herbaceous plants) have only primary growth, or this and very small amounts of secondary growth.

Woody plants such as shrubs and trees have primary growth and appreciable secondary growth. Parts I and II of Lab Exercise 7 will contrast organs lacking and having, respectively, secondary growth.

The above terminology can be summarized as follows:

- *primary growth* = primary body = apical (shoot and root) meristems and intercalary meristems ——> transitional meristems (i.e., protoderm, provascular tissue, ground meristem) ——> primary tissues (i.e., epidermis, parenchyma, collenchyma, sclerenchyma, primary xylem, primary phloem);
- *secondary growth* = secondary body = lateral meristems (i.e., vascular and cork cambia) ——> secondary tissues (i.e., secondary xylem, secondary phloem, cork, cork parenchyma).

F. Simple versus complex tissues

Simple tissues are homogeneous, consisting of mostly a single cell type, for example, epidermis, parenchyma, collenchyma, and sclerenchyma. In contrast, ***complex tissues*** are heterogeneous, consisting of two or more cell/tissue types, that is, xylem, phloem, and periderm.

II. STRUCTURAL ADAPTATIONS FOR A TERRESTRIAL LIFE

Supplement 6 gives background information on the origin and early evolution of the land plants. The first land plants needed to evolve various morphological and anatomical adaptations for life on land. These adaptations still manifest themselves today not only in the seed plants described below, but also, though to a somewhat lesser extent, in various bryophytes and especially pteridophytes. Suffice it to say, the bryophytes lack vascular tissue and most of the other specialized tissue types of the vascular plants. However, some bryophytes have a cuticle (e.g., on the leaves of some mosses and liverworts) or stomata (e.g., on the capsules of the hornworts and most mosses) like those of the vascular plants, and some specialized mosses have conducting tissue analogous to the xylem and phloem of land plants. Bryophytes and almost all extant pteridophytes lack secondary growth.

A. General—multicellularity

All land plants are multicellular structures. Multicellularity not only offers to the internally located cells protection against desiccation, but also enables the differentiation and specialization of tissues. Recall that many algae consist of parenchyma (see Lab Exercise 3, Part I-G), that is, thin-walled live cells that often have large vacuoles and that usually function in storage or photosynthesis. Comparable parenchyma (e.g., most of the flesh of apple and other fruits) occurs in all land plants, which have evolved a diversity of other tissue types.

B. Anatomical adaptations—the dermal system

The diverse functions of the dermal system include:

- protection against mechanical injury (i.e., from weather, pathogens, predators, etc.) and also desiccation (i.e., restriction of water loss);
- capability of gaseous exchange between the atmosphere and the interior of the plant (via stomata and lenticels—see below);
- absorption (primarily via root hairs of the root epidermis);
- secretion;
- even support.

The dermal system consists of the epidermis and, if developed, the periderm.

B1. Epidermis

See Figs. 6–1C, D, 6–3, 7–1A, and 7–3. The ***epidermis*** is the outermost, bounding tissue region of plant organs and usually consists of a single layer of cells. The epidermis consists of several cell types:

- ***ordinary epidermal cells*** and interspersed among these
- pairs of ***guard cells*** comprising a stoma (Figs. 6-3, 7-1A, 7-3),
- ***trichomes*** consisting of hairs, scales, and other types, and
- ***special cell types*** (e.g., silica cells).

The term ***stoma*** (singular, or stomate; plural ***stomata*** or stomates) refers to the guard cells and the ***pore*** (aperture) they enclose [The less preferable definition of "stoma" (e.g., Bold et al. 1987) is simply the hole, sans cells]. The cell walls next to the pore are usually thickened. Typically, chloroplasts occur in the epidermis only in the guard cells. Hence these are green, whereas other epidermal cells are usually transparent.

A ***cuticle,*** which consists of the waxy material ***cutin,*** covers the epidermis (sometimes cutin also occurs in and between the cell walls) and functions to prevent water loss from the plant (Fig. 7-3). The whitish (glaucous) coating found on many leaves and fruits is actually the cuticle. The cuticle is impervious to gasses and water vapor and thus protects organs from desiccation (not surprisingly, roots typically lack a cuticle). However, gases must pass into and out of organs, because carbon dioxide is needed for photosynthesis. The ***guard cells*** open and close the stomatal pores and hence regulate gas exchange between the atmosphere and the interior of the plant (Figs. 6-3, 7-3).

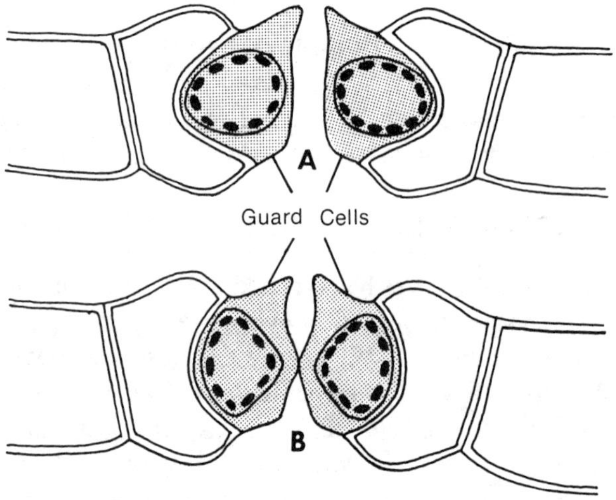

Fig. 6-3. Sections showing shape of guard cells of open (*A*) and closed (*B*) stomata of the leaf of *Phaseolus vulgaris* (bean). The term stoma refers to the guard cells and the pore (aperture) they enclose. Note the characteristically thickened cell walls next to the pore. (From Norstog & Long 1976:478; redrawn from *Botany,* 5th ed., by C. L. Wilson, W. E. Loomis & T. A. Steeves, © 1971 by Holt, Rinehart and Winston, Inc., New York. Reprinted by permission of the publisher.)

Epidermal hairs (trichomes) (Fig. 7-1A) also frequently occur in the epidermis and, among other things, may represent an additional mechanism to minimize water loss through evaporation. The epidermis also offers some mechanical support to young stems, whereas the root epidermis functions mainly in absorption via the root hairs (see Part I-D2). Some epidermal cells also function in secretion and as trigger hairs, as in carnivorous plants such as sundew (*Drosera*) or Venus fly trap (*Dionaea muscipula*) (see Lab Exercise 11, Part I-C).

The epidermis thus is the important interface layer between the plant and the environment. Not surprisingly, pteridophytes of the Lower Devonian had a well-defined epidermis with a cuticle and stomata (Fig. Sup6-1).

●●● See Fig. 7-3. Then examine a slide of a leaf transection of *Syringa vulgaris* (lilac). Identify the following structures: cuticle, epidermis, stomata, guard cells, stomatal pores (apertures), ordinary epidermal cells, and epidermal hairs (trichomes). Because the cuticle is very thin, examine the DEMO slide of a leaf transection of *Agave* (agave) and note its very conspicuous cuticle (the pinkish layer overlying the green epidermal cells). Many species of *Agave* live in the desert. Why might a thick cuticle be advantageous in such a habitat?

●●● The epidermis can be readily stripped from a leaf. Examine the DEMO slide of an epidermal peel from a leaf of *Iris* (iris). Note the stomata (identify their pores and guard cells) interspersed among the ordinary epidermal cells. Reconcile this surface view of the guard cells with the sectional view seen on the slides of *Syringa* or *Agave*.

●●● See Fig. 6–2. Finally reexamine the slide of a longisection of the root apex of *Allium* (onion) from Part I-D2. Most of the dermal system shown here is protoderm, as opposed to its differentiated epidermis.

B2. Periderm

See Figs. 7–4C and 7–5. A characteristic feature of woody plants is the presence of periderm, a secondary tissue consisting of the ***cork cambium*** (phellogen) and its two derivatives, ***cork*** (phellem) and ***cork parenchyma*** (phelloderm). The cork cambium is a lateral meristem and originates in various places, almost always external to the vascular cambium. The cork cambium is always a single layer of cells. It produces radial rows (files) of cells:

- ***cork parenchyma*** internally, in small amounts, the cells at maturity living;
- ***cork*** externally, in large amounts, the cells at maturity non-living and impregnated with ***suberin,*** a waxy, water-proofing material.

Cork is the most important component of the bark of woody plants (see Lab Exercise 7, Part II).

Functions of periderm include chiefly protection against desiccation and mechanical injury. The compact nature of cork (i.e., lack of intercellular spaces) and the presence of suberin in its cell walls provide an effective barrier to water vapor. Subsidiary functions are support and gas exchange. The latter occurs via ***lenticels,*** spongy areas running perpendicularly through the cork surface and permitting gas exchange between interior tissues and the atmosphere.

Periderm and epidermis are analogous tissues, and suberin and cutin analogous chemicals. Periderm functions in regions of secondary growth, epidermis in regions of primary growth.

Quercus suber (cork oak), the cork of commerce, is native to southern Europe and northern Africa. Every eight to ten years, especially in Portugal, its thick cork is stripped from trees older than 20 years for use in insulation, linoleum, tiles, bottle corks, floats, cork-tipped cigarettes (Mabberley 1987).

●●● Examine a slide of a transection of a three-year-old stem of *Tilia americana* (basswood, American linden). The periderm occurs at the periphery of the stem and consists of:

1. several layers of red stained, tannin-filled cork cells;
2. the cork cambium, a single layer of cells with protoplasts and nuclei located at the inner limits of the periderm;
3. cork parenchyma just internal to this cambium.

Note the radial alignment of the cells.

●●● Examine the DEMO of bottle corks. As noted in Lab Exercise 1, Part III, a typical wine bottle cork has about one billion cells. The dark streaks are the lenticels. There are two main variations:

1. The classic bottle cork is a single piece of cork, with the lenticels running perpendicular to the long axis of the cork. Why are the lenticels oriented that way?
2. Laminated bottle corks have appeared more recently, often as champagne corks. In these the lenticels usually run parallel to the long axis of the cork. Why?

A DEMO diagram (from Morey 1973:43) shows how bottle corks are cut from sheets of cork oak. How are the lenticels of cork oriented in the tree?

C. Anatomical adaptations—the ground (fundamental) system

The multifarious functions of the ground (fundamental) system include:

- assimilation (the absorption and incorporation of the nutritive products of photosynthesis);
- storage;
- protection;

- support;
- aeration;
- in the case of the endodermis, regulation of water and mineral balance.

C1. Pith, cortex, mesophyll—aeration systems

See Figs. 7–1A, B, and 7–3. Systems of ***intercellular air spaces*** facilitate the aeration of tissue by enabling a large total surface area between cells and the air in the organ. Mesophyll of leaves typically has very conspicuous intercellular air spaces (Fig. 7–3). Stems and roots also have intercellular air space systems in their pith and cortex. Aquatic plants, in particular, have modified parenchyma (***aerenchyma***) with large intercellular air spaces for buoyancy and air circulation (see Lab Exercise 11, Part II-A).

●●● See Fig. 7–3. Then reexamine the slide of a leaf transection of *Syringa* (lilac). Identify the following structures: mesophyll, intercellular air spaces.

C2. Pith, cortex, mesophyll—support

The fundamental system has two types of tissue specialized solely or mainly for support:

- ***collenchyma,*** live elongate (lengthened) cells with unevenly thickened walls found in the peripheral parts of stems, leaves, and reproductive structures and functioning for support;
- ***sclerenchyma,*** characteristically with rather evenly thickened cell walls and including two main types:
  - ***fibers,*** long pointed cells, up to a centimeter or more long, functioning mainly in support, but also in storage and protection;
  - ***sclereids,*** cells of various shapes, that is, roundish (stone cells), star-shaped (astrosclereids), columnar, etc., but not especially long, and functioning mainly for support and protection.

Sclerenchyma occurs in all organs and also in the dermal and particularly the vascular (fascicular) tissue systems.

Collenchyma is found in actively growing, that is, elongating organs. The functional cell is living, not dead. The walls have a high water content, and because of certain chemicals (i.e., pectins), the cell wall can be easily stretched; that is, it is elastic. Thus the cells can grow longer as the organ grows longer, and the collenchyma can maintain its supportive function because it stretches as the plant grows.

In contrast, sclerenchyma is functional after most organ growth is finished (Figs. 7–1A, 7–5). It also occurs in largely dead tissues such as wood, seed coats, and pits of fruits. In fact, at their functional maturity sclerenchyma cells themselves are usually dead, not living. Sclerenchyma has three functions: support (dead cells), protection (dead cells), and storage (live cells).

●●● Examine the DEMO slide of collenchyma from a petiole transection of *Apium graveolens* (celery). The collenchyma appears as groups of silvery, glistening cells in the ridges just beneath the epidermis.

C3. Endodermis—for the regulation of water and mineral balance

See Fig. 7–2. The endodermis is distinguished as follows:

- It is always a single layer of cells.
- It represents the innermost cell layer of the cortex.
- It occurs in roots of all vascular plants and in the stems of most pteridophytes.
- Its cells always bear a ***casparian strip*** consisting of suberin.

The casparian strip is present as a continuous, thickened strip or band on the four ***anticlinal*** cell walls (those perpendicular to the surface of the organ), but is absent from the two ***periclinal*** cell walls (those parallel to the surface) (Fig. 7–2D). The casparian strip is firmly attached at all its points to the protoplast of the endodermal cell (Fig. 7–2B, C) and is impenetrable due to the suberin present. Dissolved solutes can only move through the endodermal cell via its protoplast, which thus regulates what passes through the cell. In other words, the endodermis functions in the water and mineral balance of the plant.

●●● See Fig. 7–2. Then examine the DEMO models of endodermal cells, on which the red tape represents the casparian strips. Finally examine the DEMO slides of a stem transection of the pteridophyte *Psilotum nudum* (whisk fern) and a root transection of *Ranunculus acris* (buttercup). On each locate the endodermis with its casparian strips. In transections the endodermis is readily found at the junction of the cortex and vascular tissue; its casparian strips stain a glistening bright red with common histological stains (see Lab Exercise 1, Supplement B) and seem to pop readily in and out of focus with rapid focusing of the microscope. The endodermis should be more easily visible in the *Psilotum* than in *Ranunculus* because in some slides of the latter secondary wall thickening of the endodermal cells has obscured their casparian strips.

C4. Pericycle—supportive and especially meristematic and storage tissue

See Fig. 7–2A. Pericycle is generally associated with an endodermis and thus is usually absent from organs lacking an endodermis. The pericycle can be distinguished as one to several layers of typically thin-walled, often densely cytoplasmic, parenchyma cells located between the endodermis and the phloem. Pericycle has several functions: storage, support (if its cells become thick-walled), and in roots the meristematic origin of lateral roots and the vascular and cork cambia.

●●● See Fig. 7–2A. Then reexamine the DEMO slide of a root transection of *Ranunculus* (buttercup) from Part II-C3 and observe the pericycle.

D. Anatomical adaptations—the vascular (fascicular) system

The many functions of the vascular (fascicular) system include:

- conduction (water and solutes in the xylem, organic nutrients in the phloem);
- support (especially by the xylem);
- storage of nutrient materials (especially in the secondary vascular tissues).

The vascular tissues, xylem and phloem, occur internal to the cortex (including its endodermis if present) but external to the pith (if this is present). The xylem consists of thick-walled cells that stain red with common histological stains (see Lab Exercise 1, Supplement B). In roots and stems xylem is typically internal to the phloem. This stains differently and often consists entirely of thin-walled cells mostly smaller in diameter than cells of the xylem. However, some axes have thick-walled phloem fibers (e.g., see Part II-D4).

Xylem and phloem, which are almost always associated with each other, are complex tissues consisting of several cell types. The following listing gives their main cell types and their primary functions:

COMPONENTS OF XYLEM:

- *tracheids:* water (and ion) conduction, support;
- *vessel elements* (vessel members): primarily conduction, but also some support;
- *fibers:* support, sometimes also nutrient storage;
- *parenchyma cells:* nutrient storage and radial conduction of nutrients;

COMPONENTS OF PHLOEM:

- *sieve elements* (there are several types): nutrient conduction;
- *companion cells:* assistance of sieve elements in conduction;
- *sclerenchyma elements* (especially fibers): support, sometimes also nutrient storage;
- *parenchyma cells:* nutrient storage and radial conduction of nutrients.

It is the above distinctive conducting elements that actually characterize xylem and phloem. Moreover, xylem and phloem are a combination of dead and live cells at functional maturity, that is:

- dead = xylem conducting elements, and usually xylem and phloem sclerenchyma;
- live = all other things, and rarely xylem and phloem sclerenchyma.

Due to the great complexity of primary and especially secondary xylem and secondary phloem, examples will not be available for most of the details summarized below. In the following discussion no great distinction is made between primary versus secondary xylem and between primary versus secondary phloem.

D1. Xylem—conducting elements (tracheids versus vessel elements)

Tracheids and vessel elements have cell walls that are relatively thick (Figs. 7-1, 7-2A, 7-5) and differ as follows:

- ***Tracheids*** have closed end walls and are long, slender cells.
- ***Vessel elements*** (vessel members) have open (perforate) end walls called ***perforation plates*** and exhibit two main subtypes:
  - long, slender cells with ***scalariform perforation plates,*** that is, with many openings or perforations on each end wall;
  - shorter, broader cells with ***simple perforation plates,*** that is, with only one opening or perforation on each end wall.

The openings (perforations) in the first subtype are typically elongate and arranged parallel to each other in a ladderlike or scalariform pattern. These three cell types are dead when they conduct water. They represent a series from the least to the most efficient in water conduction. Essentially, the very large opening in a simple perforation plate permits the fastest movement of water.

D2. Xylem—supportive elements

The xylem effects its function of support in several ways: chemicals, secondary wall patterns, and sclerenchyma.

a. *Lignin and the submicroscopic structure of cell walls:* The cell wall consists of several chemicals, of which the two most important are ***cellulose,*** a long-chain glucose polymer, and ***lignin,*** a long-chain phenol polymer. Cellulose and especially lignin have various functions, but essentially in tandem they provide considerable strength in a manner similar to reinforced concrete. That is, the cellulose molecules are embedded in a matrix of lignin and other compounds just as in reinforced concrete the steel is embedded in a matrix of concrete.

●●● The basic organization of the cell wall appears to have evolved early in the history of the land flora and appears not to have changed significantly since Devonian times. Electron microscopic studies of wood of *Callixylon* (= *Archaeopteris*), an Upper Devonian (ca. 350 million years ago) pteridophyte and progymnosperm (see Lab Exercise 5, Part III-B), yielded patterns of wall layering similar to those in extant plants. Compare the DEMO photographs (from Schmid 1967:722) of cell wall layering in wood of *Callixylon* with the DEMO photographs (from Côté et al. 1966:434) of similar cell wall layering in wood of extant conifers. *Note:* Ignore the terminology indicated in the captions to these photographs.

b. *Secondary wall patterns of conducting elements:* The thickened cell walls of the conducting elements of xylem provide support. There are two main variations:

- ***pitted walls*** with ***pits*** (thin spots in the wall) and extensive amounts of thick cell wall;
- ***non-pitted walls*** lacking pits and thus with relatively small amounts of thick cell wall and considerable amounts of thin cell wall, two common and important types being:
  - ***annular thickenings*** consisting of rings (donuts) of thickened wall material, and
  - ***helical thickenings*** consisting of helices of thickened wall material.

The significance of the non-pitted types of wall thickenings is that they provide support to elongating structures because the cells can be extended. In contrast, the cell walls of the pitted types of cells can not be extended and hence provide support to structures that have already completed their elongation.

●●● Take part of a leaf (ca. 5 cm, or 2 in.) from the live material of *Agapanthus africanus* (lily-of-the-Nile; blue African lily), fold the leaf in half, and then gently pull the two halves apart a very small amount. You should see some white "threads," which actually are the helical wall thickenings of the conducting elements of the xylem (i.e., tracheids and especially vessel elements). Verify this by examining the leaf halves under a dissecting microscope or, preferably, a compound microscope (for the latter simply hold the leaf halves under the 10X objective; do *not* make a water mount slide preparation).

c. *Sclerenchyma:* Conducting elements, specifically tracheids, were the initial supportive elements in the extinct land plants lacking secondary growth. Then there was specialization, a division of labor so to speak, into xylem that had vessel elements functioning chiefly in conduction, and fibers (sclerenchyma) functioning primarily in support.

D3. Phloem—conducting elements

Most cells of the phloem are rather delicate (Figs. 7-1, 7-2A), consisting of thin-walled cells lacking lignin. Therefore, phloem has been rarely preserved as a fossil. The characteristic conducting elements of the flowering plants are ***sieve tube elements,*** which have distinctive end walls called ***sieve plates.*** A sieve plate has highly differentiated pores. The protoplast extends from cell to cell through the pores and hence conducts nutrients. Some pores may be occluded by a protein (so-called P-protein or slime).

●●● Examine the DEMO photographs (from Esau 1965:668) of occluded sieve plates.

D4. Phloem—supportive elements

While most phloem cells are thin-walled, phloem does provide some support from sclerenchyma cells such as fibers, which are thick-walled (Figs. 7-1A, 7-5).

●●● See Fig. 7-5. Then reexamine the slide of a transection of a three-year-old stem of *Tilia* (basswood, American linden). Note at the outermost part of the phloem the very thick-walled cells (in longitudinal view these appear elongate). These sclerenchyma cells are the type called ***fiber.*** Fibers are also evident in the DEMO slide of *Agave* (agave) examined in Part II-B1; note the bundles of thick-walled, light red-staining cells.

D5. Secondary vascular tissue

The above remarks (in Part II-D) apply largely to both the primary and secondary vascular tissues, although the latter (Figs. 7-4, 7-5) are more complex than the former. Secondary xylem and secondary phloem have all the cell types of their primary tissue counterparts, and then some additional types. However, secondary xylem (wood) has only pitted types of wall patterns, whereas the non-pitted plus pitted types occur in primary xylem. The component cells of primary and secondary vascular tissue have comparable functions. Secondary vascular tissue will be examined in Lab Exercise 7, Part II.

Es ist dafür gesorgt, dass die Bäume nicht in den Himmel wachsen.
It is so arranged that the trees do not grow into the heavens.
—Johann Wolfgang von Goethe, *Dichtung und Wahrheit,* 1811–22

NAME ______________________________ LAB SECTION # ______________ DATE ____

TITLE ______________________________

__________________ TOTAL MAG ____

TITLE ______________________________

__________________ TOTAL MAG ______

TITLE ______________________________

__________________ TOTAL MAG ____

TITLE ______________________________

__________________ TOTAL MAG ______

LAB EXERCISE **7**

Basic Morphology and Anatomy of Vascular Plants II

(Anatomy of Organs; Arrangement of Spores in a Tetrad)

OBJECTIVE

To examine the salient features of vegetative morphology and anatomy of representative examples of gymnosperms and especially angiosperms, particularly in the context of anatomy of organs.

Note: Read this material *before* coming to the lab; there is insufficient time in the lab to read the material from scratch. Due to the large amount of background information presented here, a ●●● is used to mark those paragraphs calling for actual observations or diagrams in the lab. Most unmarked paragraphs reiterate information presented in the lecture. A bullet (●) signifies an important term or concept.

Suggested Lab Diagrams (see part noted for suggested labels):
Medicago sativa (alfalfa, lucerne, an angiosperm and dicotyledon) OR *Trifolium pratense* (red clover, an angiosperm and dicotyledon) young stem x.s. (Part I-A)
Zea mays (corn, maize, an angiosperm and monocotyledon) stem x.s. (Part I-A)
Ranunculus acris (buttercup, an angiosperm and dicotyledon) root x.s. (Part I-B)
Syringa vulgaris (lilac, an angiosperm and dicotyledon) leaf x.s. (Part I-C)

PERSPECTIVE

See the "Perspective" in Lab Exercise 6 and especially its Part I-E contrasting primary versus secondary growth. Note that the chief focus of Lab Exercise 7 is on the main tissue regions of organs, *not* on individual components of, for example, the epidermis or vascular bundles. Part IV on arrangement of spores in a tetrad deals with reproductive morphology of the land plants.

Part I-E of Lab Exercise 6 distinguished between:

- ***primary growth*** resulting especially in increase in length of organs due to the activity of the apical meristems and, if present, intercalary meristems;
- ***secondary growth*** resulting mainly in increase in width (girth, thickening) of organs due to the activity of the two lateral meristems:
  - the ***vascular cambium*** producing ***secondary xylem*** (wood) internally and ***secondary phloem*** externally (these comprise the ***secondary vascular tissue***);
  - the ***cork cambium*** (phellogen) producing ***cork parenchyma*** (phelloderm) internally and ***cork*** (phellem) externally (these tissues comprise the ***periderm***).

It follows that plants can be divided into ***herbaceous*** and ***woody***. Herbs or herbaceous plants have no wood at all, or else very small amounts of wood, that is, only primary growth, or this and very small amounts of secondary growth. In contrast, woody plants such as shrubs and trees have considerable amounts of wood, that is, primary growth and appreciable secondary growth. In other words, all plants undergo primary growth, but only some plants undergo secondary growth.

Note: Before beginning this lab exercise, you should review Supplement 2 on basic descriptive terminology of morphology and anatomy.

I. ANATOMY OF THE MATURE PRIMARY BODY

This section reprises Lab Exercise 6 by focusing on the "typical" transectional (cross sectional) appearance of stems, roots, and leaves that have mainly or exclusively primary growth. Actually, these examples are "typical" only in the sense that they are the ones inevitably used in basic biology and botany courses.

••• Reread Lab Exercise 6, Part I-C, on the basic tissue regions of stems, roots, and leaves. Then examine the slides of a stem, root, and leaf in conjunction with their equivalent models. As you examine the following transections note that xylem cells typically appear larger and thicker-walled than phloem cells and stain red with common histological stains (see Lab Exercise 1, Supplement B).

A. Stem anatomy

See Fig. 6–1C, D, and especially Figs. 7–1 and 7–4A, B. A transection of a typical stem exhibits two main structural variations, proceeding centrifugally (these two types stereotype the two main classes of angiosperms, respectively, the dicotyledons and the monocotyledons—see Supplement 10):

- pith, ring of vascular bundles (xylem internal, phloem external), cortex, and epidermis;
- vascular bundles (xylem internal, phloem external) dispersed in ground tissue (no definite pith or cortex), and epidermis.

The pith is usually parenchymatous, but the cortex may contain collenchyma and sclerenchyma besides parenchyma. The areas between vascular bundles (interfascicular regions) are parenchymatous or parenchymatous and sclerenchymatous (Fig. 7–1).

Stems have undergone numerous morphological and anatomical modifications. The main functions of stems include:

- production and support of leaves and reproductive structures (cones, flowers, etc.);
- the connection between the leaves and roots for long-distance transport of water and nutrients;
- storage of water and nutrients;
- manufacture of nutrients, especially by young stems.

If the ground tissue of stems is mainly parenchyma, photosynthesis and especially storage are important functions. Collenchyma and sclerenchyma are common in stems and function for support. In specialized cases stems function to:

- overtide the organisms in times of winter or summer stress, as by rhizomes, bulbs, and other underground stem parts;
- produce adventitious roots, as the supporting prop roots of corn or maize (*Zea mays*);
- produce new individuals by asexual reproduction (see Lab Exercise 1, Part V);
- disperse the organism, as *Opuntia* (cholla cactus), where the spiny jointed stems easily fall apart;
- defend the organism against predators and pathogens;
- house ants and other organisms that live inside the stems;
- climb.

Lab Exercise 11 deals with various stem modifications of the flowering plants.

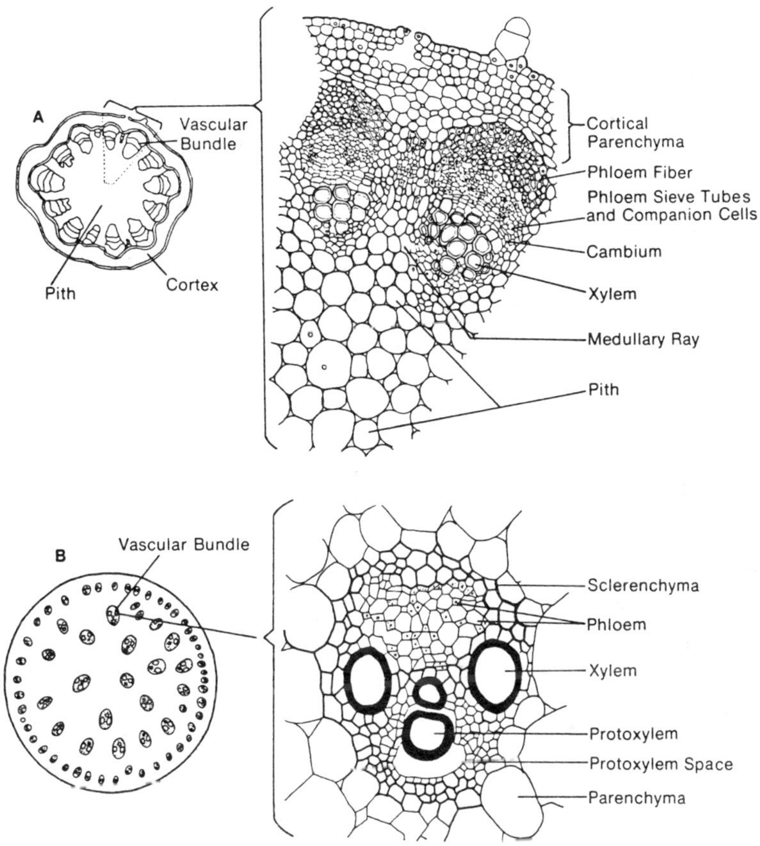

Fig. 7–1. Transections of herbaceous stems. *A,* the dicotyledon *Trifolium* (clover); *B,* the monocotyledon *Zea mays* (corn, maize) ("protoxylem" is the early, first-formed xylem). (From Norstog & Long 1976:460.)

●●● See Fig. 7–1A. Then examine, as available, a slide of a stem transection of *Medicago sativa* (alfalfa, lucerne) OR *Trifolium pratense* (red clover) and *focus only* on the *youngest* (smaller diameter) transection. Identify the following structures, as seen centrifugally:

1. central pith of large parenchyma cells;
2. ring of vascular bundles, each with primary xylem internally and primary phloem externally;
3. parenchymatous region between the vascular bundles, which may be initiating secondary growth (this would be evident as radially aligned cells);
4. cortex of small parenchyma cells;
5. epidermis bearing guard cells and occasional hairs (trichomes).

Is any sclerenchyma present? Does the epidermis have a cuticle?

●●● *Suggested diagram and labels: Medicago sativa* (alfalfa, lucerne, an angiosperm and dicotyledon) OR *Trifolium pratense* (red clover, an angiosperm and dicotyledon) young stem x.s.: pith, vascular bundles, primary xylem, primary phloem, phloem fibers (sclerenchyma), cortex, epidermis, guard cells (several pairs), epidermal hair (trichome).

••• See Fig. 7–1B. Then examine a slide of a stem transection of *Zea mays* (corn, maize). Identify the following structures, as seen centrifugally:

1. parenchymatous ground tissue;
2. vascular bundles, each with primary xylem internally and primary phloem externally, both surrounded by fibers;
3. epidermis.

A distinct pith and cortex are absent. Although the vascular bundles appear "scattered" in transection, they have an orderly arrangement over the lengthwise course of the stem. What function might the concentration of vascular bundles at the periphery of the stem relative to its middle have? What is the likely function of the fibers surrounding the vascular bundles?

••• *Suggested diagram and labels: Zea mays* (corn, maize, an angiosperm and monocotyledon) stem x.s.: vascular bundles, primary xylem, primary phloem, fibers around vascular bundle, ground tissue, sclerenchyma (fibers) in ground tissue beneath epidermis, epidermis, guard cells (several pairs). Note that a distinct pith and cortex are lacking!

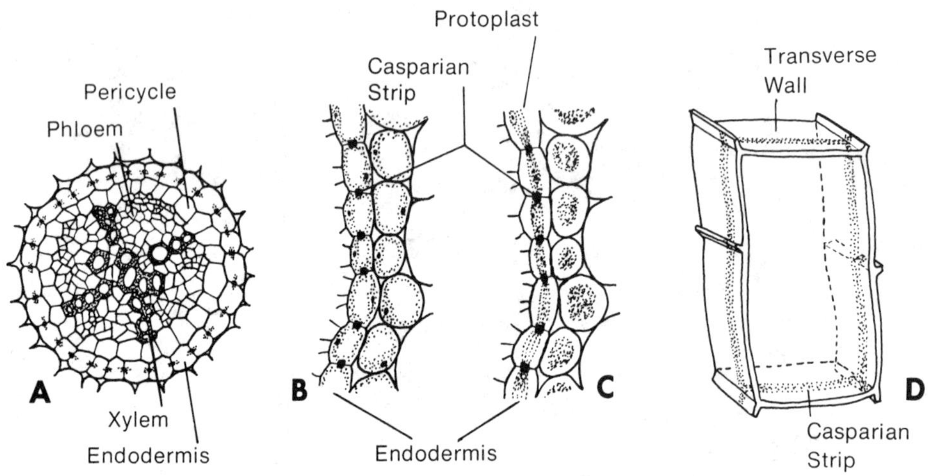

Fig. 7–2. Endodermis of root of *Convolvulus* (morning glory). *A,* transection of vascular cylinder and endodermis showing arrangement of casparian strips (the epidermis and most of cortex are not shown); *B, C,* transections of endodermis showing normal relationship of protoplasts to cell walls (*B*) and, after alcohol treatment, shrunken protoplasts adhering only to casparian strips (*C*); *D,* three-dimensional view of endodermal cell showing casparian strip on four of the six cell walls. (From Norstog & Long 1976:483; redrawn and modified from *Plant anatomy,* 2nd ed., by K. Esau, © 1965 by John Wiley and Sons, Inc., New York. Reprinted by permission of the publisher.)

B. Root anatomy

See Fig. 6–2 and especially Fig. 7–2. A transection of a typical root exhibits the following structures, proceeding centrifugally through the phloem:

- primary xylem, primary phloem alternating with the xylem points, pericycle (one or more layers of cells), endodermis (a single layer of cells), cortex, and epidermis.

Contrast this relationship of xylem and phloem with that of a stem (see Part I-A). Roots typically lack a pith and have an extensive parenchymatous cortex containing much storage starch but usually no chlorophyll. Typical aerial features of the epidermis such as a cuticle and guard cells are also lacking. Root hairs are common; they are single cells, actually outgrowths of epidermal cells.

Roots also have many important functions:

- anchorage;
- uptake of water and minerals via root hairs;
- storage of water and nutrients;

- regulation of the water and mineral balance of the plant via the endodermis (see Lab Exercise 6, Part II-C3);
- hormone production by the apical meristem and then its transport to the aerial parts.

In specialized cases roots function to:

- absorb nutrients from other organisms, as in parasites (see Supplement 11);
- climb.

Lab Exercise 11 deals with various root modifications of the flowering plants.

●●● See Fig. 7-2. Then reexamine the DEMO models of endodermal cells (the red tape represents the casparian strips) and the DEMO slide of a root transection of *Ranunculus acris* (buttercup) examined in Lab Exercise 6 (Parts II-C3 and II-C4). On the slide identify the following structures, as seen centrifugally (Fig. 7-2A):

1. central, continuous mass of primary xylem consisting of larger diameter cells centrally and several (3-5) lobes of smaller diameter cells peripherally;
2. discontinuous patches (3-5) of primary phloem located in the bays of the xylem arms;
3. single layer of thin-walled parenchyma cells comprising the pericycle;
4. endodermis with its casparian strips on the radial cell walls;
5. broad cortex consisting of parenchyma cells often containing much storage starch;
6. epidermis.

A pith is lacking. Does the epidermis have guard cells or a cuticle? Are root hairs present?

●●● *Suggested diagram and labels: Ranunculus acris* (buttercup, an angiosperm and dicotyledon) root x.s.: vascular tissue, primary xylem, primary phloem, pericycle, endodermis, casparian strip, cortex, epidermis. Note that guard cells, a cuticle, and a pith are lacking!

C. Leaf anatomy

See Fig. 6-3 and especially Fig. 7-3. The ground tissue of leaves is called ***mesophyll*** rather than cortex or pith. Mesophyll is often differentiated into:

- ***palisade tissue,*** involving columns of elongate cells, compactly arranged, thus with small air spaces between the cells (intercellular spaces);
- ***spongy tissue,*** involving irregular shaped and positioned cells, loosely arranged, hence with very large intercellular spaces.

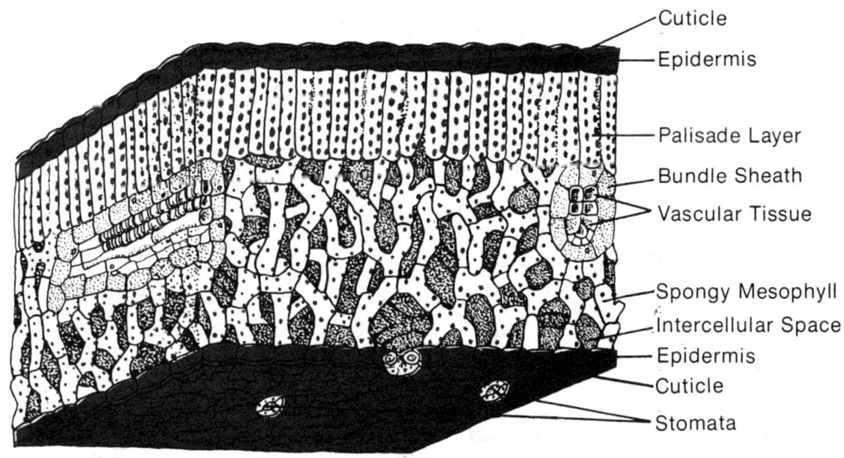

Fig. 7-3. Three-dimensional reconstruction of a leaf of a dicotyledon. Note the many chloroplasts represented by the dark spots. (From Norstog & Long 1976:477; redrawn from *Botany,* 5th ed., by C. L. Wilson, W. E. Loomis & T. A. Steeves, © 1971 by Holt, Rinehart and Winston, Inc., New York. Reprinted by permission of the publisher.)

Both palisade and spongy mesophyll contain many chloroplasts (Fig. 7-3).

The vascular bundles of leaves are called ***veins.*** A leaf usually has a large middle vascular bundle, the ***midrib,*** plus various lateral veins. Typically, xylem is *ad*axial, phloem *ab*axial, the palisade tissue also *ad*axial, the spongy tissue also *ab*axial. However, many exceptions occur.

A transection of the middle part (midrib) of a typical leaf exhibits the following structures, proceeding from the adaxial to the abaxial surface:

- cuticle, upper epidermis, palisade mesophyll, primary xylem, primary phloem, spongy mesophyll, lower epidermis, and cuticle.

The mesophyll and vascular bundles may contain much sclerenchyma.

Leaves have undergone even more evolutionary modifications than stems or roots. This is the topic of a marvelous book by Ghillean T. Prance (text) and Kjell B. Sandved (photographs) called *Leaves: The formation, characteristics, and uses of hundreds of leaves found in all parts of the world* (1985). The two most important functions of leaves are:

- storage of water and nutrients in the mesophyll (cotyledons are leaves in seeds that store nutrients for the embryo and seedling—see Lab Exercise 8, Part I-A, and Lab Exercise 10, Part II);
- manufacture of nutrients (the leaf is essentially a "photosynthetic machine").

In specialized cases leaves function to:

- protect developing vegetative and reproductive shoots;
- produce new individuals by asexual reproduction (see Lab Exercise 1, Part V);
- defend the organism against predators and pathogens;
- capture insects and other animals, as in carnivorous plants;
- house ants and other organisms that live inside the leaves;
- give buoyancy to aquatic plants;
- climb.

Lab Exercise 11 deals with various leaf modifications of the flowering plants.

••• See Fig. 7-3. Then reexamine the slide of a leaf transection of *Syringa vulgaris* (lilac) examined in Lab Exercise 6 (Parts II-B1 and II-C1). Identify the following structures:

1. upper epidermis (adaxial);
2. mesophyll consisting of the palisade tissue (adaxial) and spongy tissue (abaxial);
3. main (midrib) vascular bundle consisting of primary xylem (adaxial) and primary phloem (abaxial) surrounded by thick-walled collenchyma;
4. lower epidermis (abaxial).

Is any sclerenchyma present? Does the epidermis have a cuticle? Do stomata and hairs (trichomes) occur more commonly in the upper or the lower epidermis?

••• *Suggested diagram and labels: Syringa vulgaris* (lilac, an angiosperm and dicotyledon) leaf x.s.: cuticle, upper epidermis (adaxial), guard cells (several pairs), epidermal hair, mesophyll (palisade and spongy), air spaces between cells, vascular bundles, primary xylem, primary phloem, lower epidermis (abaxial).

D. Differences between stems, roots, and leaves

The following table contrasts the typical stem, root, and leaf of the angiosperms.

CHARACTER	STEMS (Fig. 7-1)	ROOTS (Fig. 7-2)	LEAVES (Fig. 7-3)
Pith	present*	usually absent	absent
Cortex	present	present	absent*
Mesophyll	absent	absent	present*
Chloroplasts	present	absent*	present
Endodermis	usually absent	present*	usually absent
Cuticle/stomata	present	absent*	present
Root hairs	absent	present*	absent

The asterisks distinguish one organ from the other two organs. Note that reproductive organs, for instance, cones, flowers, and their subparts, are anatomically like stems and leaves.

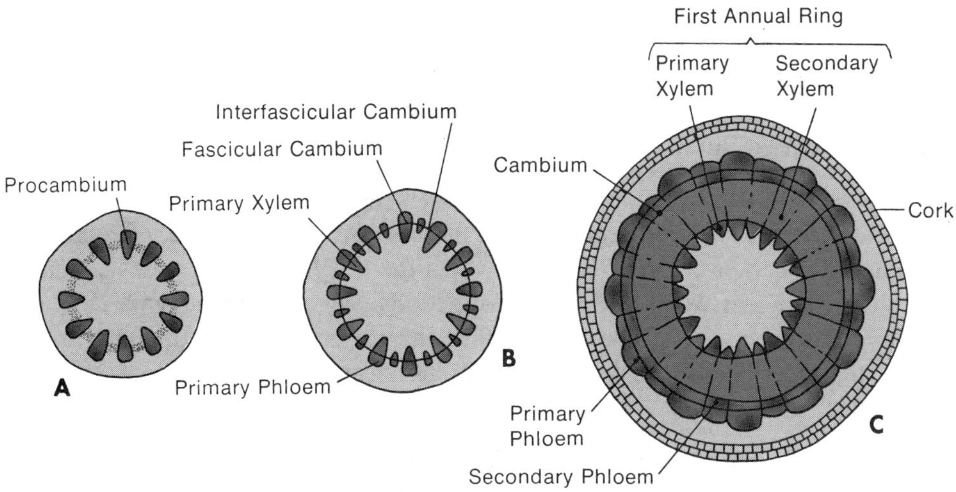

Fig. 7–4. Transections showing development of the secondary body in stems. *A,* young stem (just below shoot apex) consisting of only primary tissues; *B,* development of vascular cambium within and between the vascular bundles (respectively, fascicular and interfascicular cambia) (compare with Fig. 6–1D); *C,* one-year-old stem showing relationship of secondary to primary tissues. (From Norstog & Long 1976:464; redrawn from *Botany,* 5th ed., by C. L. Wilson, W. E. Loomis & T. A. Steeves, © 1971 by Holt, Rinehart and Winston, Inc., New York. Reprinted by permission of the publisher.)

II. ANATOMY OF A STEM WITH PRIMARY AND SECONDARY TISSUE

See Figs. 7-4 and 7-5. This section of the lab exercise also reprises some parts of Lab Exercise 6 by focusing on the "typical" (again used in the same sense as in Part I) transectional appearance of a stem that has undergone some secondary growth.

The vascular cambium originates in and between vascular bundles and eventually forms a continuous ring (Fig. 7 4B). The vascular cambium produces much more secondary xylem internally than secondary phloem externally (Figs. 7-4C, 7-5). In contrast, the cork cambium produces much more cork externally than cork parenchyma internally (Fig. 7-5). As emphasized in Lab Exercise 6, Part I-E, both vascular and cork cambia are a single layer of cells. Due to the activity of these meristems, the resultant secondary tissues usually exhibit (as seen in transection) cells nicely aligned in a radial direction from the center of the axis.

Periderm originates in various places. Because periderm is part of the dermal system and replaces the epidermis, initially the periderm originates in the cortex just beneath the epidermis. Later periderm is initiated more internally as the outer parts of the stem are disrupted by the accumulation of secondary vascular tissue internally.

The activity of the lateral meristems, the vascular and cork cambia, have definite physical effects on the primary tissues and on some of the secondary tissues they produce. Because the activity of the lateral meristems begins inside the primary body, this secondary activity will mainly affect the position and appearance of any tissues more external to the cambia, namely:

- usually unaffected = pith, primary xylem, and secondary xylem, because they are internal to the lateral meristems;
- usually affected = all other tissues, because they are external to the lateral meristems, that is:
  - the external secondary tissues—secondary phloem and periderm;
  - the primary tissues—primary phloem, cortex, epidermis.

A characteristic feature of woody plants is the presence of massive amounts of ***wood*** or secondary xylem plus varying amounts of bark. The wood typically contains ***growth rings*** (growth layers), a layer of growth as seen in transection (see Lab Exercise 7, Supplement). Growth rings are commonly called "annual rings" on the assumption that one growth ring is produced per year. However, several growth rings may be produced in one year.

The wood (secondary xylem) of large stems and roots often shows two parts:

- ***sapwood,*** an outer, lighter-colored region that is the area of active water conduction;
- ***heartwood,*** an inner, darker-colored region in which there is little or no water conduction.

The conducting cells of the heartwood become non-functional, and its parenchyma cells die. The heartwood is darker colored because pigments and other chemicals tend to accumulate in it. As the stem or root grows, the inner layers of sapwood change into heartwood. In old trees the heartwood may be a meter or more wide compared to a sapwood only a few centimeters wide.

Bark represents all the tissues external to the vascular cambium and includes:

- the secondary phloem, the parts closest to the vascular cambium functioning to conduct nutrients, plus the parts further from the cambium *not* conducting but functioning mainly in support;
- the periderm, its cork component being functionally the most important part of bark (see Lab Exercise 6, Part II-B2, for functions of cork);
- any remnants of external primary tissues (primary phloem, cortex, epidermis), although these are usually sloughed off with age of the plant.

Actually, many cork cambia usually become active in trees so that pockets of cork (and cork parenchyma) are produced repeatedly in the secondary phloem in a complicated mosaic.

Bark is essentially all the tissues of a tree external to the vascular cambium. Stripping bark from a tree, so-called girdling, will eventually kill the tree because the conducting phloem is removed. Because bark includes all tissues of a tree from the vascular cambium outward, there are two possibilities for the composition of bark, depending on the age of the tree:

Young Stem, Little Wood	***Old Stem (a Tree), Much Wood***
Secondary phloem	Younger (i.e., most recently formed) secondary phloem
Primary phloem	Later formed periderms intermixed with older secondary phloem
Cortex	
Initial periderm	*Thus:* oldest periderms and oldest secondary phloem as well as primary tissues displaced and sloughed off.
Epidermis	

It is hence a matter of degree between a young stem with rather limited secondary growth (Figs. 7–4C, 7–5) and an old tree stem with extensive secondary growth. In the tree all primary tissues will have been sloughed off, except the pith and primary xylem, which persist in the center of the stem. The bulk of the tree trunk will be secondary xylem (wood), but the bark component consisting of secondary phloem intermixed with periderm can be significant, as in trees of *Sequoia sempervirens* (redwood, coast redwood) or especially *Sequoiadendron giganteum* (big tree, Sierra redwood, giant redwood). The extremely tough and fibrous bark of *Sequoia,* which must be removed in order to saw uniform lumber, is used as insulation and garden mulch. The massive, non-resinous bark of *Sequoiadendron* makes it very resistant to fire.

Secondary growth in gymnosperms and dicotyledons is similar, but that of monocotyledons is different (see Supplement 10 contrasting these large groups of angiosperms). Secondary growth via a vascular cambium occurs not only in stems, but also frequently in roots and occasionally in leaves.

●●● See Figs. 7–4 and 7–5. Then reexamine the slide of a transection of a three-year-old stem of *Tilia americana* (basswood, American linden) examined in Lab Exercise 6 (Parts II-B2 and II-D4).

Identify the following structures, as seen centrifugally (this description also applies to Fig. 7–5, from its top to bottom):

1. pith consisting of parenchyma cells, some modified into tannin cells and mucilage cells;
2. several discontinuous patches of primary xylem;
3. three continuous increments (growth rings), one for each year, of secondary xylem containing vessel elements (the widest cells), tracheids, fibers (the narrow cells with rather thick walls), regular parenchyma cells (cells with protoplasts), and narrow and wide rays comprised of ray parenchyma cells;
4. the zone of the vascular cambium separating the secondary xylem and secondary phloem;
5. the continuous increments of secondary phloem, with narrow and conspicuous dilated vascular rays;
6. several discontinuous parts of primary phloem, with phloem fibers at the outermost parts of the bundles;
7. the cortex composed of inner parenchyma and outer collenchyma;
8. the periderm consisting of internal cork parenchyma, central cork cambium, and extensive external cork (for details see Lab Exercise 6, Part II-B2).

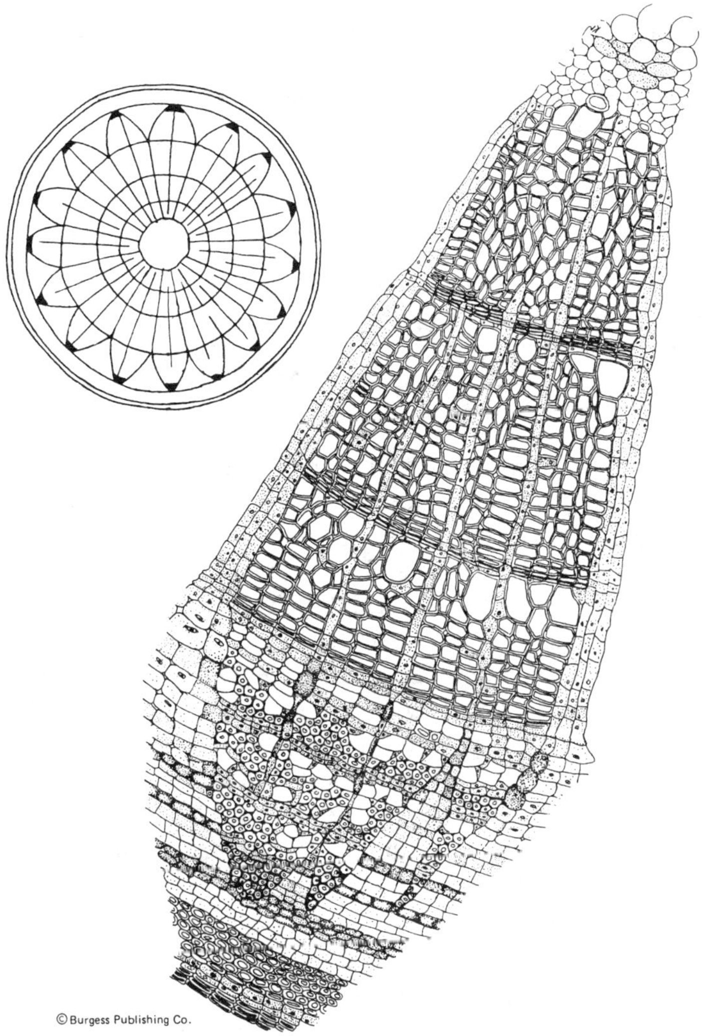

Fig. 7–5. Transection of a three-year-old stem of *Tilia americana* (basswood, American linden). See adjacent text for description.

The epidermis in the slide material will probably have been sloughed off by the development of the periderm in the outer part of the cortex (there is no epidermis evident in Fig. 7–5).

For addition information on wood, especially its growth rings and their use in dating objects, see Lab Exercise 7, Supplement.

●●● Examine the DEMO transection of a small tree trunk and the DEMO transection wedge from a large tree, for example, *Pseudotsuga menziesii* (Douglas fir) (also on DEMO is a live branch with female cones). Identify the following structures, as seen centrifugally:

1. bark, on the *Pseudotsuga* consisting of a complicated mosaic of periderm (lighter colored areas) and older secondary phloem (darker colored areas);
2. the younger, conducting secondary phloem, evident on the *Pseudotsuga* as a broad band of dark-colored tissue but not especially evident on the other, smaller specimen;

3. the vascular cambium;
4. the secondary xylem (wood) with its growth layers. Are any sapwood and heartwood evident?

III. DICOTYLEDONS VERSUS MONOCOTYLEDONS

See Supplement 10 giving the characteristics of di- versus monocotyledons, the two main groups (classes) of angiosperms.

●●● From your observations in Parts I and II are *Medicago* (or *Trifolium*), *Zea*, and *Tilia* di- or monocotyledons?

IV. MORPHOLOGICAL ADAPTATIONS—ARRANGEMENT OF SPORES IN A TETRAD

This brief section on reproductive morphology provides a link of Lab Exercises 4 and 5 with Lab Exercises 8 to 10, all of which treat mainly reproductive morphology of the land plants.

A ***sporocyte*** or ***meiocyte*** (spore mother cell) is the special cell undergoing meiosis. Thus a $2n$ sporocyte (meiocyte) produces by meiosis four $1n$ spores (meiospores). These four spores, which naturally are part of the GPTic generation (phase), constitute a ***tetrad*** (Figs. 4–3E, 8–2G, 9–4c, C). Tetrads also occur in algae and fungi, indeed always as the product of meiosis (see Supplement 4, Part II, and Figs. 3–3H, 12–2D, 12–8D, 12–9G, 13–1E). Spores of land plants can assume various arrangements in the tetrad, there being three common ones:

- In the ***tetrahedral tetrad*** each of the four spores contacts an inner side (face) of the other three spores; a three-pronged ***trilete*** (triradiate) ***mark*** or ***scar*** usually occurs on each spore (Figs. 4–1D, 5–4C, Sup6–1B), and the spores are called ***trilete spores.*** Trilete spores can also lack marks or scars (Fig. 5–5C), and then the spores are called ***alete spores.***
- In the ***tetragonal tetrad*** each of the four spores contacts an inner side (face) of only two spores and a mere edge of the third spore; a ***monolete mark*** or ***scar*** then occurs on each spore, and the spores are called ***monolete spores.***
- In the ***linear tetrad*** the four spores are arrayed end to end (Figs. 8–2G, 9–4C); no definable marks result, and the spores are called ***alete spores.***

This terminology applies both to the spores produced by homosporous plants and to the male and female spores (micro- and megaspores) produced by heterosporous plants. In addition, monolete spores tend to be more elongate than trilete spores. The tetrahedral arrangement, which has been regarded as the most efficient way to arrange four units in a tetrad, is basic to the land plants and is characteristic of most groups. Trilete spores, in fact, occurred in the first land plants (Fig. Sup6–1B). However, monolete spores characterize the ferns, whereas the linear tetrad characterizes the female spores of the seed plants.

●●● To recognize these distinctions, examine the DEMO clay models of:

1. a tetrahedral tetrad with trilete spores;
2. a tetragonal tetrad with monolete spores;
3. a linear tetrad with alete spores.

Carefully remove one of the spores from the clay tetrahedral model to be sure that you understand the configuration of this type of tetrad. Note the inked trilete mark on the inner side of each spore.

●●● Finally examine the DEMO slide of a median longisection of a cone of the lycopod *Lycopodium* (Fig. 5–3A) and observe the spores in a sporangium. Are the spores trilete, monolete, or alete? On the slide observed, why is not a scar evident on all spores of the sporangium? On which face (i.e., inner or outer) of the spore does the reticulate marking (sculpturing) occur?

NAME ______________________________ LAB SECTION # ______________ DATE ____

TITLE ______________________________

__________________ TOTAL MAG _____

TITLE ______________________________

__________________ TOTAL MAG ______

TITLE ______________________________

__________________ TOTAL MAG _____

TITLE ______________________________

__________________ TOTAL MAG ______

LAB EXERCISE SUPPLEMENT **7**

Dendrochronology

(Using the Growth Rings in the Wood of Bristlecone Pine)

OBJECTIVE

To observe the basis for the use of bristlecone pines in dendrochronology.

Note: Read this material *before* coming to the lab; there is insufficient time in the lab to read the material from scratch. Due to the large amount of background information presented here, a ●●● is used to mark those paragraphs calling for actual observations or diagrams in the lab. Most unmarked paragraphs reiterate information presented in the lecture. A bullet (●) signifies an important term or concept.

PERSPECTIVE

Dendrochronology (meaning "tree" plus "time") is the science of using trees, especially their growth rings, to date things. There are various ways to date things, but dendrochronology using bristlecone pine (*Pinus longaeva*) is a particularly accurate way. Bristlecone pines have received great notoriety because of their impressive beauty, extreme age, and considerable research potential (Ferguson & Graybill 1983; Johnson & Johnson 1978; Schmid & Schmid 1975).

I. THE TREE

Bristlecone pines are the oldest living individuals and occur in six western states of The United States:

Pinus longaeva (intermountain bristlecone pine) in California, Nevada, and Utah;

Pinus aristata (Rocky mountain bristlecone pine) in Arizona, Colorado, and New Mexico.

Pinus longaeva is much longer lived than *P. aristata* and thus is of much greater interest. Its life-span may reach 5,000 years. The oldest known living bristlecone pine tree is more than 4,600 years old and is still growing vigorously in Schulman Memorial Grove of the Ancient Bristlecone Pine Forest located east of Bishop, California. The oldest bristlecone pine grew on Wheeler Peak, Nevada. The tree was felled for research purposes and its age determined as nearly 4,900 years.

●●● The DEMO booklet (Johnson & Johnson 1978) shows typical old bristlecone pines on its covers. With die-back and weathering from sand and ice in middle age, a tree becomes greatly twisted, and its exposed wood becomes polished and etched. Very little of the tree is alive, 10% or less, resulting in a balance between the crown and non-photosynthetic bark and xylem; that is, less biomass has to be supported. This results in ***slab growth,*** in which a narrow living strip of bark is maintained on a massive dead trunk. Instead of the whole tree becoming sickly with age, it dies gradually in a succession of adjustments.

II. GROWTH RINGS IN WOOD

The wood of trees in temperate zones usually has growth rings reflecting the ages of the trees. Large, thin-walled xylem cells form during the rapid growth period of spring and early summer; these contrast vividly with the small-diameter, thick-walled cells produced later in the growing season, when moisture conditions are less favorable. The structural dissimilarity between the late wood of one season and the early wood of the next results in a ***growth ring*** (growth layer). This is often called an "annual ring," but this is improper because more than one growth ring may form in one year.

Because bristlecone pines grow very slowly, they have extremely narrow growth rings—from 100 to 200 per 2.5 cm (1 in.) of wood. In other words, a tree may add less than 5 cm (2 in.) of wood to its diameter every 200 years. In bristlecone pine the year-by-year sequence of wide and narrow growth rings forms a unique, non-repeating pattern. The narrowness and climatic sensitivity of the annual growth rings, together with the great longevity of the species and the persistence of dead wood, make the bristlecone pine invaluable in dendrochronology.

••• A DEMO compares growth layers in the wood of bristlecone pine and eastern white pine (*Pinus strobus*), a typical rapidly growing tree species:

- bristlecone pine, with very narrow growth layers of irregular thickness, about 150 growth layers occurring per 2.5 cm of wood;
- eastern white pine, with very broad growth layers of irregular thickness, about four growth layers occurring per 2.5 cm of wood.

In both species note the resin canals (the large cavities surrounded by dark-staining resin cells) and the very homogeneous nature of the wood (i.e., only tracheid water-conducting cells and vascular ray parenchyma cells). Other than the width of growth layers and, concomitantly, the width of cells, the wood of both species is very similar.

Bristlecone pines can be dated by taking a thin core of wood from the trunk using a device known as an increment borer, which is similar to a cork borer. The tree is not hurt because it secretes resin to close up the small hole the borer made. It is very helpful that bristlecone pine produces exactly one growth ring per year. Thus counting the growth rings can precisely determine the age of the tree. Longevity requires adversity. The oldest, soundest trees grow in the harshest, most exposed sites. Trees growing under more equable (hospitable) conditions do not reach venerable ages because increased moisture encourages rapid growth and the production of less dense, less resinous wood more susceptible to decay. Indicative of fast growth in a relatively benign environment is the occurrence of growth rings that are similar in width. Dendrochronologists favor trees in the exposed sites, not only because they are older, but more importantly because they are climatically sensitive and have growth rings that form a unique, non-repeating pattern [This pattern is like the line widths and spaces of commercial bar codes found on so many products today]. A core will show such unique patterns. Due to the irregular, twisted shape of old trees, many cores must be taken of one tree. These cores are overlapped and statistically correlated, a process known as ***cross dating***. In addition, there are statistical correlations with wood of other living trees as well as with dead wood lying on the ground.

••• Examine the DEMO booklet by Johnson & Johnson (1978) showing cores and the methodology of dating bristlecone pine. An actual core of this or another species may also be available on DEMO.

Although the oldest dated living bristlecone pines are less than 5,000 years old, the chronology has been extended back nearly 8,700 years (Ferguson & Graybill 1983) by using specimens from long-dead wood. Large remnants of fallen trees may persist for 4,000 years because they are destroyed, not by decay and disintegration, but only by gradual erosion. The gathering of fallen wood by souvenir collectors is prohibited in the Ancient Bristlecone Pine Forest because such wood has the potential for further extending the bristlecone pine master chronology.

The bristlecone pine chronology is highly accurate. For example, two chronologies were compiled at different times by two research groups. The two chronologies overlapped 5,395 years but disagreed by only two years. The missing rings were identified by detailed comparison of the two series, and two years were added in the chronology from the upper elevation. The two chronologies then matched perfectly! Incidentally, the bristlecone pine chronology was almost entirely worked out by the mid-1970s without the aid of computers.

III. SOME APPLICATIONS OF DENDROCHRONOLOGY

There are several applications of dendrochronology, particularly using the chronology derived from the bristlecone pine:

Radiocarbon dating

Perhaps the most important application of bristlecone pine chronology involves radiocarbon dating. The accuracy of radiocarbon dates lies both in the technology involved and the validity of certain assumptions made concerning the concentration, through time and space, of radioactive carbon in the atmosphere. Careful radiocarbon dating of ten-year segments of bristlecone pine wood of known calendar age has shown that the concentration of radioactive carbon in the atmosphere has changed over time instead of being constant as previously assumed. Scientists have now developed a detailed calibration curve that correlates the real age, based on the bristlecone pine master chronology, with the apparent age determined by radiocarbon dating. Radiocarbon dates are thus corrected by means of the more accurate bristlecone pine chronology that extends back nearly 8,700 years.

Reappraisals of many assumptions and theories in paleometeorology, archaeology, and geology have resulted. Many archaeological sites have been shown by dendrochronology to be 200 to 1,000 years older than previously believed on the basis of radiocarbon dating.

Archaeology

Wood from archaeological sites, for instance, wooden beams of an ancient building, within 1,000 miles to the east and south and 300 miles to the north of the White Mountains, California and Nevada, can be dated using the bristlecone pine master chronology. If the wood used in construction came from the same general climatic area as the bristlecone pines, it may be cross-dated with the master chronology to determine the age of the building. Fortunately, early peoples tended to use young, tall, straight trees for building, rather than trimming down large, gnarled old trees. The outer rings of the beam thus reflect the actual time of construction, and the entire ring series of the beam makes cross-dating possible. A sequence of about 40 years is necessary for an accurate comparison. Dating of archaeological sites by cross-dating, however, has had only limited application thus far.

Other

The data obtainable from bristlecone pines can be combined with other climatically sensitive variables such as pollen frequency, thickness of annual layers of silt, and isotope ratios of glacial ice to provide more information for determining and comparing local and global climates of the past.

LITERATURE CITED

Ferguson, C. W. & D. A. Graybill. 1983. Dendrochronology of bristlecone pine: A progress report. *Radiocarbon* 25:287–288.

Johnson, R. & A. Johnson. 1978. *The ancient bristlecone pine forest.* Rev. ed. Bishop, California: Chalfant Press.

Schmid, R. & M. J. Schmid. 1975. Living links with the past. *Nat. Hist.* 84(3):38–45, 84. [On bristlecone pines.]

SUPPLEMENT **8**

Introduction to the Seed Plants

(Basic Structure of Pollen Grains, Ovules, and Seeds)

God made seed plants to show what He could do with sporangia.—Schmid

I. PERSPECTIVE

The fundamental distinction between the pteridophytes and the seed plants is:

- ***Pteridophytes*** (nonseed plants, cryptogams) lack ovules and seeds, and their concomitant structures, pollen grains and pollen tubes, and seedlings.
- ***Seed plants*** (spermatophytes, phanerogams), namely, gymno- and angiosperms, have these structures; that is, each seed plant has:
  - a ***pollen grain,*** the microGPT (male GPT);
  - a ***pollen tube,*** the structure conveying sperm directly (except in cycads and *Ginkgo*) to the egg;
  - an ***ovule,*** a combination structure of a parent SPT and a megaGPT (female GPT);
  - a ***seed,*** a fertilized ovule, that is, a combination of the old parent SPT, the megaGPT, a new SPT (***embryo***), and in angiosperms the ***endosperm*** (see Lab Exercise 9, Part III-D) resulting from double fertilization;
  - a ***seedling,*** the young vegetative SPT emerging from a germinated seed.

The GPTs of seed plants are much reduced in size compared to those of pteridophytes. The largest GPT of a seed plant is the megaGPT of *Ginkgo,* which has about 8,000 cells (versus ca. 2,000 in *Pinus*) and several archegonia. The microGPTs of angiosperms are only two or three cells, and the megaGPTs of most angiosperms are only seven cells. All pteridophytes have antheridia and archegonia. Archegonia occur in most gymnosperms (except some gnetophytes) but *not* in angiosperms. Antheridia do *not* occur in any seed plant.

Note: You should be familiar with the following concepts and terminology before embarking on Lab Exercises 8 to 10 dealing with the reproductive morphology of gymno- and angiosperms.

II. POLLINATION, SYNGAMY (FERTILIZATION), AND DISPERSAL COMPARED

Pollination, syngamy (fertilization), and dispersal are events vital to the overall functioning of the reproductive structures of all seed plants.

- ***Pollination*** is the transfer of ***pollen*** (the collective term for pollen grains) from its source to a receptive surface by various agents (wind, water, animals, the plant itself, etc.). Thus pollen is transferred from a male ***source*** to a female ***target,*** that is (for definitions of terms see Lab Exercises 8 and 9):

- in gymnosperms from microsporangia *to* micropylar regions (nucellus or pollination droplet) of ovules (or to the cone scales), but
- in angiosperms from microsporangia (pollen sacs) of anthers *to* stigmas of carpels.
- ***Syngamy*** (fertilization) in seed plants, as in pteridophytes, involves the fusion of an egg and a sperm, but unlike pteridophytes, most seed plants lack swimming sperm, which are conveyed directly to the egg by pollen tubes.
- ***Dispersal*** in seed plants is the spreading of ***dispersal units*** (diaspores), that is, fruits, seeds, etc., away from the parent plant to (usually) new habitats by various agents (wind, water, animals, gravity, the plant itself, etc.).

Suffice it to say here that to form fruits and seeds, pollination and syngamy (fertilization) generally are necessary (***parthenocarpic fruits*** usually lack seeds and usually develop without syngamy). Syngamy is interpolated between pollination and dispersal.

III. ORIGIN OF THE SEED PLANTS

The evolution of the seed habit was one of the most significant evolutionary events in the history of the plant kingdom. The great adaptiveness of the seed (and ovule) and of the concomitantly developed microGPT or pollen grain of the seed plants undoubtedly explains the dominance and the considerable evolutionary success of this group. In other words, the spread of pollen (pollination) and seeds and fruits (dispersal) are great advantages that the seed plants have over the pteridophytes.

It has long been evident that some group of pteridophytes gave rise to the seed plants. However, only since 1960 has appreciable evidence been amassed to demonstrate that the strictly Devonian progymnosperms (Progymnospermophyta; see Lab Exercise 5, Part III-B) most likely were ancestral to the seed plants, specifically to the gymnosperms, some contingent of which in turn gave rise to the angiosperms (flowering plants).

IV. BASIC STRUCTURE OF POLLEN GRAINS

A pollen grain typically contains several cells and is usually an immature microGPT because further mitoses occur after the pollen grain germinates and a pollen tube forms. Sometimes a pollen grain is a mature microGPT if all mitoses occur before germination. Pollen grains contain relatively few cells (2–5, sometimes 1 or up to 43). The most important cells are:

- the ***tube cell,*** a large vegetative cell that on germination of the pollen grain emerges via the aperture to become the ***pollen tube;***
- the ***generative cell,*** a small reproductive (fertile) cell that usually after pollen germination either indirectly (in gymnosperms) or directly (in angiosperms) forms or "generates" two ***sperm,*** which are distinct cells.

A resistant wall envelops the cells of a pollen grain. The walls of both pollen grains and spores consist of:

- an ***intine,*** the inner, usually thin wall layer composed of cellulose, the usual material of plant cell walls;
- an ***exine,*** the outer, usually thick wall layer composed of the highly resistant, waxy chemical ***sporopollenin.***

Pollen grains (but *not* spores) usually also have:

- ***apertures,*** thin areas in the exine through which a pollen protoplast on germination emerges as a pollen tube.

A pollen grain of angiosperms may have zero to many apertures; in contrast, a pollen grain of gymnosperms has one aperture, or none. Megaspores of seed plants and all spores of pteridophytes lack apertures. The pollen walls of angiosperms are very highly modified compared to those of the gymnosperms and have a column-space differentiation (see Lab Exercise 9, Part III-A).

After pollination, that is, after pollen lands on a female target, the pollen grain germinates and the cytoplasm of the tube cell emerges to form a pollen tube (this is still the microGPT), which grows toward the ovule. The emergent cytoplasm synthesizes a new cell wall. Because all pollen tubes contain sperm (two produced by each pollen grain), and because the sperm of gymnosperms (except cycads and *Ginkgo*) and angiosperms are non-flagellate, the pollen tube is effectively a gamete conveyer to the megaGPT in the ovule. A ***seed*** is simply a fertilized ovule.

V. WIND-DISPERSED VERSUS ANIMAL-DISPERSED POLLEN

The following generalizations about dispersal of pollen have many exceptions:

Wind-Dispersed Pollen	*Animal-Dispersed Pollen*
Smooth-walled	Heavily ornamented or sculptured
Dry, not oily	Sticky-surfaced due to presence of oily compounds (Pollenkitt)
Individual pollen grains not adherent	Individual pollen grains frequently clumped
More pollen grains produced per plant as pollen transfer less efficient	Fewer pollen grains produced per plant as pollen transfer more efficient

[Note: Water-dispersed pollen of angiosperms may have little or no exine.]

VI. THE POLLEN GRAIN OF SEED PLANTS VERSUS THE SPORE OF PTERIDOPHYTES AND BRYOPHYTES

This table focuses on some important differences between pollen grains and spores:

Spore of Pteridophytes and Bryophytes	*Pollen Grain of Seed Plants*
Can be either a male or a female structure	Always a male structure*
Nearly always a unicellular GPT,* i.e., a single cell, the first stage of the GPT	Usually a multicellular GPT, i.e., a partly or completely developed microGPT
Usually dispersed by water and especially wind	Usually dispersed by wind and especially animals
Can land and germinate anywhere there is moisture	Must land on or near a specialized structure (ovule or stigma) in order to germinate
Germinates by wall breaking at a trilete or monolete mark to form just a filament (see Lab Exercise 7, Part IV)	Germinates via aperture(s) to form a pollen tube
*Some spores of bryophytes are multicellular.	*Seed plants, being heterosporous, have both male and female spores; the former develop into pollen grains.

VII. BASIC STRUCTURE OF OVULES

The ovule of seed plants is a specialized structure consisting of several parts (Figs. 8–2F to H, 9–1, 9–4B to F):

- 2*n* SPTic tissue consisting of:
  - the megasporangium (female sporangium) or ***nucellus,*** which is the special name for the megasporangium of seed plants;
  - one or two enclosing protective layers, the ***integument*** or ***integuments;***
- 1*n* GPTic tissue consisting of a megaspore (usually three megaspores of a tetrad abort) or else the megaGPT (***embryo sac*** is the special name for the megaGPT of angiosperms) that developed from the megaspore(s).

The ovule thus is an integumented megasporangium. The integument(s), which are structures unique to seed plants, completely cover the nucellus except for the ***micropyle,*** a small porelike opening at the apex of the ovule to permit passage of the pollen tube to the megaGPT (Fig. 9–1). In the mature seed the micropyle may be completely obliterated or it may persist in the seed coat as a small closed pore (Fig. Sup8–1).

To obtain a proper three-dimensional concept of the structure of an ovule and young seed, note that the integument is a continuous layer of tissue (Fig. 9–1). However, note that in *median* longisectional view, the integument appears as halves due to the micropyle present. If an ovule has one integument, then two halves appear (Fig. 8–2F to H, J), but if an ovule has two integuments, then four halves appear (Fig. 10–4C to H).

VIII. BASIC STRUCTURE OF SEEDS AND SEEDLINGS

Syngamy (fertilization) induces hormonal and other physiological effects that trigger:

- the new development of the ***embryo*** (very young SPT) by mitosis of the zygote;
- the concomitant transformation of the ovule into the seed.

The mature seed (Figs. Sup8–1, 8–2K, 9–4H, 10–3) thus is a composite of three generations (phases), namely:

- the old 2*n* SPT, that is:
  - the nucellus (megasporangium), which may be break down or persist to various degrees;
  - integument(s), which develop into the ***seed coat*** (testa), the protective surface layer;
- the 1*n* megaGPT, which may be completely used up or may variously persist (remnants of the microGPT are also present but usually are not identifiable as such);
- the new 2*n* SPT which is the embryo (very young SPT) before the onset of rapid growth (or before germination of the seed in seed plants), which at maturity consists of:
  - ***cotyledons*** (seed leaves), 0–18;
  - a ***shoot apex;***

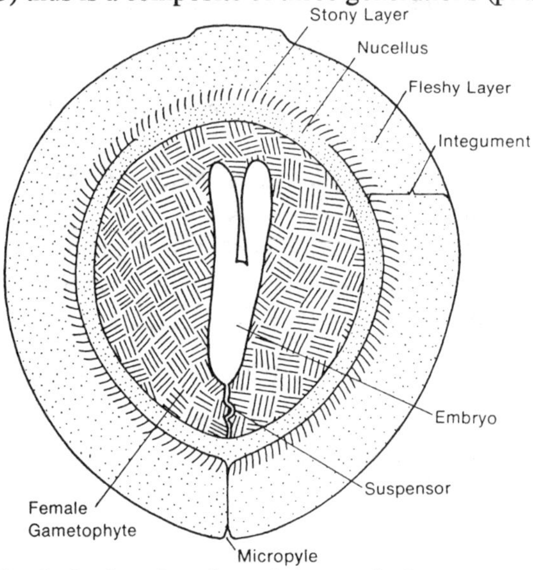

Fig. Sup8–1. Longisection of the seed of a gymnosperm. The massive megaGPT (female GPT) will provide nutrition to the developing embryo (very young SPT) and seedling. Note occluded micropyle and remnants of the nucellus (megasporangium, female sporangium). (From Norstog & Long 1976:371.)

- an ***epicotyl,*** the shoot region above the cotyledonary node;
- the short ***hypocotyl,*** the shoot region between the cotyledonary node and the radicle;
- a ***radicle*** (embryonic root), the young root;
- a ***suspensor*** (see below);

- plus in angiosperms, a usually 3*n* tissue, the ***endosperm*** (see Lab Exercise 9, Part III-D).

A developing embryo usually also has a ***suspensor*** (Fig. Sup8–1), a filamentous structure functioning to push the terminal part of the developing embryo into the nutritive tissue of the seed, that is, megaGPTic tissue in gymnosperms or endosperm in angiosperms (embryos of many pteridophytes also have suspensors—Fig. 5–4). The suspensor probably also absorbs nutrients.

As the embryo develops, the rest of the seed (i.e., the parts of the former ovule) undergoes assorted structural modifications. The transformation of the ovule into the seed thus involves various changes:

- an often appreciable increase in size;
- development of the integument(s) into the seed coat, which may become hard or fleshy;
- closure of the micropyle;
- usually partial degeneration of the nucellus (megasporangium);
- partial to complete degeneration of the megaGPT;
- development of the embryo (very young SPT) from the zygote;
- in angiosperms *only,* formation of the endosperm and usually the complete degeneration of the megaGPT.

The ovules of angiosperms are simpler in structure and usually smaller than those of the gymnosperms (compare Figs. Sup8–1 and 8–2K with Fig. 9–4H). Supplement 9 contrasts the ovules and seeds of these taxa.

Seeds often have ***dormancy,*** an inactive period, after which they germinate. The embryo emerges as the young SPT (Figs. 8–2L, 10–4).

- A ***seedling*** is the young vegetative SPT emerging from a germinated seed.

Although seedlings are actually vegetative structures, not reproductive ones, seedlings are usually treated under reproductive morphology because of their close relationship to seeds.

IX. COMMON REGIONS OF OVULES AND SEEDS

A precise boundary between the nucellus (megasporangium) and the adjacent integument can not be distinguished, except, of course, in the apical region of the ovule or seed, the ***micropylar end,*** where the integument and nucellus (megasporangium) are not fused (free). The ***chalaza*** is the lower part of the ovule where these parts are firmly joined. It is convenient for descriptive purposes to refer to this end of the ovule as the ***chalazal end.*** The ***funiculus*** is the stalk (Fig. 9–1) of an ovule or seed attaching it to ovarian tissue of angiosperms or to non-ovarian tissue of gymnosperms.

X. SUMMARY COMMENTS

The structure of pollen grains, ovules, and seeds falls into the realm of embryology. ***Embryology*** is the study of the embryo and the pre- and post-fertilization events associated with embryo formation and development, namely, (1) microsporogenesis and microgametogenesis, including development of the microsporangium and growth of the pollen tube, (2) megasporogenesis and megagametogenesis, including structure and development of ovules and seeds, and (3) embryogenesis.

- ***Microsporogenesis*** is the production of microspores (meiospores) by the meiotic division of the microsporocytes (micromeiocytes, microspore mother cells, pollen mother cells) and occurs in the microsporangia or ***pollen sacs.***

- ***Microgametogenesis*** is the development of the microspore into the mature male GPT or micro-GPT and includes events involving pollen germination and pollen tube formation.
- ***Megasporogenesis*** is the production of megaspores (meiospores) by the meiotic division of the megasporocytes (megameiocytes, megaspore mother cells) and occurs in the ovules.
- ***Megagametogenesis*** is the development of the megaspore into the mature female GPT or mega-GPT.
- ***Embryogenesis*** is the development of the embryo from its zygote and also the development of the endosperm.

It is important to reiterate: A pollen grain is the microGPT, usually an immature microGPT because further mitoses occur after the pollen grain germinates and a pollen tube forms. In other cases, a pollen grain is sometimes a mature microGPT because all mitoses occurred before germination. An ovule is a combination structure of a parent SPT and a megaGPT, whereas a seed is a fertilized ovule and thus a combination of the old parent SPT, the megaGPT, a new SPT (embryo), and in angiosperms also the endosperm, namely:

- in the ovule:
  - old SPT ($2n$), the integument(s) and nucellus (megasporangium);
  - GPT ($1n$), the megaGPT (called ***embryo sac*** in angiosperms);
- and in the seed:
  - old SPT ($2n$), the seed coat derived from the integument(s), also small amounts of nucellus (megasporangium) present;
  - GPT ($1n$), the megaGPT (massive in gymnosperms, negligible in angiosperms) (remnants of the microGPT are also present but usually are not identifiable as such);
  - new SPT ($2n$), the embryo;
  - plus endosperm (usually $3n$) in angiosperms (absent in gymnosperms).

I've heard nonsense compared with which that would be as sensible as a dictionary.—The Red Queen

LAB EXERCISE **8**

Conifers (Especially Pines) and Other Gymnosperms

OBJECTIVE

To examine the salient features of reproductive morphology of conifers (especially pine) and other gymnosperms.

Note: Read this material *before* coming to the lab; there is insufficient time in the lab to read the material from scratch. Due to the large amount of background information presented here, a ●●● is used to mark those paragraphs calling for actual observations or diagrams in the lab. Most unmarked paragraphs reiterate information presented in the lecture. A bullet (●) signifies an important term or concept.

Suggested Lab Diagrams (see part noted for suggested labels):
Pinus (pine, a gymnosperm and conifer) male cone (microstrobilus) l.s. (Part I-A)
Pinus (pine, a gymnosperm and conifer) ovule l.s. (Part I-A)

PERSPECTIVE

The gymnosperms have two divisions of extinct plants and four divisions of extant plants, with about 750 extant species in 83 genera (Kubitzki 1990). The divisions of extant taxa are:

1. cycads (Cycadophyta)—137 species, 11 genera;
2. *Ginkgo biloba* (maidenhair tree) (Ginkgophyta)—1 species;
3. conifers (Coniferophyta)—536 species, 68 genera;
4. gnetophytes (Gnetophyta)—76 species, 3 genera.

The gymnosperms were much better represented in the past, notably in the Paleozoic and particularly the Mesozoic. The gymnosperms originated from the Devonian progymnosperms (Progymnospermophyta) (see Lab Exercise 5, Part III-B). The conifers are the most familiar gymnosperms and include pines, redwoods, and junipers (see Part I-B for other examples). The gymnosperms are economically important as timber trees in the northern hemisphere, which has extensive conifer forests.

Recall from Supplement 5, Part IV, that the heterosporous haploid-diploid ($1n$-$2n$) life history characterizing *all* seed plants has male "micro-" and female "mega-" structures (Fig. Sup5–5), namely:

- ***micro-*** and ***megasporophylls,*** which bear
- ***micro-*** and ***megasporangia,*** which produce
- ***micro-*** and ***megasporocytes,*** which produce by meiosis
- ***micro-*** and ***megaspores,*** which develop into
- ***micro-*** and ***megaGPTs.***

The cones consist of sporophylls and one or more bracts. A ***nucellus*** is another name for a megasporangium. A ***bract*** is a modified, usually reduced leaf generally associated with a reproductive structure. All extant gymnosperms have unisexual cones, that is:

- ***male cones*** (microstrobili, microsporangiate strobili, pollen cones);
- ***female cones*** (megastrobili, megasporangiate strobili, ovule cones, ovulate cones, seed cones).

The male and female cones may be borne on separate, unisexual plants or on the same bisexual plant, that is:

- the ***dioecious*** condition, with unisexual, male and female plants, or micro- and megaSPTs;
- the ***monoecious*** condition, with bisexual plants, the more common situation, as in *Pinus*.

Most of this male "micro-" and female "mega-" terminology is standard usage, except that "male cone" and "female cone" are more common and, in fact, are simpler to use.

Note: Before beginning this lab exercise, you should review:

- Supplement 5 on the main types of life histories;
- Supplement 7 contrasting pteridophytes and seed plants;
- Supplement 8 on pollen grains, ovules, and seeds, and their terminology;
- Supplement 9 contrasting gymno- and angiosperms.

I. CONIFERS (CONIFEROPHYTA)

The extant conifers are characterized by the following features:

- highly branched, very woody stems;
- simple (undivided) leaves, usually small and needlelike or scalelike;
- females cones morphologically a modified compound branch system;
- non-flagellate sperm conveyed directly to the archegonium by the pollen tube.

The cones of all conifers are unisexual, that is, male and female, and are borne either on separate, unisexual plants (i.e., the dioecious condition) or, in most species, on the same, bisexual plant (i.e., the monoecious condition). Male cones of a species are smaller, simpler in structure, and more short-lived than the female cones.

A. *Pinus* (pine)

Morphologically and anatomically, *Pinus* (pine) is the best known genus of gymnosperms because of its economic importance and widespread occurrence. Although the heterosporous life history of *Pinus* is emphasized below, it should *not* be regarded as typical of all gymnosperms. Indeed, *Pinus* is atypical among conifers because its reproductive events extend over three years rather than two years, the period in most other conifers. *Pinus* is unusual among conifers in that 12 to 14 months intervene between pollination and syngamy (fertilization).

●●● Consult Figs. 8–1 and 8–2 of the heterosporous haploid-diploid ($1n$-$2n$) life history (see also Fig. Sup5–5) of *Pinus* as you examine material of this genus.

●●● The unisexual, male and female cones of *Pinus* are borne on the same plant (i.e., the monoecious condition). In common parlance, the female cones of *Pinus* are the "pine cones." See Fig. 8–2B, E. Then examine the DEMO live and dried reproductive material of *Pinus* and identify the male and female cones. Note the relative sizes of the cones and that their subunits (microsporophylls of the male cones, cone scales of the female cones—see below) are spirally arranged. Depending on their age, female cones are black-purple, green, or brown (Fig. 8–1). Do not expect to see black-purple female cones in the fall. Why?

●●● See Figs. 8–2B, C. Then examine a slide of a longisection of a male cone of *Pinus*. Identify the following structures: microsporophylls, microsporangia, and pollen grains. Note that the microsporangia are borne abaxially on the microsporophylls. Each sporophyll bears two sporangia, but this is not evident from the slide. Note the pollen grains in the microsporangia. A single bract occurs at the base of the cone but is not evident on the microscope slide.

●●● *Suggested diagram and labels: Pinus* (pine, a gymnosperm and conifer) male cone (microstrobilus) l.s.: microsporophyll, microsporangium, pollen grains. Note that there is only one bract at the base of the cone.

Each pollen grain consists of the following structures (in sequence) when it is shed from the microsporangium (Fig. 8–2D):

- two wings between which is an ***aperture,*** a thin spot in the wall of the pollen grain;
- the ***tube cell,*** a large vegetative cell that on germination of the pollen grain emerges via the aperture to become the ***pollen tube,*** which conveys the sperm to the egg;
- the ***generative cell,*** a small reproductive (fertile) cell that after germination divides twice to form (or generate) the two sperm just prior to fertilization (see below);
- two crushed ***vegetative cells*** (prothallial cells) that are of no further significance.

The football-shaped generative cell is mostly nucleus surrounded by a small halo of cytoplasm. A mature pollen grain (but still immature microGPT) consists of only four cells and lacks antheridia (Fig. 8–2D). Before the stage shown in Fig. 8–2D, each of the many microsporocytes in the microsporangium had divided meiotically to form four microspores. Each microspore of a tetrad then developed into a pollen grain, which is an immature microGPT (male GPT).

●●● See Fig. 8–2D. Then examine the DEMO slide of a longisection of a male cone of *Pinus* showing sporangia with mature pollen grains. Focus the microscope only; do *not* move the slide. Try to locate all the parts of a pollen grain noted above.

In *Pinus* and other conifers the structures supporting the ovules are called ***cone scales*** (ovuliferous scales) and *not* megasporophylls because they are morphologically more complex than mere sporophylls (i.e., each cone scale is interpreted as an evolutionarily modified shoot). Each ovule-cone scale unit (Fig. 8–2F) has on its abaxial side (i.e., subtending) an associated abaxial bract. Two ovules occur on the adaxial surface of each cone scale.

The single megasporocyte in each ovule (Fig. 8–2F) divides meiotically to form a linear tetrad (see Lab Exercise 7, Part IV) of four megaspores. The three megaspores closest to the micropyle abort (ovules are pro-choice), whereas the fourth develops into the megaGPT (female GPT) (Fig. 8–2G).

●●● See Fig. 8–2F. Then examine the DEMO slide of a longisection of an immature female cone of *Pinus* with sporocytes and/or pollen grains (i.e., pollination). Pick out a medianly sectioned ovule and identify the following structures: cone axis, bracts, cone scale, ovule, integument, nucellus (megasporangium), micropyle, and megasporocyte (the last two structures are not evident in all ovules). The female cones here represent stages just before or after pollination (Figs. 8–1, 8–2F).

Ovules of *Pinus* are pollinated by wind in spring. Pollination involves pollen transfer from the microsporangia to the ovules. The cone scales of the female cones separate slightly and pollen rests on the cone scales or sifts down between them. The tips of the ovules exude a sticky fluid (the pollination drop) that draws the pollen into the ovule through the micropyle. The cone scales then become tightly appressed again and remain so until the seeds are dispersed.

After a pollen grain has landed on the ovule, it germinates via the aperture between the two wings and forms a pollen tube, as shown in Fig. 8–2I. The pollen tube grows through the nucellus (megasporangium) of the ovule toward its archegonia (Fig. 8–2H). About a week before fertilization occurs, and a year or more after pollination had occurred, the generative cell in the pollen tube divides into two sperm that lack flagella and are of unequal size. This is the mature microGPT (Fig. 8–2I). The tip of the pollen tube enters the archegonium and ruptures, discharging its contents into the cyto-

1990 *summer:* male cones initiated; female cones initiated a bit later
1990–91 *winter:* resting period
1991 *spring:* meiosis in male and female cones
1991 *spring:* micro- and megaGPTs initiated
1991 *spring:* pollination (female cones are black-purple)
1991–92 *winter:* resting period
1992 *spring:* micro- and megaGPTs resume development
1992 *late spring or early summer:* syngamy (fertilization) (female cones are green)
1992 *early summer:* embryo development begins
1992 *late summer:* embryo development ends; seeds have matured
1992 *summer or fall:* seeds shed (female cones are brown)

Fig. 8-1. Temporal sequence of reproductive events in *Pinus* (pine). Various species, and even individuals of a species, vary appreciably from this generalized scheme because of the influence of latitude, altitude, and weather. (Modified from Gifford & Foster 1989:433.)

plasm of the egg. The larger of the two sperm fuses with the egg nucleus, and fertilization, by definition, has occurred. The remaining nuclei of the microGPT and archegonium eventually degenerate.

Each ovule of *Pinus* contains a megaGPT, which at maturity consists of about 2,000 cells and one to six archegonia (Fig. 8–2H). Each archegonium produces a large egg and several other cells. Concomitantly, the microGPT is also developing (Fig. 8–2I); at maturity it will be only six cells.

●●● See Fig. 8–2H, I. Then examine a slide of a longisection of an ovule of *Pinus* at the time of fertilization. Identify the following structures: integument, micropyle, nucellus (megasporangium), the large, massive, cellular megaGPT, and the archegonia. The thin strip that surrounds the megaGPT is the enlarged megaspore wall. Student slides of ovules are usually not median and thus may not show all of the aforenoted features. Hence, if necessary, examine the DEMO slide of a median longisection of an ovule of pine.

●●● *Suggested diagram and labels:* *Pinus* (pine, a gymnosperm and conifer) ovule l.s.: integument, micropyle, nucellus (megasporangium), megaGPT, archegonia. Note that this is a mature ovule at the time of fertilization!

After fertilization the ovule develops into a seed, and the zygote becomes an embryo (Fig. 8–2J, K).

- A ***seed*** is a combination of three generations (phases): the embryo or new SPT, a megaGPT in gymnosperms or endosperm in angiosperms enclosing the embryo, and an outer seed coat (testa) representing the modified integument or integuments, that is, old SPT.

Some nucellus (megasporangium), which is also the old SPT, may persist in gymnosperms, especially at the micropylar region. Remnants of the microGPT (pollen tube) are also present but often are not identifiable as such. What is the ploidy level of these structures?

●●● See Figs. 8–2K and Sup8–1. Then examine a slide of a seed longisection of *Pinus* showing a mature embryo. Identify the following parts of the seed:

- the ***seed coat*** (testa), the protective surface layer derived from the integument; the micropyle has closed, and much sclerenchyma often develops (some seed coats may also be fleshy) [Note: To more easily section the seed, the hardened seed coat was removed; the slide thus shows only the more internal parts of the seed listed next];
- the ***nucellus*** (megasporangium), which has largely disappeared, generally persisting as a papery layer at the micropylar end;
- the massive cellular ***megaGPT,*** which is rich in fats, carbohydrates, and proteins, and hence is a nutritive tissue for the embryo and seedling;

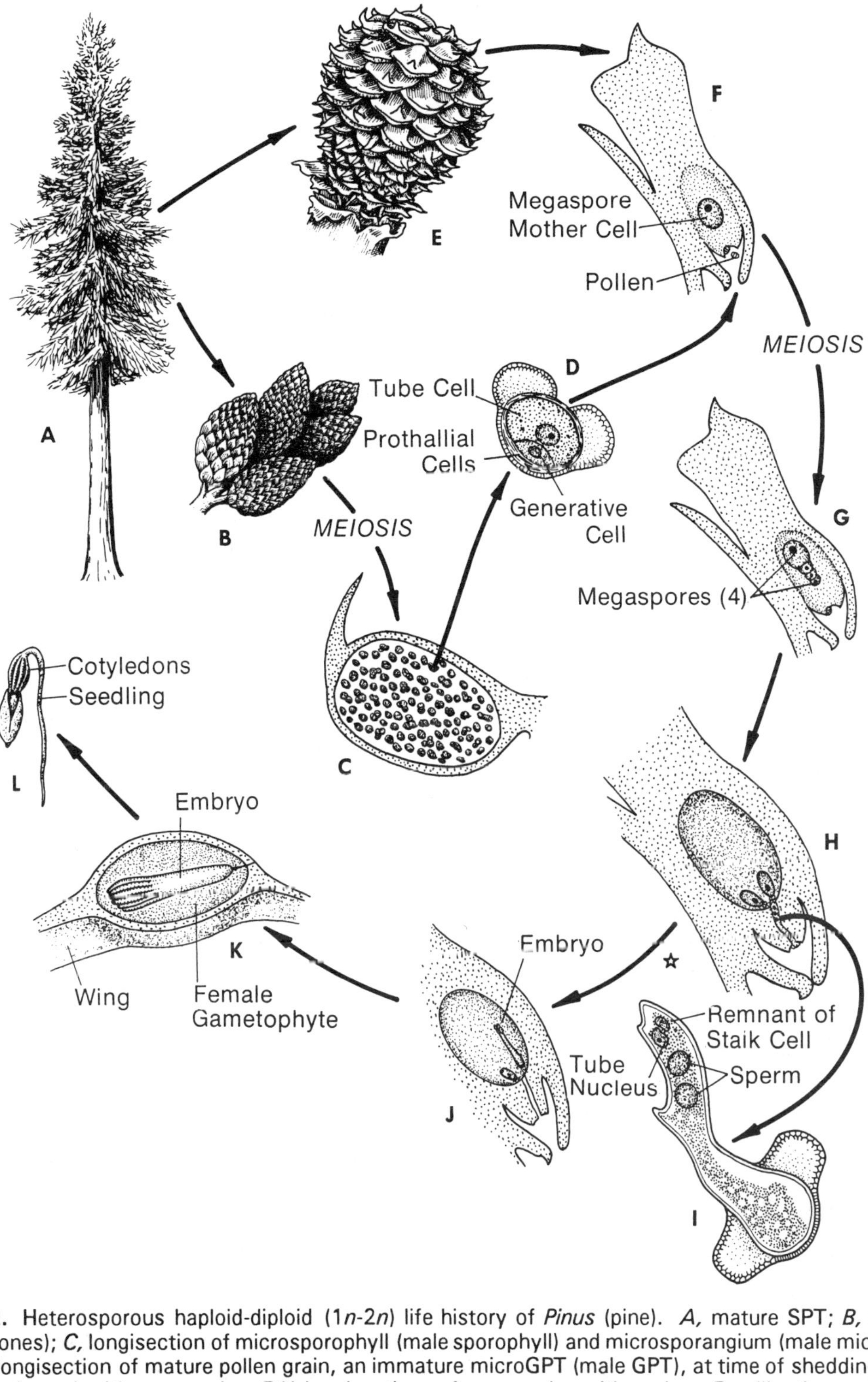

Fig. 8–2. Heterosporous haploid-diploid (1*n*-2*n*) life history of *Pinus* (pine). *A,* mature SPT; *B,* male cones (pollen cones); *C,* longisection of microsporophyll (male sporophyll) and microsporangium (male microsporangium); *D,* longisection of mature pollen grain, an immature microGPT (male GPT), at time of shedding; *E,* female cone (seed cone) with cone scales; *F-H,* longisections of cone scales with ovules: *F,* pollination and megasporocyte (megaspore mother cell); *G,* linear tetrad of four megaspores (three aborting) after meiosis; *H,* mature megaGPT (female GPT); *I,* longisection of mature microGPT, i.e., a germinated pollen grain with a pollen tube; ☆, syngamy (fertilization); *J,* longisection of seed with developing embryo (very young SPT); *K,* longisection of mature seed with embryo and nutritive megaGPT; *L,* seedling, i.e., germinated seed (note old seed coat still attached at end of cotyledons). Note the alternation of morphologically dissimilar GPTic and SPTic generations (phases), with the SPT dominant over the GPT, which is reduced to six and about 2,000 cells in the microGPT and megaGPT, respectively (compare with Fig. 9–4). (From Norstog & Long 1976:412.)

- the mature ***embryo*** (very young SPT) and its parts:
  - ***cotyledons*** (seed leaves), 3–18, averaging 8.1 in *Pinus;*
  - ***shoot apex;***
  - ***epicotyl,*** the shoot region above the cotyledonary node;
  - the short ***hypocotyl,*** the shoot region between the cotyledonary node and the radicle;
  - ***radicle*** (embryonic root), the young root.

As elaborated in Supplement 8, Part VIII, a seed is a combination of the old parent SPT, the megaGPT, and a new very young SPT (embryo).

●●● *Pinus edulis* (pinyon pine) has edible seeds, which laypersons incorrectly call "nuts." The embryos and megaGPTs of the seeds are eaten whereas the hard seed coat derived from the integument is discarded. Pinyon pine seeds are available commercially in Chinese markets, among others; usually the hard seed coat is removed. Examine the DEMO seeds of pinyon pine; some of these may be available for masticatory sampling. Note the fleshy megaGPT enclosing the embryo and identify the parts of the latter.

●●● See Fig. 8-2L. Then examine the DEMO live or pickled seedlings of *Pinus*. The many cotyledons are active in photosynthesis and initially form a cage because the old seed coat holds the cotyledons together at their tips. In this juvenile stage the leaves are borne directly on the main axis.

B. Other conifers

Other representatives of the conifers include such well known genera as *Abies* (fir), *Cedrus* (cedar), *Larix* (larch), *Pseudotsuga menziesii* (Douglas fir), *Cryptomeria japonica* (Japanese cedar), *Sequoia sempervirens* (redwood), *Sequoiadendron giganteum* (big tree, giant redwood), *Chamaecyparis* (false cypress), *Cupressus* (cypress), *Juniperus* (juniper), *Araucaria* (monkey puzzle tree, Norfolk Island pine, Chilean pine), *Podocarpus* (podocarpus), and *Taxus* (yew). All of these have unisexual male and female cones. Although most conifers have rather small male cones, those of *Araucaria,* notably *A. bidwilli* (bunya-bunya pine), may attain lengths of 10-12 cm (3.9-4.7 in.). The largest female cones occur in *Pinus* and *Araucaria.* The longest female cone is that of *P. lambertiana* (sugar pine), which may be up to 60 cm (23.6 in.) and visible from 0.4 km (0.25 mile) away. The female cones of *P. coulteri* (Coulter pine) and *A. bidwilli* are shorter, respectively, 25–35 cm (9.8-13.8 in.) and 30 cm (11.8 in.), but wider and heavier (Chamberlain 1935:290–291).

Conifers differ appreciably in the bract morphology of their female cones. The bracts of *Pinus* are small and, while obvious in sections on microscope slides, are not evident in mature cones. In contrast, the bracts of *Pseudotsuga* are very long, a distinctive feature of the genus. In *Juniperus* and many other conifers the bract and cone scale are largely fused and appear as a single structure.

●●● Examine the DEMO female cones of *Pseudotsuga menziesii* (Douglas fir), also a member of the pine family (Pinaceae), and an important timber tree in the Pacific Northwest. A tree felled in 1895 in British Columbia was 133 m (436 ft.) tall (Mabberley 1987). Note in mature cones the distinctive three-pointed bracts. These are not fused to the cone scales and so can be rather easily removed from the cone with forceps.

Although most conifers have hard, woody female cones, a few genera have fleshy reproductive structures, some of which may be dispersed by animals. There are two main variations:

- *Juniperus* and some species of *Podocarpus* have fleshy female cones.
- Other genera such as *Taxus* and some species of *Podocarpus* lack female cones but instead have fleshy ovules of varied morphology.

●●● Examine the DEMO female cones of *Juniperus* (juniper). Since the 1600s oil extracted from the fleshy cones (improperly called "berries") of *J. communis* has been used to flavor types of gin.

Many conifers are grown on university campus. As you walk about campus, take time to examine these conifers and compare their leaf and cone morphology to that of *Pinus*.

II. OTHER GYMNOSPERMS

Most extant species of gymnosperms are conifers (see "Perspective"). The three other divisions with extant representatives are the cycads (Cycadophyta), the ginkgoes (Ginkgophyta), and the gnetophytes (Gnetophyta). Various technical characteristics, of course, separate these divisions. The cycads and conifers share the gymnospermous features enumerated in the "Perspective," but they differ in many other features:

Cycads (Cycadophyta)	*Conifers (Coniferophyta)*
Stems with weakly branched to unbranched habit, with short conical or columnar trunk	Stems with highly branched habit, often with tall main, central trunk and with the branches regularly ticred around it
Stems with much pith and cortex but little secondary xylem	Stems with little pith and cortex but much secondary xylem
Leaves relatively large, usually divided	Leaves relatively small, simple, variously needlelike, scalelike, or flattened and broad in form
Sperm flagellate, swimming	Sperm non-flagellate
Cones simple, not compound*	Male cones simple, female cones compound*
Cones borne on separate plants (i.e., the dioecious condition)	Cones borne on separate plants (dioecious) or mostly on the same plant (i.e., monoecious)

* This important difference in cone morphology is *not* elaborated here!

Interestingly, the cycads, *Ginkgo*, and, presumably, some fossil gymnospermous divisions are the only gymnosperms with flagellate, swimming sperm.

DEMO examples of representatives of divisions of the extant taxa are available below for *brief* perusal.

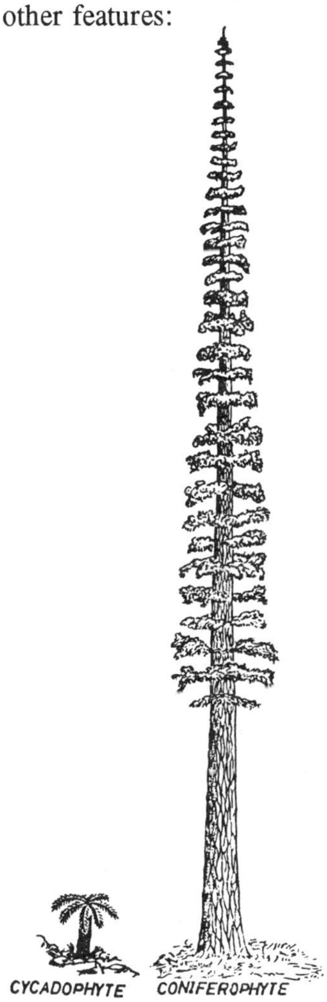

Fig. 8-3. Comparison of the habit and maximum height attained by cycads and conifers. (From Norstog & Long 1976:385; redrawn from *Structure and evolution,* by C. J. Chamberlain, © 1965 (published 1935) by The Johnson Reprint Corp., New York. Reprinted by permission of the publisher.)

A. Cycads (Cycadophyta)

The cycads have 137 extant species characterized by:

- unbranched or very sparsely branched fleshy stems;
- large compound (divided) leaves;
- flagellate, swimming sperm (also found in *Ginkgo biloba*).

The cycads bear their micro- and megasporophylls in, respectively, compact male and female cones that are borne on separate plants (i.e., the dioecious conditions). However, in *Cycas* (sego palm) the megasporophylls are not aggregated into cones.

●●● Examine the DEMO live plants of *Cycas* and *Zamia*. Note the compound leaves (one genus, *Bowenia*, has twice compound leaves). Some taxa have leaves up to 1.5 m (4.9 ft.) long. Are cones evident on any of the DEMO plants? The extant cycads have the largest cones that ever existed.

Some taxa have female cones up to 100 cm (39.4 in.) long and weighing up to 42 kg (92.6 lb.) (Chamberlain 1935:100). Details of reproductive morphology are similar to those of the conifers.

B. Ginkgophytes (Ginkgophyta)

The ginkgophytes, once widespread throughout the world, now contain only one extant, relic species, *Ginkgo biloba* (ginkgo, maidenhair tree). It is native to China, but probably does not exist outside of cultivation. Distinctive features, based mainly on the sole extant taxon, include the following:

- woody, highly branched shoots;
- fan-shaped, simple leaves with dichotomous venation;
- flagellate, swimming sperm;
- ovules and seeds borne terminally on specialized, elongate stalks (i.e., not in cones).

Ginkgo is dioecious, that is, with male and female plants. Male trees are preferred in cultivation because the undesirable females drop fleshy seeds at maturity. These contain butyric acid and not only are very smelly (butyric acid is also found in rancid butter), but also make a slimy mess on sidewalks due to their fleshiness (the effect is like slipping on a banana peel). By the way, *Ginkgo* is one of the trees most resistant to urban pollution and thus is preferred as a "street tree."

●●● Examine the DEMO live and/or pickled material of *Ginkgo,* including the distinctive features mentioned in the preceding paragraph. Note the male and female reproductive structures. The fleshy seeds of *Ginkgo,* called "nuts," are considered an edible, Oriental delicacy. Observe the DEMO can of ginkgo "nuts" and the DEMO longisection cut through one of them. What is the edible ginkgo "nut" morphologically?

Nuts!—Anthony Clement McAuliffe, reply on 23 Dec. 1944 to a German demand for surrender

C. Gnetophytes (Gnetophyta)

The gnetophytes are a singular group occupying an isolated position among the seed plants. Only fossil pollen has been found; the lack of a megafossil record certainly has not been helpful in elucidating the relationships of this group. Unlike other gymnosperms, the gnetophytes have vessel elements in their xylem; these also occur in most angiosperms and in a few pteridophytes. The gnetophytes have simple leaves and very distinctive reproductive morphology.

●●● The three genera (76 species) of gnetophytes have very divergent vegetative morphology. Examine the representatives that are on DEMO:

1. *Ephedra* (joint fir, Mormon tea, Mexican tea) grows in xeric regions of the American southwest (and other areas) and has slender stems with scalelike leaves, a habit reminiscent of *Equisetum.*
2. Species of *Gnetum* are generally vines that occur in the tropics. Their expanded foliage leaves are very reminiscent of those of the angiosperms in both form and venation.
3. The sole species in the genus, *Welwitschia mirabilis,* as the specific epithet "mirabilis" suggests, is among the most bizarre of all plants. *Welwitschia* is restricted to the extremely dry deserts of southwestern Africa (Namibia). The shoot system of *Welwitschia* produces in its lifetime only three pairs of leaves:
    - the initial two cotyledons or seed leaves;
    - the two enlarged foliage leaves that are very conspicuous, as on the DEMO;
    - a final pair of white scale bodies representing the last leaf pair.

The large foliage leaves function for up to 2,000 years, the life span of the plant (Bornman 1978). The two leaves die off at their tips but are renewed from their bases by continuous (intercalary) meristematic activity. Because their tips die, the leaves become greatly frayed and split. Old plants thus appear to have many leaves, although only two leaves are actually present.

NAME ______________________________ LAB SECTION # ______________ DATE ____

TITLE ______________________________

__________________ TOTAL MAG _____

TITLE ______________________________

__________________ TOTAL MAG _____

TITLE ______________________________

__________________ TOTAL MAG _____

TITLE ______________________________

__________________ TOTAL MAG _____

SUPPLEMENT **9**

Characteristics of Seed Plants

(Gymnosperms versus Angiosperms)

I. PERSPECTIVE

The seed plants comprise two main groups that are fundamentally separated on the basis of the enclosure of the ovule or seed:

- ***gymnosperms,*** with exposed or naked ovules/seeds—750 species, 83 genera (Kubitzki 1990);
- ***angiosperms*** (flowering plants), with enclosed ovules/seeds—about 240,000 species, 13,725 genera (Mabberley 1987).

There are, however, numerous other differences between these two groups.

II. DISTINGUISHING CHARACTERISTICS OF GYMNOSPERMS AND ANGIOSPERMS

The unifying characters of *all* extant gymnosperms (see Lab Exercise 8) and hence their differences from the extant angiosperms (see Lab Exercises 9 to 11) are:

Extant Gymnosperms	*Extant Angiosperms*
BASIC LIFE HISTORY (Figs. 8–1, 8–2, Sup5–5)	BASIC LIFE HISTORY (Fig. 9–4, Sup5–5)
HABIT AND SOME GENERAL FEATURES	HABIT AND SOME GENERAL FEATURES
All woody (mostly trees)	Woody (trees, shrubs, etc.) or herbaceous
Vessels mostly absent (i.e., tracheids only)	Vessels mostly present (tracheids also present)
Cones usually present (inflorescences, flowers, and fruits absent, but some cones fleshy and fruitlike)	Inflorescences, flowers, and fruits present (cones absent)
Ovules/seeds naked	Ovules/seeds enclosed
Pollination mostly by wind	Pollination mostly by animals, especially insects
GPTs greatly reduced	GPTs very greatly reduced
POLLEN AND MICROGPTS	POLLEN AND MICROGPTS
Pollen walls (exine) homogeneous	Pollen walls (exine) heterogeneous, differentiated into a column-space system
Pollen grain with 1 aperture, sometimes none	Pollen grain with zero to many apertures
Pollen grain always a partly developed (immature) microGPT	Pollen grain a partly or completely developed microGPT, i.e., the bi- and trinucleate conditions

Extant Gymnosperms, continued	***Extant Angiosperms, continued***
Pollen grains with 2–5 cells, sometimes 1 cell or up to 43 cells	Pollen grains with 2 or 3 cells
Sperm mostly not flagellate (multiflagellate in cycads and *Ginkgo*)	Sperm never flagellate
OVULES/SEEDS (Figs. 8–2F, G, H, K, Sup8–1)	OVULES/SEEDS (Figs. 9–4H, 10–3)
Usually larger	Usually smaller
Orientation (axis of ovule) straight	Orientation (axis of ovule) straight or especially curved
More complex in structure	Simpler in structure
Integument only 1	Integument 1 or, more commonly, integuments 2 (rarely none)
Nucellus (megasporangium) always massive	Nucellus (megasporangium) reduced, often greatly so
Pollen chamber often present	Pollen chamber never present
MegaGPT formed by 1 megaspore	MegaGPT (embryo sac) formed by 1–4 megaspores
Ovule megaGPT massive (up to 8,000 cells)	Ovule megaGPT very reduced, usually 7 cells (8 nuclei)
Archegonia mostly present (absent in *Gnetum* and *Welwitschia*)	Archegonia always absent
Seed megaGPT massive	Seed megaGPT negligible in amount
Endosperm absent	Endosperm present (typically $3n$)
Nutritive tissue of embryo/seedling the megaGPT ($1n$)	Nutritive tissue of embryo/seedling the endosperm, occasionally perisperm ($2n$)
Seed dispersal mostly by wind or gravity	Fruit/seed dispersal especially by animals
To reiterate, the ovule is a combination of: SPT ($2n$), the integument and nucellus (megasporangium) GPT ($1n$), the massive megaGPT	To reiterate, the ovule is a combination of: SPT ($2n$), the integument(s) and nucellus (megasporangium) GPT ($1n$), the very reduced megaGPT (called embryo sac)
To reiterate, the seed is a combination of: old SPT ($2n$), the seed coat derived from the integument, also small amounts of nucellus (megasporangium) present GPT ($1n$), a massive megaGPT, also negligible amounts of microGPT present new SPT ($2n$), the embryo	To reiterate, the seed is a combination of: old SPT ($2n$), the seed coat derived from integument(s), also negligible amounts of nucellus (megasporangium) present GPT ($1n$), a negligible megaGPT, also negligible amounts of microGPT present new SPT ($2n$), the embryo endosperm (usually $3n$)

For definitions of terms see Supplement 8.

SUPPLEMENT **10**

Characteristics of Angiosperms

(Dicotyledons versus Monocotyledons)

I. PERSPECTIVE

The angiosperms are divided into two classes, the dicotyledons (Dicotyledones, also called Magnoliopsida) and the monocotyledons (Monocotyledones, also called Liliopsida). These groups are often vulgarly called "dicots" and "monocots."

II. DISTINGUISHING CHARACTERISTICS OF DICOTYLEDONS AND MONOCOTYLEDONS

Dicotyledons and monocotyledons are contrasted below. The most important distinguishing characteristic is the first one on number of cotyledons (Fig. Sup10–1); it is the basis for the classificatory subdivision of angiosperms [The first and second features (the second on sieve-elements is beyond the scope of this course manual) are the most important according to Dahlgren (1983:127)].

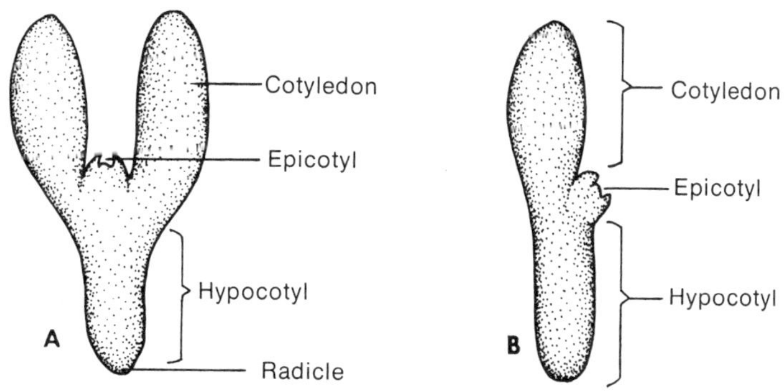

Fig. Sup10–1. Comparison of embryos of a dicotyledon (*A*) and a monocotyledon (*B*). The radicle of monocotyledons typically develops late. (From Norstog & Long 1976:454.)

Dicotyledons

Examples: magnolia, roses, oaks, sunflowers, eucalyptus

Numbers: about 10,907 genera, 188,000 species (Mabberley 1987)

Cotyledons 2 (rarely 1, 3, or 4) (Figs. 10–3B, C, Sup10–1A)

Monocotyledons

Examples: orchids, lilies, grasses, sedges, palms

Numbers: about 2,818 genera, 52,000 species (Mabberley 1987)

Cotyledons 1 (or embryo sometimes undifferentiated) (Figs. 10–3A, Sup10–1B)

Dicotyledons, continued

Sieve-element plastids with variously shaped protein bodies

As seen in transection, stem vascular bundles usually borne in a ring enclosing a pith and surrounded by a distinct cortex (Fig. 7–1A)

Often with secondary growth via a vascular cambium; hence commonly trees or shrubs, but also herbs if with entirely primary growth, or with primary growth and very little secondary growth

Leaves mostly net-veined

Mature root system either primary and/or adventitious

Floral parts, when of definite number, typically borne in sets of 5, less often 4, seldom 3 (carpels often fewer than 3)

Pollen typically with three apertures per pollen grain except in a few of the more primitive families

Seeds either endospermous (albuminous) or nonendospermous (exalbuminous) (Fig. 10–3B, C)

Monocotyledons, continued

Sieve-element plastids with wedge-shaped (cuneate) protein bodies

As seen in transection, stem vascular bundles generally "scattered" (or in 2 or more rings), with no distinct pith or cortex (Fig. 7–1B)

No secondary growth via a vascular cambium; hence usually herbs (palm and yucca "trees" achieve their size by other modes of secondary growth; a banana "tree" is entirely primary growth)

Leaves mostly parallel-veined

Mature root system entirely adventitious

Floral parts, when of definite number, typically borne in sets of 3, seldom 4, never 5 (carpels often fewer than 3)

Pollen typically with only one aperture per pollen grain

Seeds usually endospermous (albuminous) (Fig. 10–3A)

LAB EXERCISE 9

Flowering Plants (Angiosperms) I

(Flowers, Pollen Grains, and Ovules)

OBJECTIVE

To examine the salient features of reproductive morphology of flowering plants (angiosperms), namely, flowers, pollen grains, and ovules.

Note: Read this material *before* coming to the lab; there is insufficient time in the lab to read the material from scratch. Due to the large amount of background information presented here, a ●●● is used to mark those paragraphs calling for actual observations or diagrams in the lab. Most unmarked paragraphs reiterate information presented in the lecture. A bullet (●) signifies an important term or concept.

Suggested Lab Diagrams (see part noted for suggested labels):
Flowers from *two* species of angiosperms (Part I)
Agapanthus (lily-of-the-Nile, blue African lily, an angiosperm and monocotyledon) OR *Lilium* (lily, an angiosperm and monocotyledon) anther (with pollen grains) x.s. (Part III-A)

●●● *Important note:* Begin Part III-B on pollen germination of *Impatiens* early in the lab. Examine the pollen at 20 minute intervals to monitor its germination.

PERSPECTIVE

The division Anthophyta represents the angiosperms or flowering plants. These involve some 240,000 extant species and thus constitute the dominant, most ubiquitous group of land plants and the division with the greatest amount of variability. The angiosperms are divided into two classes:

- dicotyledons, with about 10,907 genera and 188,000 species;
- monocotyledons, with about 2,818 genera and 52,000 species (Mabberley 1987).

Supplement 10 contrasts these classes. In numbers and variation the angiosperms are like the insects, which have the most species of all the animal groups, that is, over 750,000 described species and many more undescribed ones. Actually, the fact that the angiosperms and insects have the largest numbers of species among, respectively, plants and animals is indicative that angiosperms and insects evolved together in response to changes they promoted in each other; that is, they coevolved.

- ***Coevolution*** is reciprocal evolutionary change in ecologically interacting species.

The story of pollination of angiosperms by insects is an obvious story, but there are various other ecological interactions between angiosperms and insects (see Lab Exercise 11, Parts I-C and I-D).

Compared to the pteridophytes and gymnosperms, the angiosperms are of rather recent origin, with fossils dating to the Cretaceous. Although there are no spectacularly promising candidates in the fossil record for an angiospermous ancestor, it is generally thought that they must have evolved from some gymnospermous precursor. The oldest clearly angiospermous fossils date from the early Cretaceous some 130 million years ago (mya). By approximately 100 mya, some 35 mya before the dinosaurs became extinct, the angiosperms had become a diverse and successful group. Unfortunately there are relatively few fossil flowers because of their ephemeral and delicate nature. As a result, the evolutionary history of flowering plants is obscure, and our understanding of the evolution of the flower is also poorly understood.

> Evolution has followed lines of opportunism, not of logic.
> —Frederick O. Bower, *The ferns,* 1923

The most distinctive feature of the angiosperms (the term literally means "vessel seed," where "vessel" means "container") is the enclosure of the ovules or potential seeds in a hollow ovary (Fig. 9–1). The gymnosperms, of course, have "naked seeds" (this is the literal meaning of "gymnosperm"). The enclosed ovules/seeds of the angiosperms and many other adaptations of the reproductive and vegetative systems apparently conferred to the early angiosperms such a great selective advantage that the group rapidly became dominant after its origin about 130 million years ago and now comprise some 240,000 species. The selective advantage undoubtedly derived from a host of factors, including:

- 1. better resistance to drought and cold due to various new structural and physiological mechanisms (Raven et al. 1992:406–407 inexplicably focus on this factor);
- 2. better resistance to pathogens (bacteria, fungi, other plants) and predators (animal), the "biochemical coevolution" of Raven et al. (1992:435–438), due to various new toxic or obnoxious chemicals produced in all organs;
- 3. more efficient water conduction due to vessels comprising vessel elements;
- 4. more efficient nutrient conduction due to sieve tubes comprising sieve tube elements;
- 5. more efficient modes of pollination due to the inflorescence (see definition below) and flower;
- 6. more efficient pollen dispersal and also incompatibility mechanisms to prevent inbreeding due to modifications of the pollen grain wall;
- 7. a more rapid life history due to the extremely reduced micro- and megaGPT;
- 8. a more rapid life history due to double fertilization and more efficient nutrition of the embryo (very young SPT before seed germination) and seedling due to the concomitant endosperm formation;
- 9. more efficient modes of fruit and/or seed dispersal due to the fruit and to modifications of the seed itself.

In contrast, and significantly, the pteridophytes and gymnosperms have rather few of the nine distinctive features of the angiosperms just enumerated. The key phrases in this discussion, and these aspects undoubtedly were very interrelated, are (1) "evolutionary reduction," (2) "greater efficiency," and (3) "more rapid completion of reproductive events." In essence, if progeny are produced more rapidly and more efficiently, greater numbers will survive for natural selection to act upon in subsequent generations.

Strictly speaking, ***inflorescences*** (clusters or aggregations of flowers), "flowers," "fruits," and their attendant structures except pollen grains, ovules, and seeds are reproductive structures unique to the angiosperms. Although comparable (analogous) structures may occur in other groups of land plants (e.g., fleshy cones in some gymnosperms—see Lab Exercise 8, Part I-B), angiosperm terminology should *not* be used to describe such structures.

Recall from Supplement 5, Part IV, that the heterosporous haploid-diploid ($1n$-$2n$) life history characterizing *all* seed plants has male "micro-" structures and female "mega-" structures (Fig. Sup5–5), namely:

- ***micro-*** and ***megasporophylls,*** which bear
- ***micro-*** and ***megasporangia,*** which produce
- ***micro-*** and ***megasporocytes,*** which produce by meiosis
- ***micro-*** and ***megaspores,*** which develop into
- ***micro-*** and ***megaGPTs.***

A ***nucellus*** is another name for a megasporangium. A ***bract*** is a modified, usually reduced leaf generally associated with a reproductive structure. The angiosperms completely lack cones and instead have either unisexual or bisexual flowers. The male and female flowers may be borne on separate, unisexual plants or on the same, bisexual plant, that is:

- the ***dioecious*** condition, with unisexual, male and female plants, or micro- and megaSPTs;
- the ***monoecious*** condition, with bisexual plants.

Note: Before beginning this lab exercise, you should review:

- Supplement 5 on the main types of life histories;
- Supplement 8 on pollen grains, ovules, and seeds, and their terminology;
- Supplement 9 contrasting gymno- and angiosperms.

I. BASIC FLORAL STRUCTURE

Because there are some 240,000 extant species of angiosperms, their reproductive structure is diverse and very complicated. Floral size varies from 0.1 mm to 1 m (0.004 in. to 39.4 in.), respectively, *Wolffia* (the whole plant is ca. 1 mm) and *Rafflesia* (see the figure on the title page and the caption on its verso). A typical flower consists of sterile (*-marked items) and fertile (#-marked items) parts (Figs. 9–1, 9–2, 9–4A) that are arranged in the following sequence (from base toward tip, and from outside toward inside of the flower):

- ***receptacle,**** the often expanded axis part of the flower bearing the various floral parts;
- ***sepals,**** the outer, often green sterile appendages, which collectively comprise the ***calyx*** and have a largely protective function;
- ***petals,**** the inner, often colored sterile appendages, which collectively comprise the ***corolla*** and usually attract animal pollinators to the flower;
- ***stamens,#*** the male sex organs or pollen-bearing structures, which collectively comprise the ***androecium,*** each stamen consisting of:
  - an ***anther,*** the pollen-containing part;
  - a ***filament,*** the stalk supporting the anther;
- ***carpels,#*** the female sex organs or ovule-bearing structures, which collectively comprise the ***gynoecium,*** each carpel consisting of:
  - an ***ovary,*** the enlarged hollow basal part (its cavity is the ***locule***) containing one to four million or so ovules;
  - a ***style,*** the slender, often hollow upper part;
  - a ***stigma,*** the expanded distal part functioning to receive the pollen.

To clarify the terminology of the gynoecium (Fig. 9–2), think of "ovary," "style," and "stigma" in reference to a side view compared to "carpel" in reference to a top-down cross sectional view. That is, a gynoecium with ten "wedges" as in a grapefruit consists of ten fused carpels, each theoretically with ovary, style, and stigma, although only one collective ovary, style, and stigma is usually designated. [*Note:* "Carpel" is more of an evolutionary concept, whereas the other terms are more common in identification manuals. Another term is "pistil" but I'll not use this because I believe in gun control☺]

Some important related floral terminology is (Fig. 9–1):

- ***whorl,*** a group of two or more *similar* floral parts (also three or more leaves) attached at the same node;
- ***pedicel,*** the stem or stalk supporting the flower in an inflorescence;
- ***peduncle,*** the stem or stalk supporting a solitary flower, that is, one not in some sort of cluster (inflorescence);
- ***perianth,*** the collective term for sepals and petals;
- ***tepals,*** perianth parts *not* clearly differentiated into sepals and petals;
- ***nectary*** (nectar gland), a structure producing ***nectar,*** a mainly sugary solution that attracts pollinating insects and other small animals;
- ***placenta,*** the part of the ovary wall where ovules or seeds attach.

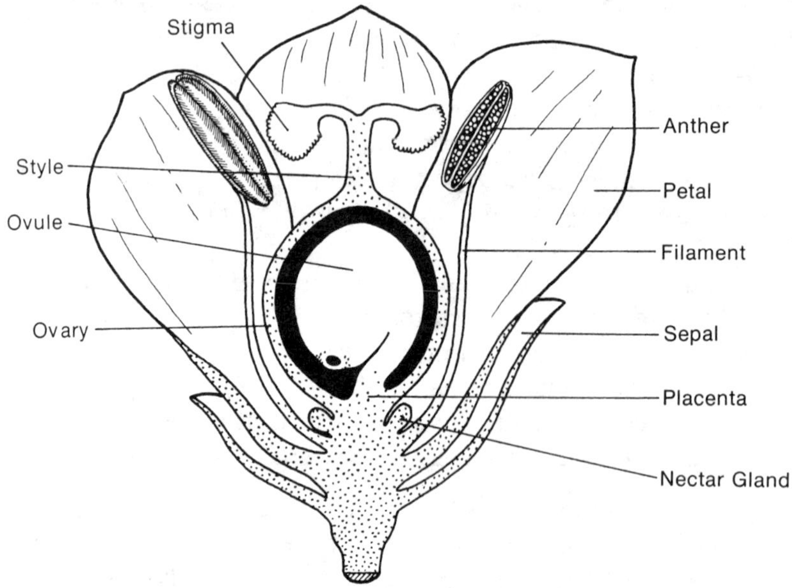

Fig. 9-1. Structure of a flower in longisection. Note the micropyle of the ovule. (From Norstog & Long 1976:425.)

The typical flower (Figs. 9–1, 9–4A) thus consists of four whorls: a calyx of sepals, a corolla of petals, an androecium of stamens, and a gynoecium of carpels. Only the fertile parts, the stamens and carpels, produce spores. Stamens and carpels are the only ***essential parts*** because only they are necessary for the flower to function sexually.

Floral parts may be completely ***free*** (separate, distinct) from each other or they may be variously ***fused*** (united) with similar or dissimilar floral parts into ***compound structures*** [Note that this usage is opposite that for vegetative parts, where compound leaves are divided into leaflets—see Lab Exercise 6, Part I-A]. For example, petals may be fused with each other to form a corolla tube. Stamens may be similarly fused or may be fused to the petals. A flower may have a single carpel (Fig. 9–2B), that is, a simple ovary, several to many separate carpels, or, mostly, a group of carpels fused together into a compound ovary (Fig. 9–2D). Sepals, petals, and stamens can fuse into a ***floral tube*** (see Lab Exercise 10, Part I). Moreover, the floral parts listed above are variously present or absent in flowers. Thus, some flowers lack a style (or styles), and unisexual flowers lack either stamens (in female or pistillate flowers) or carpels (in male or staminate flowers).

Note: There is much additional floral terminology, which is used especially by systematists. Unless instructed to the contrary, you are responsible only for the floral structures mentioned above in bold italics.

●●● See Figs. 9–1, 9–2, and 9–4A. Then dissect the live flowers available and identify the various floral parts indicated above. Note whether floral parts are free (separate) or fused. It would be profitable to use a dissecting microscope to see the finer floral details. The flowers available will depend on the season and whether sufficient quantities can be procured for the class; material in limited supply may be on DEMO. Some sample descriptions of flowers are given:

1. *Agapanthus africanus* (lily-of-the-Nile, blue African lily): Each flower has six, basally fused tepals, six stamens attached to the tepals, and one ovary consisting of three carpels, each with a cavity. The fruit (a DEMO, if available) consists of only dried carpels. Although the three outer tepals could be regarded as sepals, and the three inner as petals, these perianth parts are usually called tepals because of their very similar appearance.
2. *Aloe* (aloe): The flowers are similar to the preceding. The perianth consists of six lobes in two whorls. Depending on the species, the outer tepals are free or united for part of their length; the inner tepals are variously joined to the outer tepals.
3. *Rosa* (rose): A DEMO will show five conspicuous sepals, many petals, many stamens, and ten or more free carpels in a floral cup.
4. *Salvia* (sage) or other mint (Labiatae or Lamiaceae): The five sepals are nearly completely fused into a tube; just their tips may be evident. The five petals are fused into a corolla tube that is irregular in shape and two-lipped. The four stamens are fused to the corolla. The ovary is four-lobed, consisting of two carpels each deeply lobed; the four ovules in the gynoecium develop into four seeds in the fruit. The gynoecial features may be hard to see.
5. *Sedum* (stonecrop, orpine) or other member of Crassulaceae: There are four or five fused sepals, four or five free petals, usually ten free stamens, and a gynoecium of four or five carpels that are usually free but sometimes slightly fused at their bases.

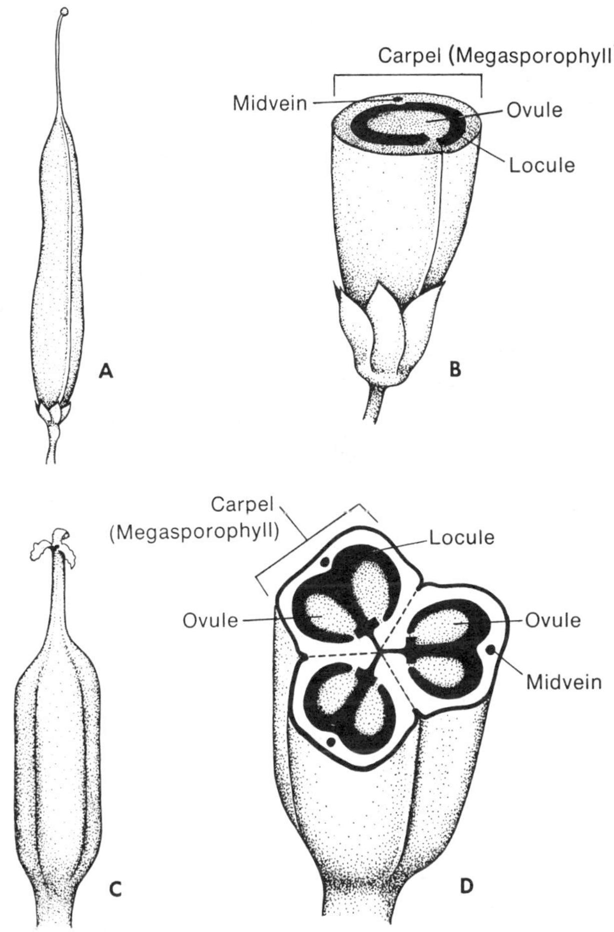

Fig. 9-2. Comparison of simple and compound ovaries in side and sectional views. *A, B,* simple ovary composed of one carpel; *C, D,* compound ovary composed of three carpels. (From Norstog & Long 1976:426.)

Flower samples 1 and 2 typify monocotyledons, whereas flower samples 3 through 5 typify dicotyledons (see Supplement 10).

●●● *Suggested diagrams and labels:* Flowers from *two* species of angiosperms: receptacle, calyx, sepals, corolla, petals (perianth and tepals if calyx and corolla undifferentiated), androecium, stamens, anther, filament, gynoecium, ovary, style, stigma, and, as relevant, carpels fused or separate.

II. SOME TRENDS OF FLORAL EVOLUTION

There are many morphological interpretations of the flower. By the traditional and classical interpretation, and still the most widely accepted one, the flower has been regarded as an evolutionarily greatly modified, determinate shoot bearing:

- sterile appendages, the sepals and petals;
- fertile appendages or sporophylls, the stamens and carpels, namely:

- ***microsporophylls*** or ***stamens,*** each of the four ***pollen sacs*** typical of a stamen considered equivalent to a microsporangium;
- ***megasporophylls*** or ***carpels,*** each ovule constituting an integumented megasporangium consisting of integument(s), nucellus (megasporangium), and megaGPT (= ***embryo sac,*** a term restricted to angiosperms).

Biologists interested in the diversity of organisms like to discuss evolutionary trends. There is some basic, more or less obvious terminology, which is used in both botany and zoology:

- ***character,*** a feature or structure of an organism, as eye or hair color;
- ***character state,*** the actual expression or trait ("distinguishing quality") of the character, as blue eyes or black hair;
- ***primitive character state,*** basically the ancestral, most basic state;
- ***specialized*** or ***derived character state,*** a state derived from the primitive one, in many but *not* all cases an advanced state.

The accompanying table contrasts primitive and derived (specialized) character states of the flower:

Character	*Primitive State**	*Specialized or Derived State*
Sexuality	Bisexual flowers, with both stamens and carpels	Unisexual flowers, with either stamens or carpels
Pollination	By beetles	By wind and by other types of animals, mainly insects
Symmetry	Radial, i.e., regular (actinomorphic) flowers	Bilateral, i.e., irregular (zygomorphic) flowers
Number of floral parts	Many per flower	Few per flower
Arrangement of floral parts	Helical or spiral	Whorled or cyclic
Fusion of floral parts	Free and distinct (e.g., petals fall off easily)	Physically fused
Size of receptacle	Large and elongate	Greatly shortened
Differentiation of perianth	Tepals present, i.e., no obvious sepals or petals	Distinct sepals and petals *or* perianth totally absent
Differentiation of stamens	Stamens leaflike, i.e., *no* distinct filament and anther present	Stamens *not* especially leaflike, distinct filament and anther present
Nuclei and aperture of pollen grain	With two nuclei (binucleate) and one furrow (monosulcate)	With three nuclei (trinucleate) and various other arrangements of furrows and/or pores
Differentiation of carpels	Carpels leaflike, stalked, without a style	Carpels *not* leaflike, with a distinct style and stigma
Position of ovary	***Superior,*** the floral parts inserted *beneath* the ovary	***Inferior,*** the floral parts inserted *above* the ovary

*As exemplified by *Degeneria vitiensis* shown in Fig. 9–3.

The relic angiosperm, *Degeneria vitiensis* (Fig. 9–3), one of two species (one described in 1942, the other in 1988) of the family Degeneriaceae, is often regarded as among the most primitive of the extant angiosperms. *Degeneria* occurs only on Fiji and is not available in cultivation. Hence *Magnolia* (magnolia), which is also primitive in many respects, is used as an approximation of the primitive flower. There are many examples of specialized flowers, for instance, those of grasses

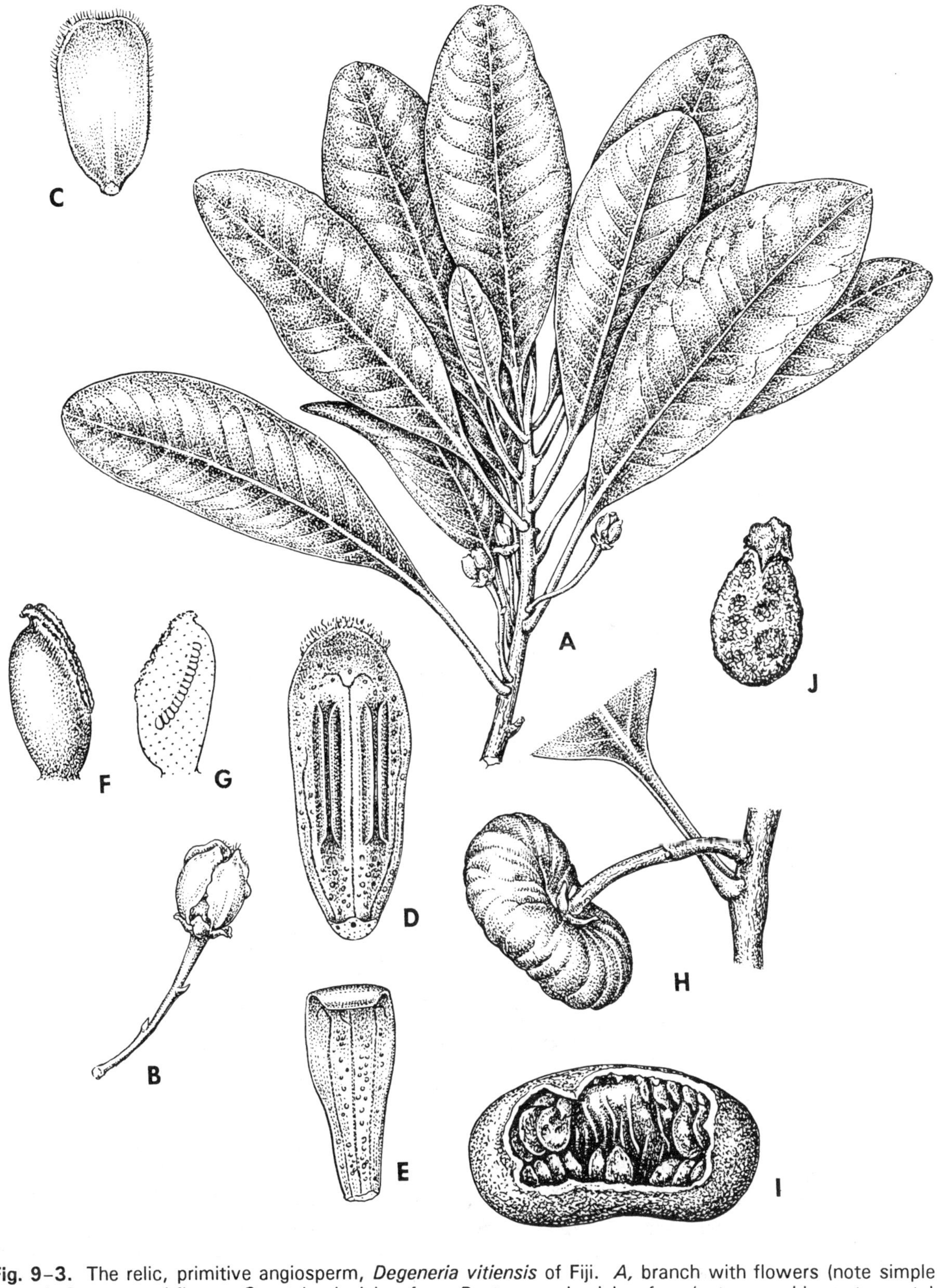

Fig. 9–3. The relic, primitive angiosperm, *Degeneria vitiensis* of Fiji. *A,* branch with flowers (note simple leaves); *B,* unopened flower; *C,* petal, adaxial surface; *D,* stamen, abaxial surface (note resemblance to a petal and the four microsporangia); *E,* sterile stamen (staminode), adaxial surface; *F,* gynoecium composed of one carpel (note the short style and conspicuous stigma); *G,* longisection of carpel showing ovules; *H,* branch with fruit; *I,* nearly mature fruit with part of wall removed to show seeds and appendages of endocarp; *J,* seed and, at top, part of attached endocarp. (Modified from Norstog & Long 1976:449; redrawn from I. W. Bailey & A. C. Smith, Degeneriaceae: A new family of flowering plants from Fiji, *J. Arnold Arbor.* 23:356–365, 1942.)

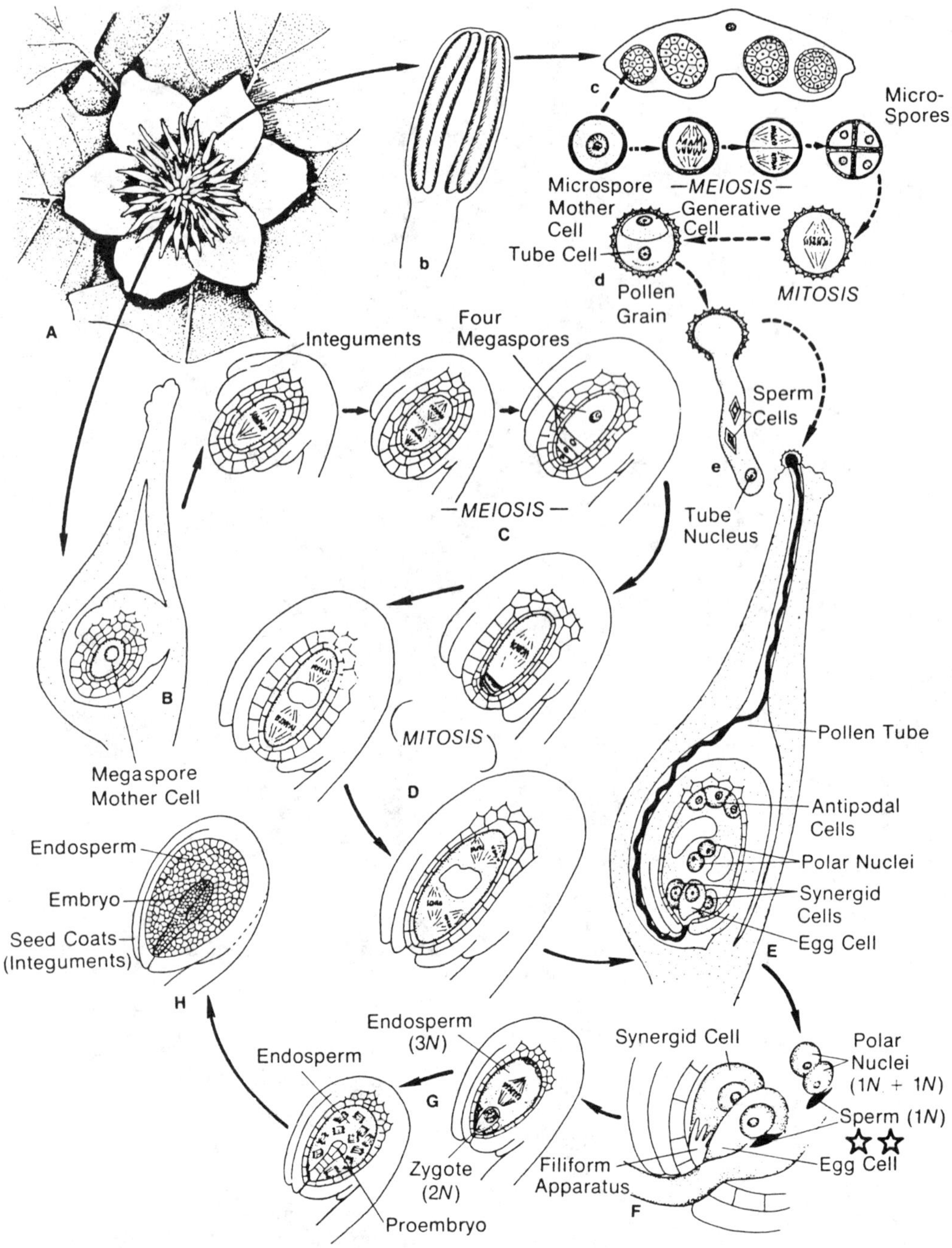

Fig. 9–4. Heterosporous haploid-diploid (1*n*-2*n*) life history of a typical flowering plant (angiosperm). *A,* flower of the SPT of *Liriodendron tulipifera* (tulip tree, tulip poplar, yellow poplar); *b,* microsporophyll or stamen with four microsporangia (pollen sacs); *c,* cross section of anther (note the four pollen sacs) showing developmental stages of microspores (microsporogenesis) (note that a tetrad of microspores results from meiosis of the microsporocyte, the "microspore mother cell"); *d,* section of mature pollen grain, an immature microGPT (male GPT), at time of shedding; *e,* longisection of mature microGPT, i.e., a germinated pollen grain with a pollen tube; *B,* longisection of a carpel (megasporophyll) with a young ovule and its megasporocyte (megaspore mother cell); *C,* longisections of ovules showing developmental stages of megaspores (megasporogenesis) (note that a tetrad of megaspores results from meiosis of the megasporocyte and that three megaspores degenerate); *D,* longisections of ovules showing development stages of megaGPT or embryo sac from three mitotic divisions; *E,* longisection of carpel showing course of pollen tube and ovule with mature seven-celled, eight-nucleate megaGPT (embryo sac) at time of syngamy (fertilization) (the megaSPT has three antipodal cells, two synergid cells, one egg cell, and one central cell with two polar nuclei); *F,* longisection of

(Gramineae or Poaceae), orchids (Orchidaceae), sunflowers (Compositae or Asteraceae), or milkweeds (Asclepiadaceae).

●●● Examine the DEMO whole and longitudinally split flowers of *Magnolia* (magnolia). Note that this relatively primitive flower has most of the primitive traits mentioned in the table. For pollen grains of *Magnolia* see the scanning electron micrograph of *Magnolia* on DEMO in Part III-A. *Magnolia* is a dicotyledon (see Supplement 10).

●●● The orchid family (Orchidaceae) is one of the most highly evolved families of angiosperms. Thus its flowers exhibit many specialized features. Which of the specialized features listed in the table apply to the live orchid flowers on DEMO? Orchids are monocotyledons (see Supplement 10).

III. EMBRYOLOGY, ESPECIALLY OF GPTs

The male and female GPTs of angiosperms have fewer component cells (respectively, at most three and seven cells—see below) than those of the gymnosperms and in this respect represent structural simplicity (Fig. 9–4). However, this clearly is evolutionary complexity because reduction has occurred through evolution. Because there about 750 extant species of gymnosperms compared to some 240,000 extant species of angiosperms, obviously there has been in the latter group greater possibility for structural diversity of the GPT.

●●● Consult Fig. 9–4 of the heterosporous haploid-diploid ($1n$-$2n$) life history (see also Fig. Sup5–5) of a typical angiosperm as you examine the material in the following sections.

A. Anthers, pollen grains, and pollination

The typical anther consists of the following parts (Fig. 9–4b, c):

- the ***epidermis,*** which may or may not persist in the mature anther;
- the ***endothecium,*** a special subepidermal layer (or layers) of cells with irregularly thickened walls, and facilitating pollen release;
- a ***mouth*** (stomium) (actually two, one on each side of the anther), a specialized region of thin-walled cells also facilitating pollen release;
- four cavities or ***pollen sacs,*** the microsporangia, which contain the pollen grains ("pollen sac" is generally restricted to the microsporangia of angiosperms);
- the ***connective,*** the mesophyll and vascular tissue located between the pollen sacs and the endothecium.

Anthers produce pollen, which is the immature microGPT (Fig. 9–4c, d). Meiosis of the microsporocyte results in the four microspores of a tetrad, which is generally of the tetragonal and especially tetrahedral types (see Lab Exercise 7, Part IV). The resultant microspores are held together by the wall of the microsporocyte. The tetrads of microspores usually break up so that the spores lie freely in the microsporangium (Fig. 9–1). Each microspore, of course, is an immature pollen grain.

The anthers of many angiosperms dehisce to release the pollen grains. This typically occurs through slits and is facilitated by the endothecium and mouth (stomium). The endothecium with its irregularly

---

Caption for Figure 9–4 continued:

micropylar end of ovule showing double fertilization (☆☆), one sperm of pollen tube uniting with the egg, the second sperm uniting with the polar nuclei; *G,* longisections showing developmental stages of embryos and endosperm in maturing seeds; *H,* longisection of mature seed with embryo (very young SPT) and endosperm. Note the alternation of morphologically dissimilar GPTic and SPTic generations (phases), with the SPT dominant over the GPT, which is reduced to three and, typically, seven cells in the microGPT and megaGPT, respectively (compare with Fig. 8–2). (From Norstog & Long 1976:436.)

thickened cell walls is comparable to the annulus of ferns and similarly involves a hygroscopic mechanism (see Lab Exercise 4, Part I). However, pollen release is gradual, *not* a violent action as is spore discharge of ferns. Water loss from the endothecium stresses the mouth regions. Their thin-walled cells constitute a weak point where the anther opens. Each anther typically has four pollen sacs (microsporangia) arranged in two pairs.

●●● See Fig. 9–4c. Then examine the DEMO model of an endothecial cell; the black ink markings represent the thickenings on five of the six walls of a cell. Finally, examine a slide of an anther transection of *Lilium* (lily) OR *Agapanthus* (lily-of-the-Nile, blue African lily). Each slide is a section of the tip of a whole flower bud; focus only on the anthers. Identify the following structures of the anther: epidermis, endothecium, mouth (stomium), connective (mesophyll, vascular bundle), pollen sacs (microsporangia), and pollen grains.

●●● *Suggested diagram and labels: Agapanthus* (lily-of-the-Nile, blue African lily, an angiosperm and monocotyledon) OR *Lilium* (lily, an angiosperm and monocotyledon) anther (with pollen grains) x.s.: epidermis, endothecium, mouth (stomium), connective (mesophyll, vascular bundle), pollen sacs (microsporangia), pollen grains (exine wall).

The walls of pollen grains of angiosperms, like those of gymnosperms, consist of several basic parts:

- ***intine,*** the inner, usually thin wall layer composed of cellulose, the usual material of plant cell walls;
- ***exine,*** the outer, usually thick wall layer composed of the waxy chemical ***sporopollenin,*** which is highly resistant, and consisting of:
  - ***columns*** (columellae) separated by spaces;
  - a ***roof*** (tectum) covering the column-space system, though a roof is often lacking;
  - ***apertures,*** thin areas that function as sites for exit of the pollen tube or pollen tubes and shaped as:
    - ***furrows,*** elongate apertures;
    - ***pores,*** circular or slightly elliptic apertures.

Furrows and pores may occur in combination in an aperture. A pollen grain may have zero to many apertures. The spaces in the exine contain enzymes that relate to compatibility and that are an evolutionary innovation for the angiosperms. That is, cross breeding is promoted, and this increases genetic variability.

Supplement 9 contrasts the pollen of gymno- and angiosperms. Pollen of gymnosperms and spores of pteridophytes lack a column-space differentiation. In addition, a pollen grain of gymnosperms has one aperture, or none, whereas spores of pteridophytes lack apertures.

Pollen grains of angiosperms have myriad features used in systematics for identification and interpretation of relationships: aggregation, polarity, symmetry, size, shape, and, especially, apertures (absence, presence, number, shape), wall layering, and surface sculpturing.

The pollen walls of angiosperms are very highly modified compared to the pollen grain walls of the gymnosperms or the spore walls of the pteridophytes. Much of this diversity in angiosperms relates to the fact that:

- spores of pteridophytes and pollen grains of gymnosperms are spread mostly by wind, whereas
- pollen grains of angiosperms are spread mostly by insects and other small animals.

Carpels of angiosperms are pollinated not only by animals (particularly insects), but also by wind, water, gravity, or by the plant itself (self-pollination).

- ***Pollination*** involves pollen transfer from the microsporangia (pollen sacs) of the anthers to the stigmas of the carpels.

Pollen grains will have their characteristic wall sculpturing when they are shed from the anther. Depending on the agent of pollination, the wall sculpturing is often modified accordingly (see Supplement 8, Part V).

●●● Reexamine the slide of an anther transection of *Agapanthus* (lily-of-the-Nile, blue African lily) OR *Lilium* (lily) and note the following features of the pollen walls: wall sculpturing of exine, the aperture.

●●● The scanning electron microscope is vastly superior to the light microscope for observing pollen. Compare the appearance of the *Agapanthus* or *Lilium* pollen just observed with the DEMO scanning electron micrographs (from Troughton & Donaldson 1972:94, 95, and Troughton & Sampson 1973:134, 135, 138, 141, 142, 145). Note on the micrographs the diversity of external wall sculpturing. In each case you might want to make a prediction as to the most likely vector (wind, water, or animal) of the pollen (see Supplement 8, Part V). The DEMO micrographs from Troughton & Donaldson (1972:94, 95) show the stereotypical pollen for dispersal by animal (for cotton pollen) and wind (for maize pollen). Maize and other members of the grass family (Gramineae or Poaceae) are pollinated mainly by wind. Finally, examine the DEMO transmission electron micrographs (from Gifford & Foster 1989:550) of sections of pollen and note the column-space differentiation of the exine. *Note:* Ignore the unfamiliar terminology in the captions of the photographs.

B. Pollen protoplasts (microGPTs) and pollen germination

A pollen grain of an angiosperm has only three types of cells (Fig. 9–4d):

- the ***tube cell,*** a large vegetative cell that on germination of the pollen grain becomes the pollen tube;
- the ***generative cell,*** a small reproductive (fertile) cell that directly forms or "generates"
- two ***sperm,*** both of which are involved in fertilization.

In the angiosperms, a pollen grain is shed from the anther as either a partly or a completely developed microGPT:

- a partly developed microGPT, that is, two-celled and thus binucleate (Fig. 9–4d), consisting of a large tube cell (vegetative cell) and a small generative cell, the latter dividing *after* pollen germination and forming two sperm in the pollen tube (Fig. 9–4e);
- a completely developed microGPT, that is, three-celled and thus trinucleate, consisting of a large tube cell (vegetative cell) and the two sperm resulting from the generative cell dividing *before* the pollen germinates.

The mature microGPT of angiosperms hence has only three cells and no antheridia (Fig. 9–4e). About 70% of the species of angiosperms have binucleate pollen, the other 30% trinucleate pollen. Among the many differences between gymno- and angiosperms is that in gymnosperms the pollen grain is always a partly developed (immature) microGPT whereas in angiosperms it may be partly developed (immature) or completely developed (mature), that is, respectively, the bi- and trinucleate conditions. The following summarizes the usual developmental sequence for the microGPT or pollen grain of angiosperms; this sequence occurs in 70% of their species (Fig. 9–4c to e):

- microsporocyte (meiocyte) in SPT ——(meiosis)——> tetrad of four microspores ——(2 mitoses per microspore)——> 3-celled mature microGPT (tube cell, 2 sperm).

●●● See Fig. 9–4d. Then reexamine the slide of an anther transection of *Agapanthus* (lily-of-the-Nile, blue African lily) OR *Lilium* (lily) from Part III-A and note the following features of the pollen protoplasts:

1. the large tube cell, which has a conspicuous nucleus and large cytoplasm;
2. the small, more densely staining generative cell, which has a conspicuous nucleus surrounded by a small halo of cytoplasm.

What is the function of these cells? Is this pollen bi- or trinucleate? When will the sperm develop?

●●● In the beginning of the lab take some pollen of *Impatiens* (snapweed, balsam) and mount it on a slide in a 10% sucrose solution with a dash of boron added. Boron plays an essential but still undetermined role in the elongation of pollen tubes; amounts from 10 to 200 parts per million are used when artificially germinating pollen (Bhojwani & Bhatnagar 1978:107). Examine the slide about every 20 minutes. How does the pollen tube emerge from the pollen grain? Note the cytoplasmic streaming in the pollen tube. Are any nuclei evident in this?

In nature, a pollen grain that has landed on the stigma germinates to produce one or more pollen tubes (Fig. 9–4E). These traverse the styles, emerge in the cavity of the ovary, and eventually enter the ovule via its micropyle (Fig. 9–4E, F).

●●● Examine the DEMO diagram (from Fahn 1990:450) of pollen on stigmas.

C. Ovules and megaGPTs

See Supplement 8, Part VII, for the basic structure of an ovule. An ***ovule*** is an integumented megasporangium consisting of several parts:

- ***integument(s)*** and the ***micropyle,*** the small porelike opening at the apex of the ovule through which the pollen tube enters;
- ***nucellus*** (***megasporangium***);
- ***megaGPT*** (***embryo sac***).

Each ovule produces an embryo sac, which is the megaGPT. Meiosis of the megasporocyte results in the four megaspores of a tetrad (Fig. 9–4B, C), which is exclusively of the linear type (see Lab Exercise 7, Part IV) in the ovules of *all* extant gymno- and angiosperms. Each ovule thus has four megaspores. In most angiosperms (70% of the known cases) the three megaspores at the micropylar end of the ovule abort (ovules are also pro-life) so that there is only one functional megaspore per ovule. This functional megaspore then undergoes three mitoses (Fig. 9–4D). Consequently, the typical mature embryo sac (megaGPT) of angiosperms consists of eight nuclei and seven cells (Fig. 9–4E):

- three ***antipodal cells*** (antipodals) at the chalazal end of the ovule;
- an ***egg cell*** and two ***synergid cells*** (synergids) at the micropylar end of the ovule;
- a large ***central cell*** containing two ***polar nuclei.***

Sometimes the polar nuclei fuse into a ***secondary nucleus*** just before fertilization. Part III-D discusses the functions of the various component parts of the embryo sac.

Considerable diversity in megasporogenesis and megagametogenesis occurs in the angiosperms. From one to four megaspores of the meiotic tetrad may participate in the formation of the embryo sac. In gymnosperms, in contrast, only a single megaspore participates in the formation of the megaGPT. Compared to the megaGPT of gymnosperms, which consists of up to 8,000 cells (versus ca. 2,000 in *Pinus*) and several archegonia, the megaGPT of angiosperms is greatly reduced, typically consisting of only eight nuclei in seven cells. There are no archegonia. This type of embryo sac occurs in 70% of the species of angiosperms. Other types of embryo sacs are equally reduced, or even more so. The following summarizes the usual developmental sequence for the megaGPT or embryo sac of angiosperms; this sequence occurs in 70% of their species (Fig. 9–4B to E):

- megasporocyte (meiocyte) in SPT ——(meiosis)——> 4 megaspores, 3 aborting ——(3 mitoses in the functional megaspore)——> 7-nucleate, 8-celled megaGPT (egg, 2 synergids, 3 antipodals, 2 polar nuclei in central cell).

D. Double fertilization and endosperm

After the pollen tube has entered the ovule via its micropyle (Fig. 9–4E), it grows through the nucellus (megasporangium) and into the mature megaGPT (embryo sac) via one of the two synergids (Fig. 9–4F). The pollen tube discharges its two sperm into that synergid. The synergids are sites of high metabolic activity and apparently secrete some chemicals to guide the pollen tube through the micropyle and the nucellus. Double fertilization now occurs.

Double fertilization is unique to the angiosperms. It involves the participation of both sperm from the pollen tube: in the embryo sac one sperm fuses with the egg to form the zygote, which develops into the embryo (very young SPT), whereas the other sperm fuses with the polar nuclei (or polar nucleus, or secondary nucleus) to form the ***primary endosperm nucleus.*** This will develop into the ***endosperm,*** the distinctive nutritive tissue of angiosperms. In summary:

- $1n$ sperm + $1n$ egg ——> $2n$ zygote ——> $2n$ embryo (very young SPT);
- $1n$ sperm + two $1n$ polar nuclei ——> $3n$ endosperm.

The endosperm is rich in fats, carbohydrates, and proteins and thus is a nutritive tissue for the embryo and, frequently, the young seedling. Endosperm is usually solid but sometimes it is a liquid, as in coconut milk.

About 70% of angiosperm species have a triploid ($3n$) endosperm because one $1n$ sperm had fused with the two $1n$ polar nuclei (or the $2n$ secondary nucleus) of the embryo sac. Actually, because an embryo sac may contain, depending on the species, anywhere from one to 14 polar nuclei that can fuse with the second sperm of the pollen tube, the ploidy level of endosperm may vary from $2n$ to $15n$. That is:

- $1n$ sperm + two $1n$ polar nuclei ——> $3n$ endosperm (in 70% of the species);
- $1n$ sperm + one $1n$ polar nucleus ——> $2n$ endosperm (in some species);
- $1n$ sperm + 14 $1n$ polar nuclei ——> $15n$ endosperm (in some species).

Endosperm formation usually precedes division of the zygote and subsequent embryo development (embryogenesis) (Fig. 9–4G). See also comments on endosperm in Lab Exercise 10, Parts II and IV.

Why, sometimes I've believed as many as six impossible things before breakfast.
—The White Queen

The antipodals usually have no function in fertilization and thus degenerate along with the synergids. Sometimes the antipodals persist and may have a nutritional role for the embryo.

Double fertilization in angiosperms induces hormonal and other physiological effects that trigger several events:

- the new development of the embryo (very young SPT) from the zygote;
- the new development of the endosperm;
- the concomitant transformation of the ovule into the seed;
- the concomitant transformation of the ovary into the fruit.

Lab Exercise 10 treats these post-fertilization topics. The other floral parts (the stamens, petals, and sepals) usually wither or shrivel, although some can persist in the fruit.

NAME ______________________________ LAB SECTION # ______________ DATE ____

TITLE ______________________________

__________________ TOTAL MAG ____

TITLE ______________________________

__________________ TOTAL MAG ______

TITLE ______________________________

__________________ TOTAL MAG ____

TITLE ______________________________

__________________ TOTAL MAG ______

LAB EXERCISE **10**

Flowering Plants (Angiosperms) II

(Fruits, Seeds, Fruit/seed Dispersal, and Seedlings)

The fruit is . . . properly speaking, the ovary brought to perfection.
—Alphonso Wood, *A class-book of botany,* 1847

OBJECTIVE

To examine the salient features of reproductive morphology of flowering plants (angiosperms), namely, fruits, seeds, fruit/seed dispersal, and seedlings.

Note: Read this material *before* coming to the lab; there is insufficient time in the lab to read the material from scratch. Due to the large amount of background information presented here, a ●●● is used to mark those paragraphs calling for actual observations or diagrams in the lab. Most unmarked paragraphs reiterate information presented in the lecture. A bullet (●) signifies an important term or concept.

Suggested Lab Diagrams (see part noted for suggested labels):
Phaseolus vulgaris (bean, an angiosperm and dicotyledon) seed l.s. (Part II)
Phaseolus vulgaris (bean, an angiosperm and dicotyledon) OR *Pisum sativum* (pea, an angiosperm and dicotyledon) seedling (Part IV)

PERSPECTIVE

As noted at the end of Lab Exercise 9, double fertilization in angiosperms induces hormonal and other physiological effects that trigger several events:

- the new development of the embryo (very young SPT) from the zygote;
- the new development of the endosperm;
- the concomitant transformation of the ovule into the seed;
- the concomitant transformation of the ovary into the fruit.

These post-fertilization topics are treated below. The other floral parts (the stamens, petals, and sepals) usually wither or shrivel, although some can persist in the fruit.

Note: Before beginning this lab exercise, you should review:

- Supplement 8 on pollen grains, ovules, and seeds, and their terminology;
- Supplement 9 contrasting gymno- and angiosperms.

I. FRUITS

After fertilization the ovary develops into the fruit, whereas the ovules develop into the seeds. Not to knock on Wood (1847), but the proper definition of "fruit" would be (Schmid 1982):

- The ***fruit*** is the ripened or matured carpel or group of carpels, with or without seeds, and with or without accessory parts that may be variously fused with the carpel(s).

Accessory parts are non-carpellary (extracarpellary) structures and include shoot parts and other floral parts, for instance, the inflorescence axis, bracts, bracteoles, pedicels, the peduncle (e.g., the fleshy part of a fig), the receptacle (in strawberry), or pedicel and receptacle (in cashew), sepals (in mulberry), petals, stamens, or ***floral tube*** derived from the fused bases of the sepals, petals, and stamens (in apple and pear). ***Fruitlets*** are the small fruits or subunits of compound (aggregate or multiple) fruits, for example, the "seeds" of a strawberry fruit.

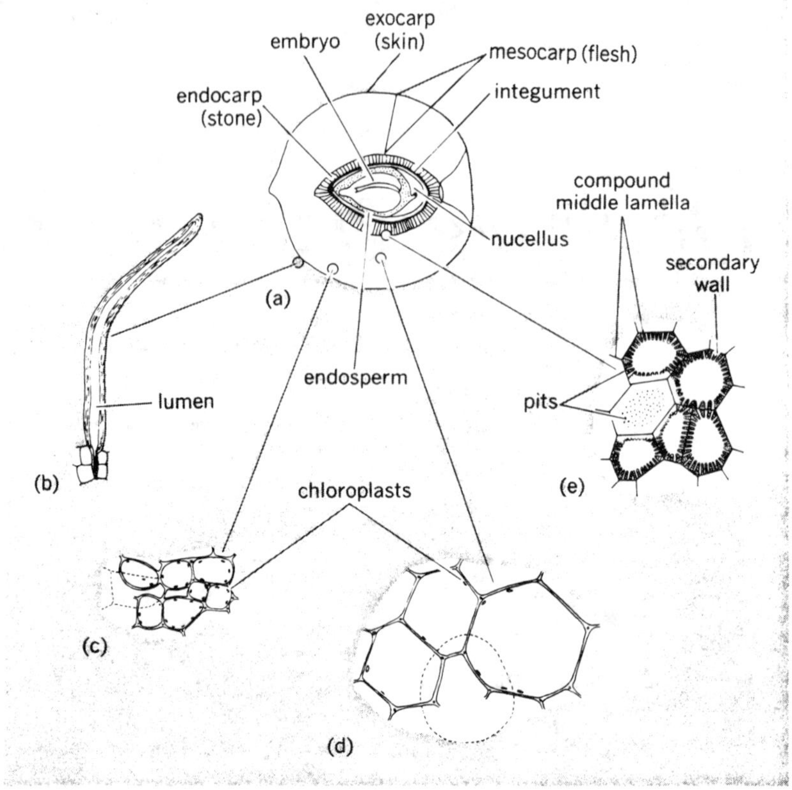

Fig. 10–1. Tissue components of the fruit (a drupe) of *Prunus persica* (peach). *(a)*, longisection of fruit about 3 cm (1.2 in.) in diameter (circles indicate positions of sections shown in *b-e*); *(b)*, epidermal hair of exocarp (skin); *(c, d)*, parenchyma of fleshy mesocarp (flesh); *(e)*, sclereids of endocarp (stone). (From R. Schmid, Fruit, *McGraw-Hill Encyclopedia of Science & Technology,* 5th ed., 5:740–746, © 1982 by McGraw-Hill Book Co., New York; redrawn from *Plant anatomy,* 2nd ed., by K. Esau, © 1965 by John Wiley and Sons, Inc., New York. Reprinted by permission of the publisher.)

Parthenocarpic fruits usually develop without fertilization and thus generally lack seeds. The following combinations are possible for parthenocarpic fruits:

no pollination and no fertilization (e.g., banana);
pollination but not fertilization (orchids);
pollination and fertilization, but the embryo aborting before fruit maturity (peach, grape).

Parthenocarpic fruits can be either a natural phenomenon or artificially induced. Because people generally do not like seeds in fruits (for obvious reasons), seeds in many fruits have been eliminated by artificial selection.

The fruit is, by definition, restricted to the angiosperms, although fruitlike structures may enclose seeds in certain gymnosperms (see Lab Exercise 8, Part I-B). The fruit is of ecological significance because of seed dispersal.

The classification of fruits is quite complicated (see Lab Exercise 10, Supplement). Suffice it to say that fruitlets and fruits *at their maturity* can be classified as dry versus fleshy and as dehiscent (splitting open) versus indehiscent (not splitting open), namely:

- ***fleshy fruits,*** the fruit wall *at maturity* partly or entirely fleshy, usually *not* splitting open (***indehiscent***);
- ***dry fruits,*** the fruit wall *at maturity* dry, usually splitting open (***dehiscent***).

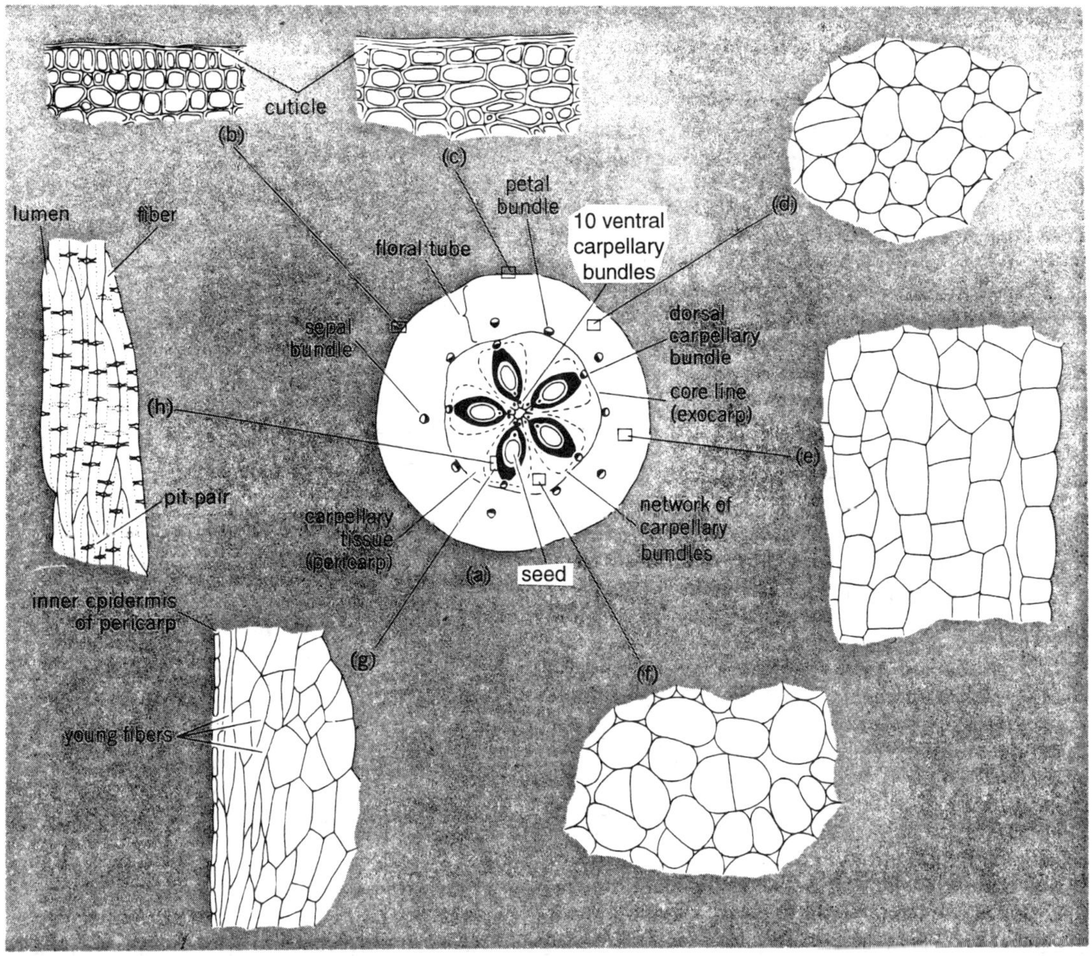

Fig. 10–2. Tissue components of the fruit (a pome) of *Malus sylvestris* (apple). *(a),* transection of fruit (rectangles indicate positions of sections shown in *b-h; b, d-g* are from a fruit 1 cm (0.4 in.) in diameter, *c* and *h* from a mature fruit); *(b, c),* epidermis and subjacent collenchyma from young and mature fruits, respectively; *(d, e),* parenchyma of floral tube part of flesh; *(f),* parenchyma of mesocarp of flesh; *(g, h),* fibers of cartilaginous endocarp from young and mature fruits, respectively. (From R. Schmid, Fruit, *McGraw-Hill Encyclopedia of Science & Technology,* 5th ed., 5:740–746, © 1982 by McGraw-Hill Book Co., New York; redrawn from *Plant anatomy,* 2nd ed., by K. Esau, © 1965 by John Wiley and Sons, Inc., New York. Reprinted by permission of the publisher.)

The dry-fleshy criterion applies to the mature fruit because immature dry fruits are fleshy to varying degrees (e.g., snow pea). All of us are familiar with such fleshy fruits as apple, peach (and other stone fruits), and pineapple. However, bear in mind that squash, tomato, eggplant, and other "vegetables" are botanically fleshy fruits. We commonly eat the seeds of dry fruits (e.g., legumes), but in some cases we eat their fleshy, hence immature fruits (snow pea, string beans; lima beans are eaten for their seeds).

The transformation of the ovary (and other parts) into the fruit involves various changes:

- an often appreciable increase in size;
- invariably changes in texture, that is, fleshiness versus dryness;
- especially in fleshy fruits, changes in chemical content, notably increased amounts of sugar, fats and decreased amounts of tannin (an astringent chemical).

Many of these changes relate to the functions of fruit/seed dispersal (see Part III). For example, fleshy fruits may become palatable to animals, and dry fruits open up to disperse their seeds.

Figures 10–1 and 10–2 show the morphology and anatomy of the fruits of peach (a drupe) and apple (a pome). Examine these diagrams for the following morphological terminology, but ignore most of the anatomical details and terminology they show. The ***ovary wall*** (pericarp) of a fruit is arbitrarily divided into three regions, which are usually much more conspicuous in fleshy fruits:

- ***exocarp*** (epicarp), the outer layer consisting of the epidermis and often subepidermal tissue, as the "skin" of some fruits;
- ***mesocarp,*** a central layer that is typically fleshy, at least partly, thus the "flesh" of some fruits;
- ***endocarp,*** the innermost layer that may be merely epidermal or, more commonly, several layers of cells, sclerenchymatous, and morphologically distinct, the "stone" or "pit" of some fruits.

For instance, peach, plum, cherry, and other stone fruits (Fig. 10–1), which are botanically "drupes," consist of a endocarp (pit or stone) of sclerenchyma fibers, a mesocarp (flesh) of largely parenchyma cells but also vascular tissue, and an exocarp (skin) of the epidermis and a few subepidermal layers. The vascular bundles deliver nutrients to the developing fruit and seed. [Note: This "carp" terminology is less often applied to dry fruits.]

●●● Obtain a "pea pod," that is, a fruit with its seeds, of *Pisum sativum* var. *sativum* (garden pea). This fruit, which is botanically a "legume," represents a single carpel and is typical of the bean or legume family (Leguminosae or Fabaceae). Slit open the fruit and note the seeds located along one wall. The part of the ovary wall where ovules or seeds attach is the ***placenta*** (Fig. 9–1). When the fruit dries out, it will split lengthwise along its two edges. Are any floral parts present other than the carpel? Save the seeds for reuse in Part II.

●●● Flowers of *Agapanthus africanus* (lily-of-the-Nile, blue African lily) were probably seen in Lab Exercise 9, Part I. Examine the DEMO, if available, of its dry fruits, which are botanically "capsules." Another fruit type, the grain (caryopsis), will be examined in Part II.

II. SEEDS

Seeds vary greatly in size, but are usually several millimeters in dimension. Orchid seeds ("dust seeds") are the most minute, weighing 0.3–14 mg and measuring 0.25–1.2 mm long and 0.09–0.27 mm wide; an orchid fruit has about 376 to 4,000,000 seeds (Arditti 1967). At the other extreme are the seeds (really one-seeded fruits) of the Seychelles palm *Lodoicea maldivica* (coco-de-mer, double coconut), which may achieve 27.2 kg (60.0 lb.) and a 50 cm (19.7 in.) length (Corner 1964; Mabberley 1987).

The transformation of the ovule into the seed involves various changes (Fig. 9–4G, H):

- an often appreciable increase in size;
- development of the integument(s) into the seed coat;
- closure of the micropyle;
- usually partial degeneration of the nucellus (megasporangium);
- usually complete degeneration of the megaGPT;
- development of the embryo (very young SPT) from the zygote;
- development of the endosperm from the second fertilization of the double fertilization.

The seed thus generally contains an embryo as the essential part and at maturity typically consists of several parts (see Figs. 9–4H and 10–3):

- the ***seed coat*** (testa), the protective surface layer(s) derived from the integument or integuments; note that the micropyle has closed (it may be evident as a small closed pore), and the seed coat becomes variously fleshy through parenchyma development and/or hardened or woody through sclerenchyma development;
- the mature ***embryo*** (very young SPT), consisting of various parts that are comparable to those in the seeds of gymnosperms (see Supplement 8, Part VIII, and Lab Exercise 8, Part I-A):

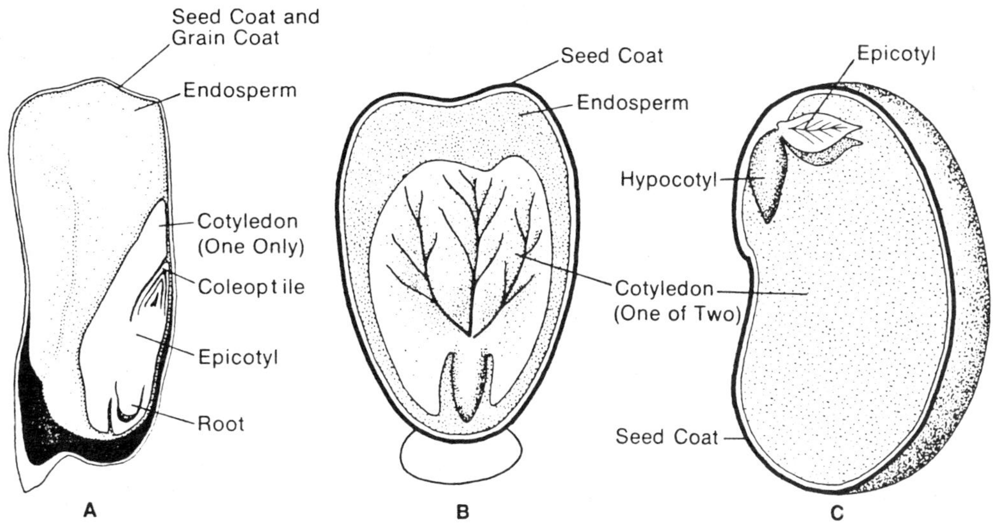

Fig. 10–3. Longisections of seeds. *A,* grain (caryopsis) of the monocotyledon *Zea mays* (corn, maize), an indehiscent dry fruit with a single seed (the coleoptile is a sheath enclosing the epicotyl); *B,* seed of *Ricinus communis* (castor bean), a dicotyledonous type with persistent endosperm; *C,* seed of *Phaseolus vulgaris* (bean), a dicotyledonous type lacking endosperm in the mature seed. (From Norstog & Long 1976:441.)

- ***cotyledons*** (seed leaves), 0–2 (see Parts IV and V and also Supplement 10, especially Fig. Sup10–1);
- ***shoot apex;***
- ***epicotyl,*** the shoot region above the cotyledonary node;
- the short ***hypocotyl,*** the shoot region between the cotyledonary node and the radicle;
- ***radicle*** (embryonic root), the young root;

- variable, often considerable amounts of ***endosperm*** (see below), the nutritive tissue of the embryo and the seedling;
- variable, usually negligible amounts of ***nucellus*** or megasporangium (this very occasionally proliferates into a nutritive perisperm—see below);
- usually negligible amounts of ***megaGPT*** (but see below), in contrast to the considerable megaGPT in gymnospermous seeds;
- remnants of the ***microGPT*** (pollen tube), which usually is not identifiable as such.

Endosperm is sometimes accompanied by a nutritive tissue derived from the nucellus, that is, ***perisperm.*** Endosperm is typically 3*n* (but see Lab Exercise 9, Part III-D), whereas perisperm is always 2*n*. Both nutritive tissues contrast with the 1*n* nutritive megaGPTic tissue of gymnospermous seeds. Although all of these nutritive tissues are functionally similar, they are quite dissimilar morphologically. The megaGPT of the angiosperms, because of its extreme reduction (see Lab Exercise 9, Part III-C), is negligible in amount in the seed. That is, those cells not directly involved in fertilization, the antipodals and synergids, typically degenerate, although the former sometimes persist and may have a nutritional role. Endosperm and perisperm may occur in the same seed (e.g., beet), or only one of these, usually the former, may be present.

Completely mature seeds (i.e., ones before germination) are of two types concerning endosperm:

- ***endospermous seeds*** (albuminous seeds), with appreciable amounts of endosperm (Figs. 9–4H, 10–3A, B);
- ***non-endospermous seeds*** (exalbuminous seeds), lacking endosperm either because of its ephemeral development in the seed or its complete or nearly complete absorption by the embryo (Fig. 10–3C).

In endospermous seeds the embryo varies greatly in size depending on the amount of endosperm persisting in the mature seed. In non-endospermous seeds the embryo almost completely fills the seed coat; the cotyledons, in particular, store nutrient reserves.

●●● See Fig. 10–3C. Then under a dissecting microscope examine a soaked seed of *Phaseolus vulgaris* (bean) and identify externally the micropyle and seed coat. Next carefully remove the seed coat and separate the two cotyledons to reveal the parts of the embryo within. Note that the embryo fills the seed coat. The seed is non-endospermous (exalbuminous). Identify the following parts of the embryo: shoot apex, epicotyl, cotyledons, hypocotyl, radicle, root apex, and root cap. What is the fate of these parts when the seed germinates?

●●● *Suggested diagram and labels: Phaseolus vulgaris* (bean, an angiosperm and dicotyledon) seed l.s.: seed coat and parts of embryo (shoot apex, epicotyl, cotyledon, hypocotyl, radicle, root apex, root cap). Note what happened to the endosperm!

Note: Wash your hands after touching the bean seeds because they were treated with a fungicide.

●●● With regard to the fresh seeds (so-called peas) of *Pisum sativum* var. *sativum* (garden pea) saved from Part I, either eat them, take them home as a souvenir, or, if you wish, under a dissecting microscope identify the structures just noted for *Phaseolus*. Pea embryos have conspicuous root caps. The young pods of snow pea used in Asian cuisines are of another variety (*macrocarpon*). Also examine the DEMO of store-bought "split peas." What does each "split" part represent?

●●● Examine the DEMO slide of a longisection of a nearly mature seed of *Capsella bursa-pastoris* (shepherd's purse). The embryo is the large central, pink-staining mass of cells. Identify the parts of the embryo noted above. The cotyledons become very large and lie parallel to the main axis of the embryo. A small amount of endosperm surrounding the embryo is still evident, but eventually it will be absorbed by the embryo. Note the developing seed coat enclosing the embryo.

The so-called seed or kernel of *Zea mays* (corn, maize) is actually a fruit, botanically a "grain" or "caryopsis" (Figs. 10–3A, LabSup10-2f). Each ovary contains one seed, and the outer parts of the seed are intimately fused with the inner parts of the ovary. A massive cotyledon (scutellum) specialized to absorb nutrients from the extensive endosperm, and a sheath (coleoptile) enclosing the epicotyl are two structures typical of embryos not only of corn but also of other members of the grass family (Gramineae or Poaceae).

●●● See Fig. 10–3A. Then examine a slide of a longisection of a kernel of corn and identify the endosperm and other structures shown on the diagram. The seed is endospermous (albuminous).

Summary regarding ovules and seeds of gymnosperms and angiosperms

To reiterate from Supplement 8, Part X, the ovule or seed is a combination structure of several generations (phases), namely:

- in the ovule:
  - old SPT (2*n*), the integument(s) and nucellus (megasporangium);
  - GPT (1*n*), the megaGPT (called ***embryo sac*** in angiosperms);
- and in the seed:
  - old SPT (2*n*), the seed coat derived from the integument(s), also small amounts of nucellus (megasporangium) present;
  - GPT (1*n*), the megaGPT (massive in gymnosperms, negligible in angiosperms) (remnants of the microGPT are also present but usually are not identifiable as such);
  - new SPT (2*n*), the embryo;
  - plus endosperm (usually 3*n*) in angiosperms (absent in gymnosperms).

The ovules and seeds of angiosperms are simpler in structure and usually smaller than those of the gymnosperms (see Supplement 9).

III. FRUIT AND SEED DISPERSAL

See also Part I for comments on dispersal. Seeds, fruits containing seeds, or even parts of branches or entire plants are dispersed by various agents—wind, water, animals (particularly mammals), gravity, and actively by the plant itself (self-dispersal). Fleshy fruits are typically dispersed by animals. Dry fruits are dispersed by animals, wind, and other agents. This emphasis on animal dispersal of fruits and seeds versus dispersal by wind and other abiotic agents parallels the situation for pollen dispersal (see Lab Exercise 9, Part III-A). In contrast, in gymnosperms chiefly wind disperses pollen and seeds (see Lab Exercise 8, Parts I-A and I-B).

IV. SEEDLINGS

Seeds are resistant structures capable of withstanding unfavorable conditions due to a dormant period. For example, in the desert, seeds will germinate only when there has been spring rain, which leaches out from the seed various inhibitor chemicals. Hence when water, temperature, oxygen, light, and other conditions are right for successful growth of the young SPT, the seed germinates and a seedling emerges.

- A *seedling* thus is the young vegetative SPT emerging from a germinated seed.

The first event of seed germination usually consists of emergence of the radicle, the embryonic root (Fig. 10–4A). This root ruptures the seed coat at the micropylar region and typically grows downward into the soil or substrate. There are then two patterns of subsequent seedling growth:

- In ***above-ground germination*** (epigeal germination), the cotyledons and the hypocotyl emerge from the seed coat and, as a result of the elongation of the hypocotyl, become elevated above the soil level; part of the seed coat is also commonly carried above the ground.

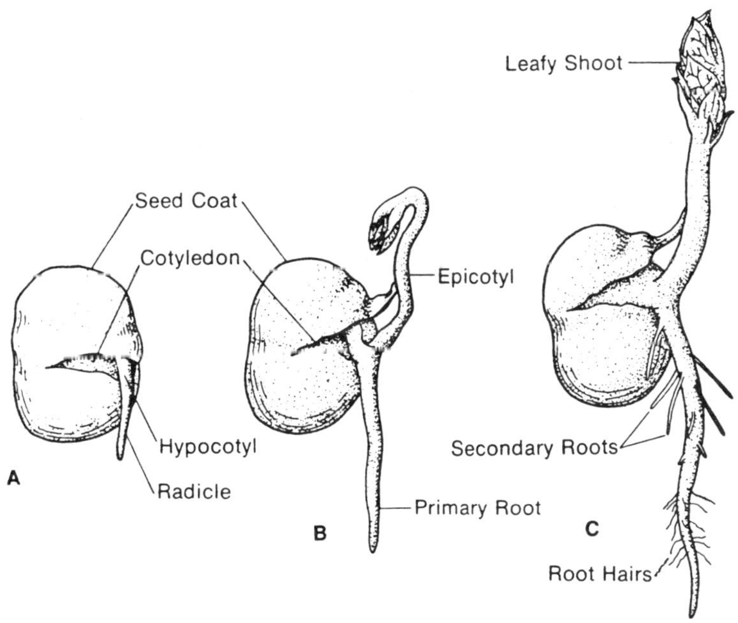

Fig. 10–4. Side views of successive stages in seed germination (*A*) and seedling development (*B, C*) of the dicotyledon *Pisum sativum* (garden pea). The germination type is below-ground (hypogeal). (From Norstog & Long 1976:455.)

- In ***below-ground germination*** (hypogeal germination), the cotyledons remain within the seed coat and thus below ground, the epicotyl alone extending above the soil level (Fig. 10–4).

Cotyledons in seedlings supply nutrients for seedling development by newly manufacturing (photosynthesizing) nutrients, by transporting nutrients previously stored in the cotyledons, or by absorbing nutrients from the endosperm.

••• See Fig. 10–4 on seed germination of *Pisum sativum* (pea). Then examine the DEMO diagrams (from Raven et al. 1992:442, 449) of seeds and seedlings. Finally examine the live seedlings of *Phaseolus vulgaris* (bean), *Pisum sativum* (pea), and *Zea mays* (corn, maize). These exemplify the above two types of germination. For each species identify the mode of germination and the following parts of the seedling:

1. cotyledons (or their scars);
2. old seed coat (if present);
3. root system;
4. stem, node, internode;
5. first pair of foliage leaves (these are simple in *Phaseolus* and *Zea,* but compound, with three leaflets, in *Pisum*);
6. second (and later) pair of foliage leaves (these are simple in *Zea,* but compound in *Phaseolus* and *Pisum,* each leaf with three leaflets);
7. the shoot tip or terminal bud, and the axillary buds (in *Zea* the shoot tip is hidden in the sheathing leaves).

Phaseolus (bean) and *Pisum* (pea) both occur in the legume family, Leguminosae, but they differ in the mode of seed germination, leaf morphology, and other features. *Note:* It is permissible to take home a seedling of each species for later observation and drawing.

●●● *Suggested diagram and labels: Phaseolus vulgaris* (bean, an angiosperm and dicotyledon) OR *Pisum sativum* (pea, an angiosperm and dicotyledon) seedling: cotyledons (or their scars), old seed coat (if present), root system, stem, node, internode, first pair of foliage leaves (simple), second (and later) pair of foliage leaves (compound, each with three leaflets), shoot tip or terminal bud, axillary buds. Show the soil level on the seedling and note whether germination is above-ground (epigeal) or below-ground (hypogeal)!

V. DICOTYLEDONS VERSUS MONOCOTYLEDONS

See Supplement 10 giving the characteristics of di- versus monocotyledons, the two main groups (classes) of angiosperms. As suggested by these names, these groups have, respectively, two versus one cotyledon per embryo (very young SPT) (Fig. Sup10–1). However, embryos of some specialized taxa may lack cotyledons, for instance, orchids, which are monocotyledons. Cotyledon number is the most fundamental difference between di- versus monocotyledons.

●●● From your observations in Part II are *Phaseolus, Pisum, Capsella,* and *Zea* di- or monocotyledons?

NAME ______________________________ LAB SECTION # ______________ DATE ____

TITLE ______________________________

__________________ TOTAL MAG ____

TITLE ______________________________

__________________ TOTAL MAG ______

TITLE ______________________________

__________________ TOTAL MAG ____

TITLE ______________________________

__________________ TOTAL MAG ______

LAB EXERCISE SUPPLEMENT **10**

Résumé of Fruit Classification

OBJECTIVE

To key out various types of fleshy and dry fruits and thus learn features of their morphology.

Note: Read this material *before* coming to the lab; there is insufficient time in the lab to read the material from scratch. Due to the large amount of background information presented here, a ●●● is used to mark those paragraphs calling for actual observations or diagrams in the lab.

Suggested Lab Diagrams (see part noted for suggested labels):
Pome, drupe, legume, *and* capsule (Part I)

PERSPECTIVE

Fruits have been classified on the basis of numerous criteria, including dehiscence versus indehiscence, dry versus fleshy texture, morphology (form), development (ontogeny), relationship of the ovary to other reproductive parts, and number of carpels and seeds. All detailed keys to or classifications of fruit types, including the key in Part II, are patently artificial for several reasons: (1) They are biased toward fruits from temperate areas. (2) They are biased toward fruits of economic importance. (3) They ignore the numerous transitional fruit types. (4) They mix rather indiscriminately morphological, histological, developmental, ecological, and functional criteria.

I. THE MORPHOLOGY OF FRUITS, ESPECIALLY USING A KEY TO DETERMINE FRUIT TYPES

For background information on fruit morphology see Lab Exercise 10, Part I and Figs. 10–1 and 10–2. Figure LabSup10–1 shows fleshy fruits, Fig. LabSup10–2 indehiscent dry fruits, and Fig. LabSup10–3 dehiscent dry fruits.

●●● Various fleshy and dry fruits are available either on DEMO or, where specifically indicated, for dissection. In conjunction with Figs. LabSup10–1 through LabSup10–3, examine these fruits and classify them as to type using the key in Part II.

●●● *Suggested diagrams and labels:* Pome, drupe, legume, *and* capsule: Label if fleshy or dry fruit and, as relevant, exocarp, mesocarp, endocarp, ovary tissue, floral tube tissue, epidermis ("skin"), "stone," seed.

Note: Do *not* attempt to memorize the following résumé of fruit classification. On the lab examination you will be given a simplified key to use (sans examples) and asked to classify a fruit according to its specific morphological type. For example, see at the end of this lab exercise the sample question for the lab final examination.

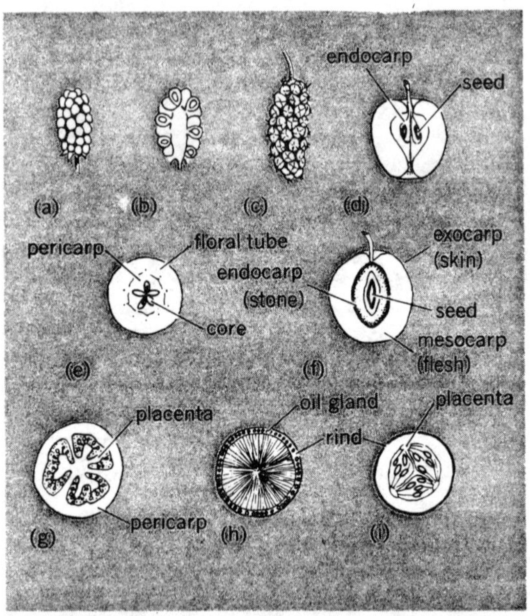

Fig. LabSup10–1. Fleshy fruits. *(a, b)*, aggregate fruit (blackberry) in surficial and sectional views, respectively; *(c)*, multiple fruit (mulberry); *(d-i)*, simple fruits: *(d, e)*, pome (apple); *(f)*, drupe (cherry); *(g)*, berry (tomato); *(h)*, hesperidium, a modified berry (orange); *(i)*, pepo, a modified berry (cucumber); *d* and *f* are longisections, *e* and *g-i* transections. (From R. Schmid, Fruit, *McGraw-Hill Encyclopedia of Science & Technology,* 5th ed., 5:740–746, © 1982 by McGraw-Hill Book Co., New York.)

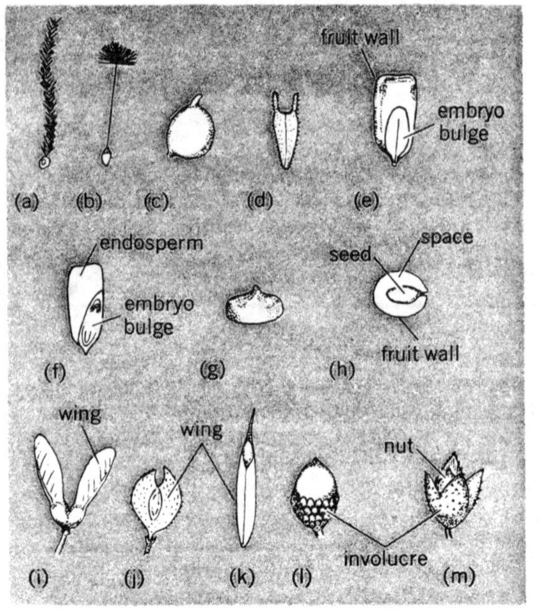

Fig. LabSup10–2. Indehiscent dry fruits. *(a-d)*, achenes: *(a)*, clematis; *(b)*, dandelion; *(c)*, buttercup; *(d)*, sticktight; *(e, f)*, caryopsis or grain (corn, maize) in surficial and sectional views, respectively; *(g, h)*, utricle (goosefoot) in surficial and sectional views, respectively; *(i-k)*, samaras: *(i)*, maple; *(j)*, elm; *(k)*, ash; *(l)*, nut and involucre (acorn) of oak; *(m)*, nut of beech. (From R. Schmid, Fruit, *McGraw-Hill Encyclopedia of Science & Technology,* 5th ed., 5:740–746, © 1982 by McGraw-Hill Book Co., New York.)

II. KEY TO FRUIT TYPES

The following key to or classification of fruit types is based on the traditional types recognized by taxonomists and botanists of mainly temperate areas and on the descriptive summary in Schmid (1982).

I. Fruit formed from several ovaries—classify the individual fruits (fruitlets) in Part II in the résumé for simple fruits.
  A. Fruits developing from one flower—***aggregate fruits.*** Examples: blackberry (Fig. LabSup10-1a, b), strawberry, raspberry, magnolia, rose, dogwood.
  B. Fruits developing from several flowers (i.e., an inflorescence)—***multiple fruits.*** Examples: pineapple, fig, mulberry (Fig. LabSup10-1c), pandan, sweet gum.

II. Fruit formed from a single ovary of one flower—***simple fruits***
  A. Fruit wall partly or entirely fleshy, usually indehiscent—***fleshy fruits***
    1. Both mesocarp and endocarp fleshy—***berry*** (may be formed from superior or inferior ovary and one to many carpels, and may have one to many seeds)
      a. Fruit mostly fleshy, except exocarp skinlike, sometimes the outer skin ("peel") fleshy and easily separable—***berry proper.*** Examples: avocado, banana, grape, kiwi, tomato (Fig. LabSup10–1g), and ugni.
      b. Fruit with thick, leathery, separable rind (outer layer)—***hesperidium,*** a modified berry. Examples: citrus fruits (Fig. LabSup10–1h).

c. Fruit with hard or tough inseparable rind (outer layer)—***pepo***, a modified berry. Examples: cucumber (Fig. LabSup10–1i) and other members of the gourd family (Cucurbitaceae).

2. Mesocarp fleshy but endocarp stony—***drupe*** or ***stone fruit*** (may be formed from superior or inferior ovary and one to many carpels, and many drupes have one to many seeds). Examples: peach (Fig. 10–1), plum, cherry (Fig. LabSup10–1f), and other stone fruits.
3. Mesocarp fleshy but endocarp thin and papery or cartilaginous—***pome*** (formed from inferior ovary of several carpels, with several seeds). Examples: apple (Figs. 10–2, LabSup10–1d, e), pear and pomegranate (the last is a leathery pome or ***balausta***).

B. Fruit wall dry—***dry fruits*** (*Note:* immature dry fruits are fleshy to various degrees, e.g., snow pea or garden pea)

1. Fruit not splitting open at maturity, or, in various nuts, only the outer part splitting open—***indehiscent fruits***
   a. Fruit bearing a wing or wings—***samara*** (fruit proper is like an achene—see below). Examples: maple, elm, and ash (respectively Fig. LabSup10–2i through k).
   b. Fruit not winged, usually single-seeded
      1. Seed entirely fused to fruit wall—***caryopsis*** or ***grain.*** Examples: corn (maize) (Fig. LabSup10–2e, f) and other members of the grass family (Gramineae or Poaceae).
      2. Seed not fused to fruit wall except at attachment point of seed
         a. Fruit small, with a thin, non-bladdery wall—***achene***
            1. Fruit formed from a superior ovary—***achene proper.*** Examples: clematis (Fig. LabSup10–2a), buttercup (Fig. LabSup10–2c), and the fruitlets of rose and strawberry.
            2. Fruit formed from an inferior ovary—***cypsela,*** a modified achene. Examples: sunflower, dandelion (Fig. LabSup10–2b), sticktight (Fig. LabSup10–2d), and other members of the sunflower family (Compositae or Asteraceae).
         b. Fruit small, with a thin and bladdery wall—***utricle.*** Examples: beet and goosefoot (Fig. LabSup10–2g, h).
         c. Fruit large, with wall hard throughout, derived from several carpels but usually one-seeded, often at least partly enclosed in an involucre—***nut.*** Examples: beech (Fig. LabSup10–2m), hazelnut, walnut, hickory, and acorn of oak (Fig. LabSup10–2l). In some nuts (e.g., hickory) there is a dehiscent leathery or fibrous husk (so-called exocarp), which is derived from the involucre and perianth surrounding the ovary.
2. Fruit splitting completely open at maturity—***dehiscent fruits***
   a. Fruit formed from single carpel
      1. Fruit splitting along one suture—***follicle.*** Examples: milkweed (Fig. LabSup10–3a) and the fruitlets of magnolia.
      2. Fruit splitting along two sutures—***legume.*** Examples: snow pea, garden pea, bean (Fig. LabSup10–3b), and many other members of the legume family (Leguminosae or Fabaceae).
      3. Fruit constricted between seeds and breaking crosswise into one-seeded segments—***loment,*** a modified legume. Example: beggar's ticks (Fig. LabSup10–3c).
   b. Fruit formed from two or more carpels
      1. Fruit formed from two carpels and with two sutures, the valves opening from below upward to expose the persistent seed-bearing partition (replum) still attached to stalk
         a. Fruit elongate, narrow—***silique.*** Examples: mustard (Fig. LabSup10–3d, e) and many other members of the mustard family (Cruciferae or Brassicaceae).

b. Fruit short and broad—***silicle.*** Examples: shepherd's-purse (Fig. LabSup10-3f, g) and other members of the mustard family (Cruciferae or Brassicaceae).

2. Fruit formed from two or more carpels, splitting variously, but into a like number of vertical valves or sutures—***capsule.*** Examples: jojoba, cotton, poplar, lily (Fig. LabSup10-3h), and yucca. The ***pyxis*** subtype (e.g., purslane, Fig. LabSup10-3i) dehisces circumferentially, the top separating as a lid.
3. Fruit splitting into one-seeded usually indehiscent units (mericarps or sometimes "cocci"), each a derivative of a carpel—***schizocarp.*** Examples: mallow and members of the umbel family (Umbelliferae or Apiaceae).

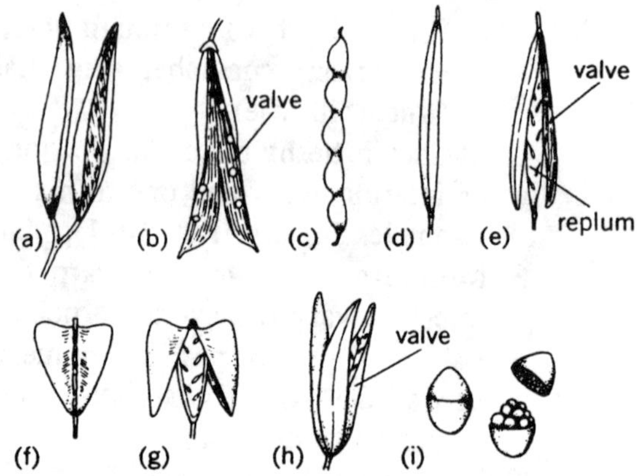

Fig. LabSup10-3. Dehiscent dry fruits. *(a),* follicles (milkweed); *(b),* legume (bean); *(c),* loment (beggar's ticks); *(d, e),* silique, closed and open, respectively (mustard); *(f, g),* silicle, closed and open, respectively (shepherd's purse); *(h),* capsule (lily); *(i),* pyxis, a modified capsule, closed and open (purslane). (From R. Schmid, Fruit, *McGraw-Hill Encyclopedia of Science & Technology,* 5th ed., 5:740–746, © 1982 by McGraw-Hill Book Co., New York.)

Sample question for the lab final examination

(1) What type of fruit is this? Use the following key.
 (A) Drupe
 (B) Pome
 (C) Follicle
 (D) Legume
 (E) Capsule.

Key to Some Fruit Types

A. Fruit wall partly or entirely fleshy, usually indehiscent—***fleshy fruits***
 1. Both mesocarp and endocarp fleshy—***berry*** (may be formed from superior or inferior ovary and one to many carpels, and may have one to many seeds)
 2. Mesocarp fleshy but endocarp stony—***drupe*** or ***stone fruit*** (may be formed from superior or inferior ovary and one to many carpels, and many drupes have one to many seeds)
 3. Mesocarp fleshy but endocarp thin and papery or cartilaginous—***pome*** (formed from inferior ovary of several carpels, with several seeds)

B. Fruit wall dry—***dry fruits***
 1. Fruit not splitting open at maturity, or, in various nuts, only the outer part splitting open—***indehiscent fruits,*** which include various types such as the samara, caryopsis or grain, achene, and nut
 2. Fruit splitting completely open at maturity—***dehiscent fruits***
  a. Fruit formed from single carpel
   1. Fruit splitting along one suture—***follicle***
   2. Fruit splitting along two sutures—***legume***
  b. Fruit formed from two or more carpels, splitting variously, but into a like number of vertical valves or sutures—***capsule***

NAME ______________________ LAB SECTION # ___________ DATE ____

TITLE ______________________

________________ TOTAL MAG ____

TITLE ______________________

________________ TOTAL MAG _____

TITLE ______________________

________________ TOTAL MAG ____

TITLE ______________________

________________ TOTAL MAG _____

SUPPLEMENT **11**

Biological Nutrition and Associations Betwixt Organisms

I. PERSPECTIVE

Most of the organisms encountered in Lab Exercises 1 to 10 undergo photosynthesis, that is, the production of their own nutrients ("food"). However, there are other ways that organisms either manufacture or obtain nutrients. In addition, dissimilar organisms may exhibit associations with each other that range from very close to quite distant. Lab Exercise 11 on organ and whole-plant modifications of angiosperms and Lab Exercises 12 and 13 on fungi amplify these aspects.

II. TYPES OF BIOLOGICAL NUTRITION

The major types of biological nutrition are as follows:

- ***autotrophic nutrition,*** able to synthesize organic compounds from inorganic materials in the environment by:
  - ***photosynthesis*** (photoautotrophy), the production of carbohydrates from water and carbon dioxide in the presence of chlorophyll by using light energy;
  - ***chemosynthesis*** (chemoautotrophy), the production of carbohydrates from water and carbon dioxide by using energy released by oxidizing inorganic molecules such as nitrogen (this is the process of nitrification, *not* nitrogen fixation), sulfur, and iron compounds;
- ***heterotrophic nutrition,*** unable to manufacture organic nutrients, the nutrients thus obtained by:
  - ***absorption,*** either by:
    - ***saprobism,*** absorption from dead and decaying organic matter (***saprobes,*** as saprobic fungi, or ***saprophytes,*** which refers to saprobic plants), or by
    - ***parasitism,*** absorption from living organisms (***parasites,*** as parasitic fungi or parasitic plants);
  - ***ingestion,*** either by:
    - ***phagotrophy,*** that is, cells ingesting or engulfing insoluble nutrient particles, or by
    - ***holozoic nutrition,*** that is, eating food and breaking it down or digesting it in internal digestive organs.

Organisms may have more than one mode of nutrition, for instance, hemiparasites (see Lab Exercise 11, Part I-A), which may be mostly autotrophic and photosynthesizing but partly heterotrophic and parasitic (absorbing). ***Haustorium*** (singular, plural ***haustoria***) is the general term for the penetrating and absorbing structure of plant and fungal parasites. A ***host*** is the organism on or in which the parasite lives. In ecological parlance, autotrophic organisms are "primary producers" whereas heterotrophic organisms are "consumers."

The above modes of nutrition occur among the major groups of organisms (see Supplement 1) as follows:

- *prokaryotes:*
  - *bacteria:* mostly saprobic, sometimes photosynthetic or chemosynthetic (Raven et al. 1992);
  - *blue-green bacteria:* mostly photosynthetic, rarely saprobic (Bold & Wynne 1985);
- *plants:*
  - *algae:* mostly photosynthetic, sometimes parasitic (e.g., some dinoflagellates and green, brown, and red algae), phagotrophic (e.g., most euglenoids, some diatoms and dinoflagellates), or saprobic (e.g., some diatoms, dinoflagellates, and cryptomonads) (Bold & Wynne 1985);
  - *bryophytes:* mostly photosynthetic, very rarely saprophytic (Schofield 1985);
  - *pteridophytes:* all photosynthetic;
  - *gymnosperms:* all photosynthetic except *Podocarpus ustus* parasitic (Schmid 1981);
  - *angiosperms:* mostly photosynthetic, sometimes parasitic, saprophytic, or holozoic (e.g., carnivorous plants);
- *fungi:* mostly saprobic (saprobic fungi, *not* "saprophytes," which refers to saprobic plants), sometimes parasitic, rarely phagotropic (e.g., slime molds), *never* photosynthetic;
- *protists:* mostly phagotropic (e.g, amoebae), sometimes parasitic;
- *animals:* mostly holozoic, sometimes parasitic.

III. TYPES OF ASSOCIATIONS BETWEEN ORGANISMS

Dissimilar organisms may exhibit close, symbiotic or loose, non-symbiotic associations with each other (effects: + = positive, − = negative, 0 = neutral):

- ***symbiotic associations*** or ***symbiosis,*** a close association of two or more dissimilar organisms (symbionts):
  - ***parasitism*** (+ −), where the association is harmful to one of the organisms (e.g., mistletoes, ergot of rye, the wheat rust pathogen, *Puccinia graminis,* parasitic on wheat and barberry, and many other fungal and non-fungal parasites);
  - ***mutualism*** (+ +), where the association is advantageous to both organisms (e.g., mycorrhizae and bacterial root nodules involved in nitrogen fixation);
  - ***commensalism*** (+ 0), where one organism is benefitted but the other is neither stimulated nor inhibited (e.g., lichens according to recent interpretations);
- ***non-symbiotic associations,*** a loose, non-obligate association of two or more dissimilar organisms, for instance, *fungal examples* as carnivorous or predaceous fungi, the fungus gardens of insects, and wood-rotting fungi, *plant examples* as carnivorous plants, herbivory, most cases of pollination and dispersal, ecological competition, and animal examples as predation.

Remember: Botanists do it in style, but there's always a stigma attached!

LAB EXERCISE **11**

Flowering Plants (Angiosperms) III

(Some Organ and Whole-plant Modifications: Specializations of the Lunatic Fringe?)

OBJECTIVE

To examine briefly some of the diverse organ and whole-plant modifications of angiosperms that are enumerated below.

Note: Read this material *before* coming to the lab; there is insufficient time in the lab to read the material from scratch. Due to the large amount of background information presented here, a ●●● is used to mark those paragraphs calling for actual observations or diagrams in the lab. Most unmarked paragraphs reiterate information presented in the lecture. A bullet (●) signifies an important term or concept.

Suggested Lab Diagrams (see part noted for suggested labels):
Mistletoe (dwarf or leafy type) (Part I-A)
Any two carnivorous plants, one with an active trap, another with a passive trap (Part I-C)
Dischidia rafflesiana (Part I-D)
Any hydrophyte (Part II-A)
Any xerophyte (Part II-B)
Any epiphyte (Part II-D)

PERSPECTIVE

All angiosperms studied up to this point have been structurally "typical" or "conventional." However, many species of angiosperms, indeed entire genera and even families, exhibit distinctive, often bizarre modifications of their organs (leaves, stems, and roots) or even of the entire plant. These modifications represent adaptations to different modes of nutrition or to different ecological niches, or sometimes even to both in the same plant. Included in these modifications are such categories as (these are defined below): parasites, saprophytes, carnivorous (insectivorous) plants, ant plants (myrmecophytes), hydrophytes (aquatic plants, including mangroves and seagrasses), xerophytes (involving succulence and other features), halophytes, epiphytes, vines and climbers, etc.

In contrast, the gymnosperms are a relatively depauperate group in numbers of species and thus are rather homogeneous. Most species are trees and shrubs and, other than extensive xerophytism for leaves, exhibit only a few of the previous adaptive modes. For instance, some species of *Gnetum* exhibit the vine habit, and two species of the conifer *Podocarpus* the parasitic and the aquatic habits. The latter two modifications occur in New Caledonia (Schmid 1981). The ferns from an evolutionary viewpoint are a moderately successful group today, and in this respect second only to the angiosperms in numbers of taxa. Perhaps not surprisingly, the ferns, have many species that are aquatic, epiphytic, and xerophytic, although none of the other aforenoted categories. Finally, the other pteridophytes (lycopods, horsetails, psilophytes) have many aquatic and epiphytic species.

The organ and whole-plant modifications exemplified below can be categorized as:

- ***specializations related mainly to nutrition*** (sometimes to supplement it), that is, ones involving parasites, saprophytes, carnivorous plants, and ant plants (myrmecophytes) (see Part I);
- ***specializations related mainly to habit or habitat*** and not especially to nutrition, that is, ones involving hydrophytes (aquatic plants), xerophytes, halophytes (salt-tolerant plants), epiphytes, and vines (and other types of climbers) (see Part II).

These categories are not mutually exclusive. There are many overlapping examples. Thus, *Dischidia* is a vining, epiphytic ant plant (see Part I-D), whereas *Cuscuta* (dodder) is a vining parasite.

Note: Before beginning this lab exercise, you should review Supplement 11 on the major types of biological nutrition and associations betwixt organisms.

I. SPECIALIZATIONS RELATED MAINLY TO NUTRITION

The following specializations involve plants with organ or whole-plant modifications functioning mainly for nutrition, sometimes to supplement it:

A. Parasites

Parasites live on or in another organism and derive some or all of their nutrition from the ***host,*** the organism on or in which the parasite lives. The parasitic angiosperms obtain water and minerals and, in a few demonstrated cases, also nutrients from a host plant, which is physically penetrated by the parasite. Water is obtained via interconnections between the xylem of the host and parasite, whereas nutrients are obtained via phloem-phloem interconnections. The part of the parasite penetrating the host is called the ***endophytic system*** and is an evolutionarily modified root. The endophytic system is the same as the ***haustorium*** (singular, plural ***haustoria***), the general term for the penetrating and absorbing structure of plant and fungal parasites. There are two types of parasites based on nutrition:

- ***Holoparasites*** are parasites entirely dependent on the host for fixed carbon as well as water and nutrients; that is, the parasite is entirely heterotrophic and lacks chlorophyll for photosynthesis, for instance, *Cuscuta* (dodder).
- ***Hemiparasites*** (or ***semiparasites***) obtain water and possibly some nutrients from the host but also capable of photosynthesis; that is, the parasite is partly heterotrophic and partly autotrophic, for example, *Arceuthobium* (dwarf mistletoe) and *Phoradendron* (mistletoe, leafy mistletoe).

In addition, there are two classes of parasites based on their organ nature:

- ***root parasites,*** their seeds germinating in the soil and then establishing root contact with the host;
- ***shoot parasites,*** their seeds dispersed to the host and then germinating there, part of the seedling subsequently penetrating the shoot system of the host and developing as an endophytic system.

●●● Examine the DEMO live material of *Arceuthobium* (dwarf mistletoe) parasitic on *Pinus* (pine) and *Phoradendron* (mistletoe, leafy mistletoe) parasitic on *Quercus* (oak). Both are examples of shoot parasites, specifically shoot hemiparasites. Identify the host and the parasite. Which parasite is similar to the mistletoe of Christmas-kissing fame? A DEMO reprint (Scharpf & Hawksworth 1974) gives brief information on *Phoradendron.*

●●● A DEMO slide of a longisection of the shoot hemiparasite *Phoradendron californicum* (mesquite mistletoe) on a stem longisection of *Prosopis juliflora* (mesquite), the host, may be available for examination. The parasite, which is blue-green colored on the slide, consists of the following parts:

- an exposed aerial part;
- an embedded endophytic system consisting of:
  - an ***axial system*** (longitudinal system) extending from the outside of the stem to the secondary vascular tissue;
  - a *sinker system* extending into the vascular tissue.

The sinker system is especially evident in the pink-staining wood of the host, where the parasite appears as the blue-green, raylike cells with conspicuous nuclei.

●●● *Suggested diagram and labels:* Mistletoe (dwarf or leafy type)—host plant and leaf, stem, and endophytic (haustorial) system of parasite.

B. Saprophytes

Saprophytes make use of dead and decaying plant and animal materials; that is, saprophytes obtain fixed carbon as well as nutrients from dead or decaying plant and animal organic material. Saprophytes generally are non-photosynthetic, or only slightly photosynthetic. Some presumed saprophytes may actually be parasites, at least partly so. For example, recent work has shown that *Monotropa* (Indian pipe) is really partly parasitic on tree roots via a fungal intermediary that forms a mycorrhizal association with the tree (see Lab Exercise 13, Part III-A). This intermediary acts as a bridge and actually passes organic materials from the tree to the "saprophyte."

●●● See the cover figure of *Monotropa* and its caption on the verso of the title page. Then examine the DEMO photograph of *Monotropa.* This plant is completely white, lacking chlorophyll for photosynthesis.

C. Carnivorous plants

Carnivorous plants, often less accurately called "insectivorous plants," capture insects and other small animal prey. These plants have received much attention ever since Charles Darwin's *Insectivorous plants* (1875, 2nd 1888 rev. by Francis Darwin). A recent contribution is B. E. Juniper et al.'s (1989) *The carnivorous plants,* a superb and most scholarly treatise. Adrian Slack's (1988) excellent *Carnivorous plants* recently received a limited revision. In addition, juvenile literature (juvenilia) on plant carnivory has not been neglected; there are in English at least ten fairly recent juvenile books (Schmid 1991). Such literature can actually be vastly superior to adult books in their pictorial content. Needless to say, carnivorous plants are one of the most popular examples of botanical humor.

Carnivorous plants have highly modified leaves that trap the animals and can be classified based on the trapping mechanism (Juniper et al. 1989; Slack 1988):

- ***passive traps:***
  - ***pitfall or pitcher traps,*** for example, *Brocchinia,** *Catopsis,** *Cephalotus* (West Australian pitcher plant), *Darlingtonia* (cobra lily), *Heliamphora* (sun pitcher plants), *Nepenthes* (tropical pitcher plants), *Sarracenia* (trumpet pitcher plants);
  - ***flypaper or adhesive traps,*** for example, *Byblis* (rainbow plants), *Drosophyllum* (Portuguese sundew), *Ibicella,** *Triphyophyllum**;
- ***active traps:***
  - ***flypaper or adhesive traps,*** for example, *Drosera* (sundews), *Pinguicula* (butterworts);
  - ***spring or snap traps,*** for example, *Aldrovanda* (waterwheel plant), *Dionaea* (Venus fly trap);
  - ***bladder or suction traps,*** for example, *Utricularia* (bladderwort), possibly *Genlisea.*

Fig. 11–1. "Sacrificed to a man-eating plant," *American weekly,* 26 Sep. 1920. (From S. Prior, Carnivorous plants and "the man-eating tree," *Field Mus. Nat. Hist. Bot. Leafl.* 23:1–20, 1939.)

The definitions of these types are obvious. The asterisked genera are recently discovered cases of carnivory. Incidentally, cultivated pitcher plants should have their pitchers periodically filled with water.

Carnivorous plants live in bogs and other nutritionally poor (nitrogen poor) habitats. The leaves in a number of cases secrete enzymes to digest the trapped animals (or these simply decay, as in some of the passive traps). In this way nitrogenous compounds are made available to the plant.

●●● Examine the exceptionally fine pictures in the two DEMO children's books: Cynthia Overbeck's *Carnivorous plants* (1982) and Jerome Wexler's *Secrets of the Venus's fly trap* (1981) intended for, respectively, ages ten through 12 and eight through 12! Examination of the DEMO cartoons is optional. Carnivorous plants, not surprisingly, often figure in cartoons. The two cartoons in Figs. 11–1 and 11–2 from 1920 and 1925 issues of *American Weekly* are classics of the genre. On DEMO are discussions by Prior (1939) and Schwartz (1974) of such examples of science fiction.

●●● Then examine the DEMO live material of various carnivorous plants, including the DEMO diagrams and text (from Slack 1988:28–29, 72–73, 77). For each sample determine what type of trap mechanism is employed, that is, whether the trap is active or passive, and the relevant subtype. Take note of any dead animals in the pitchers or on the flypaper traps. Your TA or instructor may demonstrate these mechanisms.

●●● *Suggested diagrams and labels:* Any two carnivorous plants, one with an *active* trap, another with a *passive* trap. Make some notes on the trap mechanisms!

D. Ant plants (myrmecophytes)

Angiosperms have formed various interesting associations with ants, including pollination of flowers by ants, dispersal of seeds by ants, protection from herbivory by ants, the feeding of ants, and the housing of ants (Beattie 1985). The last includes ***ant plants*** (myrmecophytes) such as tropical epi-

Fig. 11–2. "Escaped from the embrace of the man-eating tree," *American weekly,* 4 Jan. 1925. (From S. Prior, Carnivorous plants and "the man-eating tree," *Field Mus. Nat. Hist. Bot. Leafl.* 23:1–20, 1939.)

phytes housing ants in modified leaves or stems. The epiphytes usually live in nutrient-poor environments and thus obtain some nutrients from the waste products of the ant colony and/or from soil the ants bring into the nest.

●●● Examine the excellent pictures of ant plants in the DEMO article by Soepadmo (1978). Then examine the DEMO live material of *Dischidia rafflesiana* (to-take-a-dump plant). Note the inflated tubular leaves ("pitcher leaves"), one of which has been longitudinally sectioned to reveal its contents. ***Adventitious roots*** (i.e., ones arising from shoots or leaves rather than roots or the radicle) formed above the modified leaves grow into the leaves and obtain nutrients from the debris that ants produce in the leaves. The roots and the ants enter the leaf cavity by an opening located at the base of the hollow leaves. Are the ant leaves the only type produced along the length of shoot of this vining epiphyte? Might there be a difference in the degree of development of the root system between "pitcher leaves" with debris in them and those lacking such?

●●● *Suggested diagram and labels: Dischidia rafflesiana*—stem, regular leaf, hollow ant-inhabiting leaf, stem-borne (adventitious) roots in the hollow leaf.

II. SPECIALIZATIONS RELATED MAINLY TO HABIT OR HABITAT AND NOT ESPECIALLY TO NUTRITION

The problem of water economics is critical to all plants that live on land (for overview see Supplement 6, Part V, and for anatomical adaptations see Lab Exercise 6, Part II). The distinctive anatomical/morphological features related to plants that grow in different environmental extremes with regard to degree of water availability can be loosely categorized as follows:

- ***hydromorphic,*** referring to structural features typical of plants (***hydrophytes*** or ***aquatic plants***) that require a large supply of water and that may grow partly or entirely submerged in water (fresh water, sea water, or brackish water);

- ***mesomorphic,*** referring to structural features typical of plants (***mesophytes***) that require abundant available soil water and a relatively humid atmosphere;
- ***xeromorphic,*** referring to structural features typical of plants (***xerophytes***) that are adapted to dry habitats.

The Greek prefixes "hydro," "meso," and "xero" mean, respectively, "wet," "intermediate" ("middle"), and "dry." ***Halophytes*** (salt-tolerant plants) are plants growing under the influence of salt deposits or salt water. Bear in mind that these characterizations are rather loose and often overlap, for example: The xerophytes of one region may be much more mesophytic than the mesophytes of another region. Certain xeromorphic features may occur in mesophytes. Halophytes may be either hydrophytes or xerophytes. ***Epiphytes,*** non-parasitic plants growing on other plants, often have xeromorphic features.

Hydrophytes (aquatic plants) and xerophytes represent adaptations to extreme environmental conditions, which can be summarized as follows:

Feature	*Hydrophytes (aquatic plants)*	*Xerophytes*
Sunlight	Often too little	Too much
Heat	No problem (if near surface)	Too much
Water	No problem	Too little
Wind	Not relevant	Often too much
Air circulation	Too little	No problem or excessive

Because they usually occur in desert areas, xerophytes are also subject to flash floods as well as to weapons tests. Depending on whether they are hydrophytes or xerophytes, halophytes can be subjected to either of these extremes.

As a result of living under extreme environmental conditions, plants have evolved distinctive morphological and/or anatomical modifications to compensate for the above excesses and deficiencies (for some general anatomical adaptations see Lab Exercise 6, Part II). Most of the land plants encountered in Lab Exercises 1 and 4 to 10 were autotrophic (photosynthesizing) mesophytes, whereas most of the prokaryotes and algae encountered in Lab Exercises 2 and 3 were aquatic ("hydrophytes"). In the following specializations the entire organism is modified in direct response to a particular environment:

A. Hydrophytes (aquatic plants, water plants), including mangroves and seagrasses

The ***hydrophytes*** (aquatic plants, water plants) comprise plants that live partly or completely in water, including such subtypes as the ***seagrasses*** or marine angiosperms, which are strictly marine monocotyledons (e.g., eelgrass), and ***mangroves,*** which are plants inhabiting tidal swamp lands in the tropics and subtropics (e.g., red and black mangrove). There are four main types of hydrophytes:

- ***Emergent plants*** have their roots in the soil or mud beneath the water, but parts of their shoots above the water level and exposed to air. Examples include *Typha* (cattail), *Cyperus* (sedge), *Juncus* (rush), and *Oryza* (rice).
- ***Rooted floating plants*** are anchored in the mud but have leaves floating on the water surface (the stem may be a rhizome or erect but submerged). Usually the upper leaf surface is waxy, water repellent, and has all or most of the stomata. Examples include *Nymphaea* (water lily).
- ***Rooted submerged plants*** are anchored in the mud but have their entire shoot system underwater. The shoot has little lignified tissue and thus is very flexible, incapable of support outside the water. The leaves have a very thin cuticle, few stomata, and little vascular tissue. The mesophyll may be reduced to one cell layer. Examples include *Zostera* (eelgrass), *Elodea* (elodea, pondweed, waterweed), and *Myriophyllum aquaticum* (milfoil, parrot's feather, water feather), which is often cultivated in aquaria.

- ***Free-floating plants*** or ***floaters*** have a root system not in contact with the soil and a shoot system floating or rising above the water surface (the roots thus dangle freely underwater). The leaves are modified to prevent waterlogging by having a thick cuticle and/or large trichomes that trap air. Examples include *Lemna* (duckweed) and *Eichhornia* (water hyacinth).

The main structural modification consistent to all four types is increased amounts of ***aerenchyma,*** which is parenchyma with large, intercellular spaces that serve such functions as flotation and air circulation. Other features may also promote buoyancy.

●●● Various live aquatic plants are available either on DEMO or, where specifically indicated, for dissection. Examine each sample and determine the type hydrophyte represented. Then for each type make a hand section or, for bulky tissues, simply break open the plant to obtain an idea of the internal structure of aquatic plants. Also examine the DEMO potted plant and the DEMO slide of a transection of a leaf of *Nymphaea* (water lily). What are some of the apparent adaptations for life in an aquatic environment? A DEMO article by Soepadmo (1980) gives excellent pictures of mangroves, which exemplify emergent plants.

●●● *Suggested diagram and labels:* Any hydrophyte, for example, water hyacinth (*Eichhornia,* with fleshy sectioned petiole). Label the air spaces and the main parts of the plant!

One of the nice things about water plants is that they never need watering.
—Christopher Lloyd, *The well-tempered garden,* 1965

B. Xerophytes, including succulents

Xerophytes are plants adapted to arid situations where water is at a premium. Xerophytes thus have a variety of anatomical and morphological adaptations (i.e., xeromorphic adaptations) to reduce water loss, for instance, a thick cuticle, sunken stomata, and hairiness. Concomitantly, many xerophytes are ***succulents,*** that is, plants with shoots modified to store water; ***stem succulents, leaf succulents,*** and ***root succulents*** are recognized. In addition, often there are structural means to strengthen the plant, such as extra sclerenchyma.

●●● Examine the DEMO photographs (from Trager 1985:78) and the DEMO live material of stem succulents (cacti, euphorbs) and leaf succulents (*Aloe*—aloe, *Lithops*—stone plant). Many succulent xerophytes occur in the spurge or euphorb family (Euphorbiaceae) and the cactus family (Cactaceae). Note the structural (morphological) similarity. This is an example of ***convergent evolution,*** defined as structural similarity in two evolutionary lines *not* due to common ancestry, *but* rather due to adaptation to ecologically similar situations. In other words, this similarity is a long-term environmental/ecological rather than evolutionary/genetic resemblance. What anatomical features might be relevant as xeromorphic adaptations?

●●● *Suggested diagram and labels:* Any xerophyte. Label the main parts of the plant!

C. Halophytes (salt-tolerant plants)

Halophytes (salt-tolerant plants) are plants growing under the influence of salt deposits or salt water. Halophytes may be either hydrophytes, such as the seagrasses and mangroves, or xerophytes, such as many plants of salt flats, for example, *Atriplex* (saltbush) and *Salicornia* (pickleweed). Halophytes usually are succulent, generally have reduced leaves, and often are articulated (the succulent leaves make for conspicuous nodes); halophytic grasses are an exception to these statements. Halophytes are able to grow under saline conditions by *not* absorbing much salt or, if salt is absorbed, by storing it in the plant (e.g., *Salicornia*) or by secreting salt (e.g., some mangroves).

••• Examine the DEMO live material of *Salicornia* (pickleweed). Note the succulence and reduced leaves of the shoot. Try tasting part of a shoot. Boiled young shoots make a good pickle and can be put in a pickling mix.

D. Epiphytes, including epiphylls

Epiphytes are plants growing on other plants because the seeds of the epiphytes germinated on the other plants rather than in soil or water. Therefore, epiphytes grow attached to the trunks and branches of trees, shrubs, and woody vines (lianas), or sometimes even occur on surfaces of live leaves. The host plant is used only for support, and not for nutrition as is the case with parasites. Most vascular plant epiphytes (including the epiphytic ferns) have roots that never contact the soil. Hence epiphytes usually live in a constantly wet habitat, for example, tropical rain forests or montane cloud forests, or else have modifications to store and conserve water.

Stranglers such as strangler fig (*Ficus*) are usually regarded as epiphytes. The fig starts off as a true epiphyte but subsequently becomes rooted in the soil. Then the strangler undergoes rapid secondary growth and eventually engulfs and strangles the host tree. In contrast, shoot parasites like the hemiparasitic mistletoes have been considered epiphytes in view of their partial parasitism, but they are probably best *not* regarded as epiphytes.

Epiphylls are epiphytes growing specifically on leaves. Epiphylls include algae, bryophytes, and similar small organisms. In the tropics some leaves of host plants may frequently be almost completely covered with epiphylls. Why are epiphylls often more deleterious to the host than epiphytes living on bark? Why do many epiphytes also show specializations relating to nutrition (e.g., carnivory, myrmecophytism)?

The orchid and bromeliad (pineapple) families (respectively, Orchidaceae and Bromeliaceae) have mostly epiphytic representatives. Epiphytic orchids frequently have thick leaves and/or pseudobulbs (i.e., swollen internodes) for water storage and succulent roots for attachment and photosynthesis. There are two main types of epiphytic bromeliads (incidentally, *Ananas*, pineapple is terrestrial):

- ***Tank plants*** collect rain water and debris in a reservoir formed by the tightly overlapping leaf bases. The water in the tanks has a distinctive fauna containing frogs, slugs, insects, and various microorganisms. In parts of South America mosquitoes are so dependent on tank bromeliads that malaria can be significantly abated by reducing the local bromeliad population (Benzing 1980:215). Incidentally, cultivated tank bromeliads should have their cups periodically filled with water.
- ***Air plants*** are rooted to the bark and have specialized trichomes to trap water and to reduce water loss. During wet weather the trichomes flatten against the epidermis and absorb water and minerals from dew and rain runoff, whereas during dry weather the trichomes separate a bit from the epidermal surface, the intervening air spaces acting as a barrier to water loss.

••• Examine the DEMO live material of epiphytic orchids and bromeliads. Note the various modifications of these plants relevant to their epiphytic mode of life. Wet a small part of the air plant (*Tillandsia usneoides*—Spanish moss, graybeard) and note the color change from silvery gray to green. The former color is due to the air spaces between the trichomes and the other epidermal cells.

••• *Suggested diagram and labels:* Any epiphyte. Label the main parts of the plant!

E. Vines (and other types of climbers)

Many angiosperms are ***climbers***, that is, plants using objects or other plants as a support. The most familiar climbers are ***vines***, which climb by various means. ***Lianas*** are tropical, woody vines. There are various types of climbers (descriptions and examples mostly follow Fisher 1989):

- ***Twiners*** are vines or climbers that wrap around objects. Their shoots are characterized by long internodes, delayed leaf expansion, and a unique circular movement of the shoot apex. The last involves a "search movement" of the elongating internodes behind the shoot apex. This better allows the growing shoot to randomly contact a nearby object and twine around that object for support. Leaves expand only below the region of movement. Examples include *Dioscorea* (yam), *Ipomoea* (sweet potato), *Lonicera* (honeysuckle), and *Pueraria* (kudzu).
- ***Tendril climbers*** are vines or climbers that have long, cylindrical, climbing structures (***tendrils***) that in some species respond to physical contact by a rapid growth response causing the tendril to coil around the object. Roots, stems, leaves, especially leaflets, and even inflorescences may be modified into tendrils. Examples include *Clematis* (clematis), a petiole tendril climber, *Gloriosa* (glory lily, climbing lily), *Macfadyena* (*Doxantha*) (cat's claw, yellow trumpet vine), both leaf tip tendril climbers, *Parthenocissus* (Virginia creeper), an inflorescence tendril climber, *Smilax* (greenbrier), a stipule tendril climber, *Vanilla* (vanilla), a root tendril climber, and *Vitis* (grape), an inflorescence tendril climber.
- ***Root climbers*** are vines or climbers that have adventitious roots on the aerial stem attaching the shoot to its support. For example, *Hedera helix* (English ivy) has adventitious holdfast roots that grow into rough surfaces and secrete an adhesive substance that tightly binds the roots to the surface. This is evident by comparing roots on unsupported parts of the stem with adhesive roots "peeled" from concrete or bark; the latter roots will retain small amounts of substrate. Note that such adhesive roots are *not* root tendrils (see above). Other examples include *Monstera* (split leaf philodendron, Swiss cheese plant) and *Philodendron* (philodendron), which has both aerial attachment roots and feeder roots.
- ***Scramblers*** are climbers (not really vines) that climb in an irregular, sprawling manner. They tend to have long shoots that grow through and over vegetation; branches developing at an angle to the main stem prevent slipping (Herklots 1976). Examples include *Canarina campanulata,* where the main stem grows upwards and ends in a hanging flower, and *Solanum ochranthum,* where sticky leaves may help to hold the slender branches in place (Herklots 1976).

The terminology of climbers appears not to have been standardized. Nevling (1968), for example, discusses twiners, branch climbers, inflorescence climbers, leaf climbers, root climbers, and weavers, the last category being climbers lacking an obvious adaptation. Incidentally, strangler fig (see Part II-D) is often mistaken as a climber.

●●● Examine the DEMO diagrams (from Fisher 1989:87–89) of climbers. Then examine the DEMO live material of vines and determine if they are of the twining, tendril climber, or root climber type. *Dischidia* from Part I-D should also be reexamined.

SELECTED LITERATURE ON ORGAN AND WHOLE-PLANT MODIFICATIONS OF ANGIOSPERMS

Arber, A. 1920. *Water plants: A study of aquatic angiosperms.* Cambridge: University Press. [Also 1963 reprint.]

Beattie, A. J. 1985. *The evolutionary ecology of ant-plant mutualisms.* Cambridge: Cambridge University Press.

Benzing, D. H. 1980. *The biology of bromeliads.* Eureka, California: Mad River Press.

Chapman, V. J. 1976. *Mangrove vegetation.* Vaduz: J. Cramer.

Darwin, C. 1888. *Insectivorous plants.* 2nd ed. rev. by F. Darwin. New York: D. Appleton and Co. [1st ed. 1875.]

Fisher, J. B. 1989. Modifications of plant organs. Whole plant specializations. Pp. 60–93 in P. Kaufman, T. F. Carlson, P. Dayanandan, M. L. Evans, J. B. Fisher, C. Parks & J. R. Wells. *Plants: Their biology and importance.* New York: Harper & Row, Publishers.

Herklots, G. 1976. Flowering tropical climbers. Folkestone, England: Dawson, Science History Publications.

Heslop-Harrison, Y. 1978. Carnivorous plants. *Sci. Amer.* 238(2):104–108, 111–115, 162.

Huxley, C. 1980. Symbiosis between ants and epiphytes. *Biol. Rev.* 55:321–340.

Juniper, B. E., R. J. Robins & D. M. Joel. 1989. *The carnivorous plants.* London: Academic Press.

Kuijt, J. 1969. *The biology of parasitic flowering plants.* Berkeley: University of California Press.

Madison, M. 1977. *Vascular epiphytes: Their systematic occurrence and salient features.* Selbyana 2:1–13.

Nevling, L. I., Jr. 1968. Some ways plants climb. *Arnoldia* 28:53–67.

Overbeck, C. 1982. *Carnivorous plants.* Photographs by Kiyoshi Shimizu. Adapted from a translation by Chaim Uri of Kiyoshi Shimizu's Shokuchūshokubutsu (1975). Minneapolis: Lerner Publications Co.

Perry, F. 1981. *The water garden.* New York: Van Nostrand Reinhold Co.

Pietropaolo, J. & P. Pietropaolo. 1986. *Carnivorous plants of the world.* Portland: Timber Press.

Pijl, L. van der. 1955. Some remarks on myrmecophytes. *Phytomorphology* 5:190–200.

Prior, S. 1939. Carnivorous plants and "the man-eating tree." *Field Mus. Nat. Hist. Bot. Leafl.* 23:1–20.

Rauh, W. 1984. *The wonderful world of succulents: Cultivation and description of selected plants other than cacti.* 2nd ed. Translated from the German by H. L. Kendall. Washington, D.C.: Smithsonian Institution Press.

Reimold, R. J. & W. H. Queen (eds.). 1974. *Ecology of halophytes.* New York: Academic Press.

Richards, P. W. 1952. *The tropical rain forest: An ecological study.* Cambridge: The University Press.

Scharpf, R. F. & F. G. Hawksworth. 1974. Mistletoes on hardwoods in the United States. *U.S.D.A. For. Serv., For. Pest Leafl.* 147:1–7.

Schmid, R. 1991. The fascination with carnivorous plants in juvenile literature. *Carnivorous Pl. Newslett.* 19:32–40.

Schnell, D. E. 1976. *Carnivorous plants of the United States and Canada.* Winston-Salem, North Carolina: John F. Blair, Publisher.

Schwartz, R. 1974. *Carnivorous plants.* Ed. by D. Leavy. New York: Praeger Publishers.

Sculthorpe, C. D. 1967. *The biology of aquatic vascular plants.* London: Edward Arnold (Publishers).

Slack, A. 1988. *Carnivorous plants.* Rev. ed. London: Alphabooks [Original ed. by Ebury Press, London, 1979, and The MIT Press, Cambridge, Massachusetts, 1980.]

Soepadmo, E. 1978. Ant-plants. *Nature Males.* 3(4):12–19.

———. 1980. Mangroves. *Nature Males.* 5(1):14–23.

Tomlinson, P. B. 1986. *The botany of mangroves.* Cambridge: Cambridge University Press.

Trager, J. 1985. Cases of convergence. *Euphorbia J.* 3:77–79. [Pt. 2 in *Ibid.* 4:71–76, 1987.]

Visser, J. 1981. *South African parasitic flowering plants.* Cape Town: Juta.

Waisel, Y. 1972. *Biology of halophytes.* New York: Academic Press.

Warming, E. [Assisted by M. Vahl]. 1909. *Oecology of plants: an introduction to the study of plant-communities.* Translated from the 2nd (1902) German edition and edited by P. Groom & I. B. Balfour. Oxford: Clarendon Press. [Also 1930–33 4th German edition by E. Warming & P. Graebner.]

Weber, H. C. & W. F. Forstreuter (eds.). 1987. *Parasitic flowering plants.* Gladenbach: Druckerei Kempkes, Offset- + Buchdruck GmbH.

Wexler, J. 1981. *Secrets of the Venus's fly trap.* Photographs by the author. New York: Dodd, Mead & Co.

NAME ______________________________ LAB SECTION # ______________ DATE ____

TITLE ______________________________

__________________ TOTAL MAG _____

TITLE ______________________________

__________________ TOTAL MAG ______

TITLE ______________________________

__________________ TOTAL MAG _____

TITLE ______________________________

__________________ TOTAL MAG ______

NAME ______________________________ LAB SECTION # ______________ DATE ____

TITLE ______________________________

__________________ TOTAL MAG ____

TITLE ______________________________

__________________ TOTAL MAG ______

TITLE ______________________________

__________________ TOTAL MAG ____

TITLE ______________________________

__________________ TOTAL MAG ______

SUPPLEMENT **12**

The Fungi

(It's Another World)

I. PERSPECTIVE

From the viewpoint of importance to humanity, the fungi (plural, singular *fungus*) and the angiosperms are by far the two most important large groups of organisms that have historically been regarded as plants. As elaborated below in Part III and in Supplement 1, Part II, the fungi were traditionally regarded as plants but are now universally placed in a separate kingdom, the Fungi. ***Mycology*** is the study of fungi, and the persons studying fungi are ***mycologists.***

II. THE UBIQUITY AND SOCIOLOGICAL AND ECONOMIC IMPORTANCE OF FUNGI

The fungi have been, currently are, and always will be incredibly important from a sociological and especially economic viewpoint. The many morphological, ecological, pathogenic, sociological, and economic aspects of fungi have inspired numerous fascinating books, which often have colorful titles:

H. J. Brodie's *Fungi: Delight of curiosity* (1978);
C. M. Christensen's *The molds and man* (1965);
R. C. Cooke's *Fungi, man and his environment* (1977);
C. T. Ingold's *Dispersal in fungi* (1953) and *Fungal spores: Their liberation and dispersal* (1971);
E. C. Large's *The advance of the fungi* (1940);
M. K. Matossian's *Poisons of the past: Molds, epidemics, and history* (1989).

The fungi directly or indirectly can make one's day and have been variously good, bad, and very ugly (*Il Buono, il Bruto, il Cattivo*):

- *The good:* Yeasts are important in beer and bread production (see Lab Exercise 12, Part II-B). Cheese and soy production derive from fungi (see Lab Exercise 12, Part III-B). Mushrooms, truffles, and many other fungi are directly edible. The saprobic fungi, along with many saprobic bacteria, are very beneficial in disposing of dead and decaying organic matter.
- *The bad:* Yeasts exist as a normal component of mucous membranes but sometimes get out of control (e.g., in vaginal and other infections). Other fungi are serious parasites of animals and plants, causing many millions of damage each year (e.g., the wheat rust pathogen *Puccinia graminis*—see Lab Exercise 13, Part I).
- *The ugly:* Yet other yeasts may become systematic and pathogenic, sometimes fatal (e.g., candidiasis). Other fungi are rather slow-acting, lung-rotting, often fatal (coccidioidomycosis). Many fungi are deadly poisonous if eaten, as the quick-killing poisonous mushrooms that contain nerve toxins (neurotoxins). Some fungi have had really ugly effects, killing millions of people, causing mass migrations (the potato blight pathogen *Phytophthora infestans*), even striking down armies, and thus changing the course of history (the rye pathogen *Claviceps purpurea,* ergot).

The last two parasitic fungi have caused two historical pestilences (murrains). The first is a water mold (Oomycota), the second a sac fungus (Ascomycota). Incidentally, yeasts are sac fungi, whereas mushrooms and *Puccinia graminis* are club fungi (Basidiomycota).

The potato blight pathogen (*Phytophthora infestans*) has been of notable sociological significance. Potatoes (*Solanum tuberosum*) are a New World crop brought from the Andes to Europe by the Spanish in the late 1500s. Potatoes quickly became an important crop in northern Europe, where it was ideally suited to the cold, damp climate. By 1800 the peasant agriculture of Ireland had become almost entirely dependent on potato. Peasants ate mainly potatoes, 3.6–6.4 kilos (8–14 lb.) per person per day (Large 1940:23), which seems a hell of a lot of spuds to eat in one day—potato soup, potato bread, and presumably potato ice cream. In one July week in 1845 the fungus wiped out nearly the entire potato crop, which also failed in 1846. Famine ensued. The effect by 1860 was a million Irish dying, one and a half million emigrating (Large 1940:39), mostly to the northeastern United States to cities such as Boston and New York. Other European countries were much less drastically affected because of a more diversified agriculture.

The potato you eat is actually an underground stem, and the "eyes" are shoot buds. Potato can be propagated by planting parts of the potato tuber bearing the "eyes." This is nothing more than vegetative or asexual propagation of the plant (see Supplement 4, Part V). Potatoes in Ireland were of uniform genetic constitution, and once the disease entered the country, all the plants were equally susceptible to it. The epidemic was unstoppable until the disease eventually wore itself out.

Of like sociological significance has been ergot of rye (*Claviceps purpurea*), which affects the fruit (grains) of rye (*Secale cereale*). The fungus produces potent toxins of two types: (1) alkaloids causing gangrene (appendages and limbs fall off); (2) amides, including a naturally occurring counterpart of the hallucinogen LSD, effecting formication (not fornication), numbness, convulsions, even psychotic delusions. Rye bread was a staple of European peasants and of the early American immigrants; some peasants ate 0.9–1.4 kilos (2–3 lb.) of bread a day (Matossian 1989:xii). This is a heck of a lot of bread to eat. Naturally many people ate contaminated bread. This fungus causes the horrid disease called "St. Anthony's fire" or "ergotism," which is not the same as "egotism" or "the disease of conceit," the title of a 1989 Bob Dylan song. Historical consequences of ergotism include:

1. In 994 more than 40,000 people died in Europe.
2. The Black Death occurred in Europe in 1348–1350 with demographic effects lasting until 1490.
3. The witch trials in Salem and other New England communities in 1692 may have convicted persons with convulsive ergotism.
4. In 1722, the disease, and maybe also "egotism" in this case, struck down the Russian cavalry on the eve of battle for the conquest of Turkey. This changed the course of history because the Turkish Ottoman empire was saved, only to disintegrate after World War I because the Turks were allies of the Germans. Had the Turks been defeated by the Russians nearly 200 years earlier, there might be an entirely different situation in the Middle East and Yugoslavia today.
5. Mass hysteria occurred in France in 1789.

Matossian's (1989) recent book treated ergot of rye. The following review is slightly modified from the one appearing in *Taxon* 39:258 (1990):

> Matossian, Mary Kilbourne. *Poisons of the past: Molds, epidemics, and history.* Yale University Press, New Haven, 1989, xiv, [i], 190 pp., illus. [Contents—in 4 topic areas: (1) intro.: food poisoning and history; case study of Russia and its neighbors; (2) health history of Europe: a new look in the distant mirror; mycotoxins and health in early modern Europe; witch persecution in idem; the great fear of 1789; population explosion of 1750–1850; (3) health history of colonial New England: throat distemper; ergot and the Salem witchcraft affair; great awakening or great sickening?; (4) reflections: social control of mass psychosis; plant health and human health; biblio.; index.] [This fascinating and important study by a historian of Russia maintains that the 14th century black plague in Europe, the 1692 Salem, Massachusetts witch trials, mass hysteria in France

in 1789, and various other historical phenomena resulted from people eating food contaminated with microorganisms. Particularly implicated is rye bread made from grains infected with ergot, *Claviceps purpurea,* which produces a naturally occurring counterpart of the hallucinogen LSD. Such phenomena were widespread because rye bread was a dietary staple of the peasant population; some peasants ate 1–1.5 kg of rye bread a day. Linnda R. Caporael (Ergotism: The satan loosed in Salem?, *Science* 192:21–26) in 1976 had already proposed the ergotism-witchcraft link, but Matossian amasses much detailed circumstantial evidence (12 maps, 7 graphs, 11 tables) to support the proposition. This book has already received appreciable attention in the popular media. However, many laypersons will be disappointed by the sparsity of illustrations from historical works. Only two are present; of the 21 other figures two are of ergot, the rest maps and graphs. Nevertheless, this work is a true botanical-historical tour de force.—Rudolf Schmid, UC]

III. WHY FUNGI ARE A DIFFERENT KINGDOM

Traditionally, the fungi were placed in the plant kingdom and studied by botanists because only two kingdoms were recognized, namely, plants and animals. Since Whittaker's 1969 proposal of a five-kingdom classification system of living organisms, the fungi have increasingly been recognized as a fifth kingdom (Fungi) coordinate with the plants (Plantae), animals (Animalia), protists (Protista), and prokaryotes (Monera). Part II of Supplement 1 contrasts the fungal kingdom with the four other kingdoms and gives details on the late recognition of the fungi as a separate kingdom in the five-kingdom system accepted by most biologists today.

The fungi are eukaryotes, as are the plants, animals, and protists, but the fungi differ significantly from the latter kingdoms (see also Supplement 1, Part II, and Supplement 11, Part III):

- Unlike the animals and protists, which lack cell walls, most fungi have cell walls, as do prokaryotes and plants. However, the vegetative bodies of the slime molds lack cell walls.
- Unlike plants and the prokaryotes, most fungi have their cell walls composed of a matrix of ***chitin,*** a tough, resistant, nitrogen-containing polysaccharide. However, some fungal cell walls are cellulosic, for example, in water molds (Oomycota).
- Unlike plants, which are mostly autotrophic, and animals, which are mostly heterotrophic but holozoic, the fungi are distinctive in their mode of nutrition, which is heterotrophic and by absorption. However, the slime molds are phagotrophic, that is, with cells ingesting nutrients. Thus most fungi absorb nutrients either
  - from dead and decaying organic matter (***saprobes,*** saprobic fungi, *not* "saprophytes," which refers to saprobic plants) or
  - from living organisms (***parasites,*** parasitic fungi).

All fungi lack chloroplasts and thus are *never* green, photosynthetic, or autotrophic.

IV. CLASSIFICATION OF FUNGI

There is disagreement about the number of divisions of fungi. Because the main groups of fungi appear to be only remotely related, sharing nothing more than their lack of chloroplasts and a heterotrophic mode of nutrition, these main groups are recognized as divisions, with over 74,500 species according to Raven et al. (1992:729–731) (the algae have only about 22,650 species; the 18,000–20,000 species of lichens are not placed in any division):

1. cellular slime molds (Acrasiomycota)—65 species;	slime molds	"lower fungi" (3.36% species)
2. plasmodial slime molds (Myxomycota)—450 species;		
3. chytrids (Chytridiomycota)—750 species;	"true fungi"	
4. water molds (Oomycota)—475 species;		
5. bread/etc. molds (Zygomycota)—765 species;		
6. sac fungi (Ascomycota)—30,000 species;		"higher fungi" (96.64% species)
7. club fungi (Basidiomycota)—25,000 species;		
8. imperfect fungi (Deuteromycota)—17,000 species		

Because fungi are no longer regarded as plants, their divisional names end in *mycota* rather than in *phyta*. Scagel et al (1982:123) note that about 100,000 species of fungi have been described and that estimates of the total number are 250,000 or more species. Raven et al. (1992) include divisions 1 to 4 plus the algae in the Protista. Alexopoulos & Mims (1979) and Bold et al. (1987—Alexopoulos did the fungal part of this book) do not. The latter approach is favored here (similarly, I do not regard algae as protists).

Divisions 1 and 2 include the "slime molds," whereas divisions 3 to 8 include the "true fungi" ("fungi" used subsequently is employed mainly in the sense of "true fungi"). The slime molds have phagotrophic nutrition and lack cell walls (except on the spores) and hyphae, instead forming a naked, slimy, protoplasmic mass. In contrast, the "true fungi" have absorptive nutrition, cell walls, and hyphae. Multicellular fungi typically consist of microscopic threads or filaments that branch extensively over or within the substratum used for nutrients. The filaments of fungi are universally called ***hyphae*** (plural, singular ***hypha***), a term used hereafter. Hyphae may be ***septate,*** with cross walls, or ***coenocytic*** (nonseptate), mostly without cross walls (except at the sites of reproductive structures). A ***mycelium*** is the mass of hyphae constituting the vegetative body (thallus) of a fungus.

Divisions 1 to 5 are sometimes referred to as the "lower fungi" as opposed to the structurally more complex "higher fungi" representing divisions 6 to 8 (for various reasons these designations are not the best). The "higher fungi" completely lack coenocytic hyphae and flagellate cells, even in the aquatic species. These structures characterize the "lower fungi" except that the bread/etc. molds (division 5) totally lack flagellate cells (and only one species of Acrasiomycota has flagella—Alexopoulos & Mims 1979:48). [Regarding as "higher fungi" divisions 5 to 8 (e.g., Bold et al. 1987:724) rather than divisions 6 to 8 (here; Scagel et al. 1982) belies the previous statement about coenocytic hyphae.]

Division 8 is actually a "form-division" because of the artificial nature of this group. These imperfect fungi are designated "imperfect" because they lack sexual stages. Either these have not yet been discovered or else they do not exist, having been evolutionarily lost. On the basis of vegetative features, the imperfect fungi are predominantly sac fungi (Ascomycota), rarely club fungi (Basidiomycota, which would be suggested by such technical details as clamp connections and dolipore septa). The imperfect fungi reproduce only asexually, mainly by conidia (see Lab Exercise 12, Part III-B), but sometimes by budding or fission (in the imperfect yeasts) or other means (see definitions of these terms below). Imperfect fungi have distinctive scientific names, of course. If sexual reproduction is discovered in an imperfect fungus, a second name is used for the sexual stage. For example, the sexual stages *Eurotium, Sartorya,* and *Emericella* all belong to the asexual stage *Aspergillus* (Alexopoulos & Mims 1979).

V. PECULIARITIES IN THE LIFE HISTORIES OF FUNGI

Supplement 5 treats the main types of life histories, namely, haploid (1*n*), diploid (2*n*), and haploid-diploid (1*n*-2*n*), and their terminology (see especially Figs. Sup5–1 to Sup5–3). All three types occur in fungi:

a haploid (1*n*) life history in most fungi, but

diploid (2*n*) and haploid-diploid (1*n*-2*n*) life histories in only a few fungi.

An alternation of generations (phases) characterizes the haploid-diploid (1*n*-2*n*) life history. That is:

- a multicellular 2*n* ***sporophyte*** (SPT) or sporophytic (SPTic) or spore-producing phase alternates with
- a multicellular 1*n* ***gametophyte*** (GPT) or gametophytic (GPTic) or gamete-producing phase.

However, fungi are not plants, and "phyte" means "plant." Consequently, in those few fungi with haploid-diploid (1*n*-2*n*) life histories, for example, the chytrid *Allomyces* (Fig. 12–9; Lab Exercise

12, Part IV), the "SPT" is called the ***sporothallus***, the "GPT" the ***gametothallus*** (Alexopoulos & Mims 1979; Bold et al. 1987). In such alternation of generations (phases), a 2*n* spore-producing "SPT" or sporothallus typically alternates with a 1*n* gamete-producing "GPT" or gametothallus.

Syngamy (fertilization) in the fungi, like in the algae, is either isogamy, anisogamy, or, much less often, oogamy (see definitions in Fig. Sup4-4 and Supplement 4, Part III). However, the life history of the "higher fungi" has a most important variation for syngamy, namely the ***dikaryotic phase*** (dikaryophase, heterokaryotic phase) consisting of two-nucleate (binucleate) cells (Figs. 12-2, 12-6, 12-7, 13-1). As noted in Supplement 4, Part III, cell fusion (syngamy, fertilization) consists of:

- ***cytoplasmic fusion*** (plasmogamy);
- ***nuclear fusion*** (karyogamy).

[Note that "plasmogamy" is *not* "union of the protoplasts" as is so often defined (e.g., Raven et al. 1992:754, etc.) because the "protoplast" includes both the cytoplasm and the nucleus.] Usually these fusions occur very quickly after each other, so that for practical purposes there is only one fusion. In contrast, there can be appreciable delay between cytoplasmic fusion and nuclear fusion, as in some 55,000 species of mushrooms and other "higher fungi." Here there is ***cytoplasmic fusion*** (plasmogamy), which is the union of the cytoplasm of the gametes (sex cells) or cells of two mating strains, but *not* accompanied by union of their nuclei (Figs. 12-1H, 12-2B, 12-7B). A $n+n$ state (dikaryon) thus is established in each cell. Each cell is two-nucleate or binucleate ($n+n$). These cells may divide mitotically to form ***dikaryotic hyphae*** comprised of many $n+n$ cells (i.e., many dikaryons) (Figs. 12-6, 12-7B). Eventually in certain $n+n$ cells the two nuclei in the cell fuse. ***Nuclear fusion*** (karyogamy) ensues and a zygote results (Figs. 12-1I, 12-2B, 12-7E) [Note that from a terminological viewpoint, karyogamy and not plasmogamy results in the zygote]. Nuclear fusion occurs after a brief or long interval, respectively, in the sac and club fungi. The delay may be as much as eight to ten months, as in the wheat rust pathogen *Puccinia graminis* (see Lab Exercise 13, Part I).

In summary, the two-nucleate or binucleate state that persists for some time in such fungi, before nuclear fusion occurs, is designated $n+n$ and is called a ***dikaryotic phase***. From the perspective of life histories, the dikaryotic phase ($n+n$) is interpolated between the 1*n* and 2*n* phases. That is, the life history with the dikaryotic phase is usually fundamentally a haploid (1*n*) life history.

Caption for figure overleaf:

Fig. Sup12-1. 'Alice in Wonderland and the caterpillar.' Priscilla Fawcett's rendition (in Norstog and Long 1976:250) of Alice on tiptoes peering at a caterpillar hookahed down on a mushroom may be completely unknown outside botany; the artist informs me (personal communication, Jan. 1986) that she never received any comment about her drawing of Alice and the caterpillar. This drawing compares most favorably with its famous Tennielian counterpart and is decidedly superior to Harry Roundtree's 1928 version (reproduced in Ovenden 1972:28), which shows Alice in roaring 20s hairstyle and garb. Because of the excellence of Priscilla Fawcett's drawing, I reproduce here with her permission not only her drawing but also her comments on its background:

> When I was doing the drawing in spring 1973, I wanted the caterpillar to be both a caterpillar and a youngish Oxford type 'into drugs' as Lewis Carroll and his contemporaries were. It was quite respectable to experiment a little with drugs. I suppose a few got hurt though. Obviously if a caterpillar smoked, it would have to do so through its respiratory system and exhale through the spiracles [the small openings along each side of the thorax and abdomen of an insect that permit breathing], making multiple smoke rings, etc., feasible. I made the spiracles rather big because if they had been the right size they wouldn't have been so obvious.

Besides the delightful caterpillar, I particularly like the wonderful menagerie of gazing characters from Alice's *Adventures.* Of course, the botanical detail is also superb, as is to be expected from this renowned botanical artist, whose work is reproduced elsewhere in this manual. (Drawing from Norstog and Long 1976:250; text modified from Schmid 1989:5.)

PF

LAB EXERCISE **12**

Fungi I

(Cell/tissue Organization and Reproduction)

OBJECTIVE

To examine fungi for (1) their sociological and economic importance, (2) their cell/tissue organization, and (3) their reproductive morphology.

Note: Read this material *before* coming to the lab; there is insufficient time in the lab to read the material from scratch. Due to the large amount of background information presented here, a ●●● is used to mark those paragraphs calling for actual observations or diagrams in the lab. Most unmarked paragraphs reiterate information presented in the lecture. A bullet (●) signifies an important term or concept.

Suggested Lab Diagrams (see part noted for suggested labels):
Rhizopus (a bread mold) asexual sporangia (Part III-A1)
Eurotium (a sac fungus) cleistothecium OR *Sordaria* (a sac fungus) perithecium (Part III-E)
Agaricus brunnescens (cultivated white mushroom, a club fungus) basidiocarps (Part III-F)
Coprinus (inky cap mushroom, a club fungus) cap x.s. (Part III-F)

PERSPECTIVE

The fungi are so different from the other major groups of organisms that a fifth kingdom (Fungi) coordinate with the plants (Plantae), animals (Animalia), protists (Protista), and prokaryotes (Monera) is now recognized. The fungi have eight divisions (actually seven divisions plus the form-division of imperfect fungi), with over 74,500 species according to Raven et al. (1992:729–731) (the algae have only about 22,650 species; the 18,000–20,000 species of lichens are not placed in any division):

1. cellular slime molds (Acrasiomycota)—65 species;		
2. plasmodial slime molds (Myxomycota)—450 species;	slime molds	
3. chytrids (Chytridiomycota)—750 species;		"lower fungi"
4. water molds (Oomycota)—475 species;		(3.36% species)
5. bread/etc. molds (Zygomycota)—765 species;	"true fungi"	
6. sac fungi (Ascomycota)—30,000 species;		
7. club fungi (Basidiomycota)—25,000 species;		"higher fungi"
8. imperfect fungi (Deuteromycota)—17,000 species		(96.64% species)

Supplement 12, Parts III and IV, elaborates various complications about the classification of fungi, including their relationship to the other kingdoms of organisms. The "lower fungi" are structurally less complex than the "higher fungi" (for various reasons these designations are not the best). Division 8 is actually a "form-division" because of the artificial nature of this group.

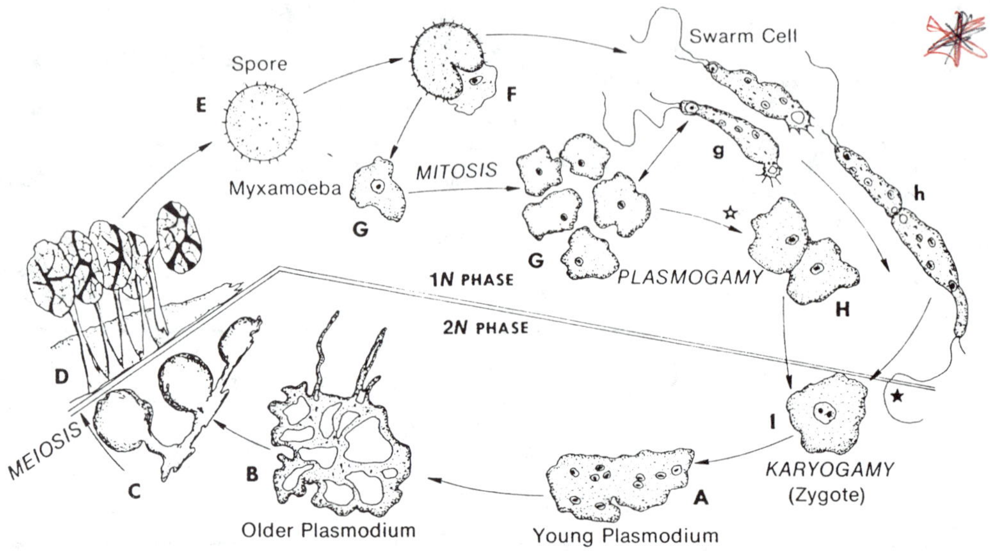

Fig. 12–1. Diploid (2*n*) life history of a plasmodial slime mold (Myxomycota). *A, B,* young and old 2*n,* multinucleate plasmodia, respectively; *C, D,* developing and mature sporangia, respectively; *E,* meiospore; *F,* germination of spore; *g, G,* flagellate cells (swarm cells) and nonflagellate cells (myxamoebae), respectively, either of which can act as gametes; *h, H,* initial syngamy (fertilization) at ☆, i.e., cytoplasmic fusion (plasmogamy) of swarm cells or myxamoebae, respectively; *I,* final syngamy at ★, i.e., nuclear fusion (karyogamy). (From Norstog & Long 1976:253; redrawn from *Introductory mycology,* 2nd ed., by C. J. Alexopoulos, © 1962 by John Wiley and Sons, Inc., New York. Reprinted by permission of the publisher.)

Note: Before beginning this lab exercise, you should review:
- Supplement 5 on the main types of life histories;
- Supplement 11 on the major types of biological nutrition and associations betwixt organisms;
- Supplement 12 on the fungi and their characteristics.

I. THE UBIQUITY AND SOCIOLOGICAL AND ECONOMIC IMPORTANCE OF FUNGI

For background information on the good, bad, and ugly aspects of fungi see Supplement 12, Part II.

●●● Numerous fungal spores occur in the air and on landing on a suitable medium may develop into mature fungi. Examine the DEMO petri dishes with the culture medium exposed, for two hours each, *one* and *two* weeks earlier. This DEMO shows the impressive number of fungal spores circulating in a normal room.

●●● Not all fungi are nice. Examine the DEMO color photographs (from Delacrétaz et al. 1976:99, 105, 131, 159) of fungal diseases of humans. These diseases are groady to the max, and some fungal diseases can be fatal. However, such diseases are very rare and are common in:

tropical areas,
areas with poor sanitation,
areas without access to decent medical care, and
persons with suppressed immune systems.

In fact, the vast majority of fungi are harmless to humans and play a beneficial role in the ecosystem.

●●● Examine the DEMO photograph (from O'Brien & O'Brien 1972:105) depicting women and children in 1846 starving due to the potato famine in Ireland that had resulted form the potato blight pathogen (*Phytophthora infestans*). Also examine the DEMO copy, if available, of Matossian's (1989) book on the ergot of rye pathogen (*Claviceps purpurea*), as well as any available DEMO reviews of the book.

●●● On a lighter note, fungi have provided many examples of humor, especially cartoon humor. In fact, mushrooms, more than any plants or any other fungi, have played a great role in botanical humor. Various cartoons are on DEMO. The DEMO papers by Emerson (1969, 1973) offer additional examples. See also Fig. Sup12-1.

●●● Fungi have also appeared in poetry and doggerel, often in a humorous context. The DEMO poem, John Updike's "Ode to rot" (from *The Atlantic monthly,* Jan. 1985, p. 83), is a good example.

●●● Fungi, particularly mushrooms, are common subjects of juvenilia. Examine the very fine pictures in the DEMO children's books: Sylvia A. Johnson's *Mushrooms* (1982), intended for ages ten through 12, and Millicent E. Selsam's *Mushrooms* (1986) and Barrie Watts's *Mushroom* (1986), intended for ages four through eight. Which of these books deals with more than mushrooms?

●●● Finally examine the dried material of *Ustilago maydis* (corn smut), a club fungus parasitic on corn, as well as any available DEMO newspaper articles about it. While this parasite causes serious economic losses (up to 25% of a crop), it was considered a delicacy by the Aztecs and is still highly esteemed by Mexicans today. Currently this delicacy sells for $8.00–10.00 a pound (0.45 kg). Incidentally, the International Maize and Wheat Improvement Center holds, biennially, a Smut Workers' Workshop (e.g., the 5th workshop was held in Apr. 1986 in Ciudad Obregón, México).

II. CELL/TISSUE ORGANIZATION IN FUNGI

Recall (see Lab Exercise 3, Part I) that any multicellular plant or fungal body *not* differentiated into roots, stems, and leaves is called a ***thallus*** (singular, plural ***thalli***). This refers to all multicellular bodies in the algae, bryophytes, and fungi, but in the vascular plants "thallus" refers only to the GPT. Moreover, in fungi "thallus" refers mainly to the vegetative body. The filaments of fungi are universally called ***hyphae*** (plural, singular ***hypha***). A ***mycelium*** is the mass of hyphae constituting the vegetative body (thallus) of a fungus. ***Haustorium*** (singular, plural ***haustoria***) is the general term for the penetrating and absorbing structure of fungal parasites. A ***host*** is the organism on or in which the parasite lives.

Also recall (see Lab Exercise 3, Part I) that the algae have the following types of cell/tissue organization, ranging from the simplest type to the most complex: unicells, colonies, filaments, coenocytes (or tubes or siphons), *membranes, pseudoparenchyma, *parenchyma, and *parenchyma plus phloem. The asterisked types do *not* occur in fungi, which in addition have pseudoplasmodia and plasmodia.

A. Slime bodies—Pseudoplasmodia and plasmodia

See Fig. 12-1. These structures occur only in the two divisions of slime molds and not in the "true fungi." Hence this is another difference between these groups (for other distinctions see Supplement 12, Part IV). ***Pseudoplasmodia*** and ***plasmodia*** (plural, singular ***pseudoplasmodium, plasmodium***) consist only of protoplasts bounded by plasma membranes. The two divisions of slime molds can be contrasted as follows:

- *cellular slime molds (Acrasiomycota):* vegetative body an amoebial stage (called myxamoeba in the slime molds), eventually forming a ***pseudoplasmodium,*** an amoeboid structure behaving as a unit, the individual amoebae retaining their cellular identity;
- *plasmodial slime molds (Myxomycota):* vegetative body an amoebial stage (myxamoeba), eventually forming a ***plasmodium,*** an amoeboid structure behaving as a unit, *but* the individual amoebae losing their cellular identity; that is, a multinucleate mass (in essence, a coenocyte—see below) surrounded by a membrane results (see Fig. 12-1).

These organisms are called "slime molds" because of the slimy, naked, advancing protoplasmic mass (Fig. 12-1B) that creeps over the substratum and ingests organic matter (bacteria, yeasts, fungal spores, and bits of decaying plant and animal matter), possibly also absorbing some material from the surrounding environment. When the nutrient supply is depleted, the protoplasmic mass migrates to a new site. When moisture and/or the nutrient supply become limiting, the protoplasmic mass differentiates into delicate and ephemeral sporangia and other types of reproductive structures bearing spores. These are usually enclosed in a rigid cell wall as in the "true fungi." Spores on herbarium specimens have germinated after 76 years in storage (Bold et al. 1987:696).

A typical slime mold has a diploid ($2n$) life history (Fig. 12–1; see also Fig. Sup5–2). The $2n$ plasmodium results from the fusion (syngamy, fertilization) of either two $1n$ flagellate cells (swarm cells) or two $1n$ amoeboid cells (myxamoebae), either of which thus can act as the gametes. The amoeboid cells feed on bacteria and yeasts, divide mitotically, and when they reach critical mass fuse in pairs.

●●● Examine the DEMO culture of the "slugs" of the cellular slime mold *Dictyostelium* and the DEMO photograph (from Carolina Biological Supply Co., 1987 calendar, or from another available source) of its reproductive body. The slugs represent the vegetative amoebial stage, specifically the pseudoplasmodium.

●●● See Fig. 12–1. Then examine the DEMO accounts (from Alexopoulos & Mims 1979:62–63 and Katsaros 1989:3) of "The Blob" that alarmed Texas in 1973. This was *Fuligo septica,* a widespread species with the largest bodies of any slime mold, up to about 30 cm (11.8 in.) in diameter and 3 cm (1.2 in.) thick (Katsaros 1989:21). Finally examine the DEMO culture of the plasmodial slime mold *Physarum polycephalum,* which is feeding on bacteria living on the oatmeal flakes in the culture medium. Only the veinlike parts of the plasmodium are alive. The plasmodium is coated with a thick sheath of slime. The plasmodium is bright yellow in contrast to the white slime tracks left by the organism on the agar surface. Plasmodia readily regenerate themselves; pieces removed from the large plasmodium thus become new individuals.

●●● Place a small piece of the plasmodium (sans water) on a petri dish or on a microscope slide but do *not* add a cover glass. Then examine the preparation with a dissecting microscope (highest magnification) and/or compound microscope (two lowest magnifications). Observe the rapid movement of the vacuoles and cytoplasmic particles in the plasmodium. This phenomenon, called ***protoplasmic streaming,*** is universal to all cells. It may be up to 1.45 mm/sec (0.06 in./sec) in *Physarum polycephalum,* the fastest protoplasmic streaming known (Bold et al. 1987:691). What is the direction of streaming? Does it change? Apparently there is no close relationship between direction of streaming and direction of plasmodial movement.

B. Unicells

See Fig. 12–2. As in the algae, ***unicells*** are completely separate cells due to the parent cell wall being cast off after cell division. Unicellular fungi include chytrids and the yeasts.

Yeasts are notorious for causing various infections (see Supplement 12, Part II) but are also invaluable in the manufacture of bread and beer (Fig. 13–3). *Saccharomyces cerevisiae* (brewer's yeast) (the generic name means "sugar fungus") can live without oxygen (anaerobic respiration) and ferment carbohydrates, enzymatically breaking down glucose (a sugar):

1 glucose + 2 NAD (a coenzyme) ——> 2 ethyl alcohol + 2 carbon dioxide + 2 NAD

The by-products of alcoholic fermentation are used as follows:

- ethyl alcohol (ethanol), in brewing (alcoholic fermentation) to make beer, wine, sake, hard liquor (fermented corn for bourbon, fermented barley for Scotch, fermented rye for rye), and
- carbon dioxide, in brewing to make beer, champagne, and crackling wine fizzy and, as a leavening agent, to make bread rise (i.e., bubbles become trapped in the dough and leaven it).

Incidentally, in bread-making the alcohol evaporates during the baking process.

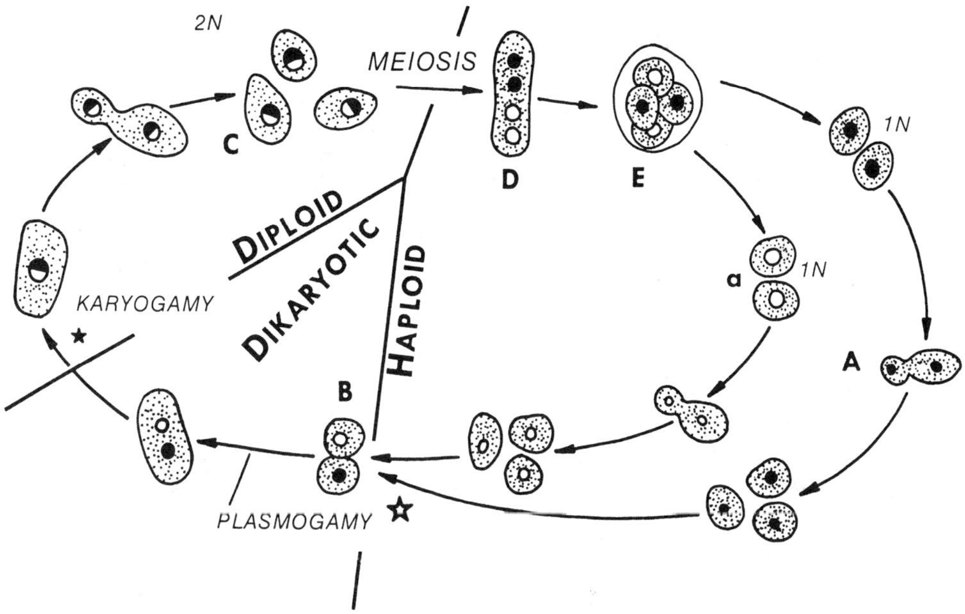

Fig. 12–2. Haploid-diploid (1*n*-2*n*) life history of *Saccharomyces cerevisiae* (brewer's yeast), a unicellular sac fungus (Ascomycota). *a, A,* budding 1*n* yeast cells of opposite mating strains; *B,* syngamy (fertilization), initially, at ☆, cytoplasmic fusion (plasmogamy), finally, at ★, nuclear fusion (karyogamy); *C,* budding 2*n* yeast cells; *D, E,* young and mature asci (sporangia), respectively, each ascus with four ascospores. Note the alternation of morphologically similar generations (phases) and the brief dikaryotic (*n* + *n*) phase. (From Norstog & Long 1976:263; redrawn from *Introductory mycology,* 2nd ed., by C. J. Alexopoulos, © 1962 by John Wiley and Sons, Inc., New York. Reprinted by permission of the publisher.)

Beer-making typically involves these processes (Weiner 1977): malt, mash, brew, ferment, store, and drink! (1) Grains of barley (*Hordeum vulgare*) are partly germinated (malted), during which enzymes (amylase) in the grain convert starch in the grain to sugar. (2) The malted grain is then mashed, that is, mixed with hot water to form a wort, a sweet brown liquid. (3) During the brew stage, hops (female inflorescence bracts of *Humulus lupulus*) are added to a boiling mix to add flavor. (4) During the alcoholic fermentation stage, which takes a week or longer, yeast is added, and the glucose is converted to alcohol and carbon dioxide. When fermentation is done, the wort is known as beer. Bubbles are added by one of three methods: (a) sealing the ferment, (b) injecting carbon dioxide under pressure, or, the best way, (c) adding sugar (priming) or wort (krausening), bottling the beer, and letting bubbles develop naturally. By the by, Budweiser has bigger bubbles and goes flat faster than quality beer because the carbon dioxide is artificially injected. (5) Beer can be optimally stored for only relatively short periods. (6) Prosit! Skoal! A votre santé! Cheers!

A loaf of bread, a jug of wine, and thou.—Omar Khayyám

Natural wine results when yeast cells on grapes mix with the juice of the crushed grapes. Under oxygenless (anaerobic) conditions, the yeast breaks down the glucose in the grape juice to alcohol. Modern vintners, however, achieve fermentation by adding pure yeast cultures to relatively sterile grape juice. On the other hand, yeasts must be excluded when making grape juice and sweet cider.

Yeasts are mainly sac fungi, but they may also be club and imperfect fungi. Most yeasts have a haploid (1*n*) life history (see also Fig. Sup5–1), but Fig. 12–2 shows the haploid-diploid (1*n*-2*n*) life history (see also Fig. Sup5–3) of *Saccharomyces cerevisiae,* which has alternation of similar generations (phases). This well-studied species was recently discovered to be filamentous under starvation conditions (Hoffman 1992). Yeasts consist of two groups, based on their asexual reproduction:

- the fission yeasts reproducing by ***fission,*** the simple splitting of a cell into two cells by constriction and formation of a cell wall;
- the budding yeasts reproducing by ***budding,*** the production of a small ***bud*** (outgrowth) from the parent cell, the nucleus dividing and one resultant nucleus moving into the bud.

Note in Fig. 12–2 that budding involves both $1n$ and $2n$ cells. Fission and budding can be incredibly quick, as those of you know who have had yeast infections or have used yeasts to make beer or bread. One cell of *Saccharomyces cerevisiae* dividing at maximal rate and if unchecked could in just two weeks form around the earth a layer 3 m (9.8 ft.) deep (Hoffman 1992).

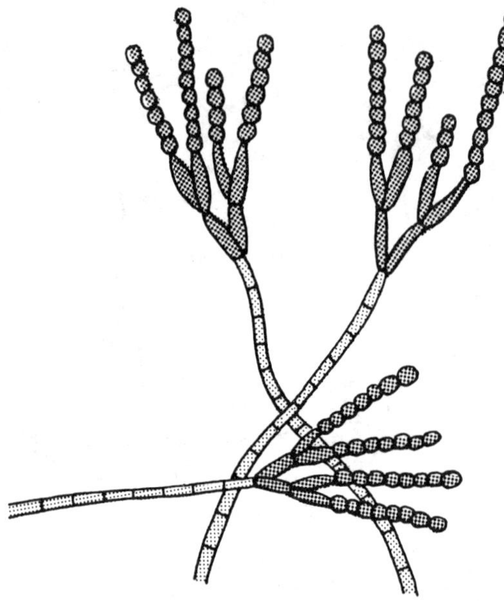

Fig. 12–3. Conidia (conidiospores), the asexual spores of *Penicillium,* an imperfect fungus (Deuteromycota) with septate hyphae. (From Norstog & Long 1976:269.)

●●● See Fig. 12–2. Then prepare a water mount slide preparation of a culture of vegetative cells of *Saccharomyces*. What type of asexual reproduction is present?

C. Colonies

Colonies in algae and blue-green bacteria are groups of loosely associated cells, from few (2, 4, 8, 16) up to about 60,000 (see Lab Exercise 3, Part I-B). However, in fungi the term ***colony*** is used rather differently, referring to a mass of individual cells living together (e.g., yeasts) or "to many hyphae growing out of a single point and forming a round or globose thallus" (Alexopoulos & Mims 1979:591).

D. Septate hyphae

See Figs. 12–3, 12–6, 12–7, 12–8, 13–1, and 13–2A. Hyphae may be ***septate,*** with cross walls, or ***coenocytic*** (nonseptate), without cross walls (see next category), and in either case can form a mycelium. Septate hyphae may occur singly (Fig. 12–3) or especially as pseudoparenchyma (see below). The individual cells of septate hyphae may each contain a single nucleus (Figs. 12–6, 12–7), two nuclei (Figs. 12–6, 12–7), the dikaryotic ($n+n$) condition (see Part III-D), or many nuclei (e.g., *Aspergillus*). The multinucleate cells can not be confused with coenocytic hyphae because of the many cross walls present. Examples of septate hyphae will be seen in Parts III-B, III-E, and III-F.

E. Coenocytic (nonseptate) hyphae (coenocytes)

See Figs. 12–4, 12–5, and 12–9. ***Coenocytic hyphae*** lack cross walls, except typically at the sites of reproductive structures. Such hyphae constitute coenocytes, which, as in the algae (see Lab Exercise 3, Part I-D), are basically gigantic multinucleate cells. The "lower fungi" have coenocytic hyphae but lack septate hyphae; the reverse situation applies to the "higher fungi." Examples of coenocytic hyphae will be seen in Part III-A.

F. Pseudoparenchyma

See Figs. 12–6, 12–8, 13–1, and 13–2A. ***Pseudoparenchyma*** (plectenchyma) results from the loose or compact interweaving of contiguous hyphae to form a more or less three-dimensional structure. Mushrooms and other examples of pseudoparenchyma will be seen in Parts III-E and III-F.

III. REPRODUCTION IN SELECTED FUNGI

Fungi reproduce asexually and sexually by various means. Asexual reproduction includes spore production:

- in sporangia, by nonmotile spores (Fig. 12–5a, A) or by motile (flagellate) spores called ***zoospores*** (Figs. 12–4A, 12–9F);
- naked at the tips or sides of hyphae (***conidiophores***) by nonmotile spores called ***conidia*** (plural, conidiospores; singular ***conidium***) (Figs. 12–3, 13–1B, C).

The nonmotile spores are typically dispersed by wind. For asexual reproduction in the yeasts see Part II-B. For sexual reproduction in the "higher fungi" see Parts III-D to III-F.

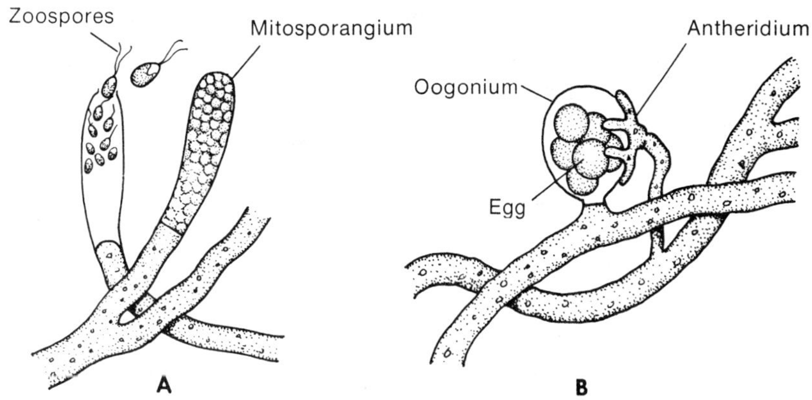

Fig. 12–4. Reproduction in *Saprolegnia,* a coenocytic (tubular, siphonous) water mold (Oomycota). *A,* asexual sporangia (zoosporangia, mitosporangia) producing zoospores (mitospores); *B,* sexual reproduction (oogamy). The gametes are the only 1*n* structures in this organism, which has a diploid (2*n*) life history. (From Norstog & Long 1976:257; redrawn from *Cryptogamic botany,* 2nd ed., vol. 1, by G. M. Smith, © 1955 by McGraw-Hill Book Co., Inc., New York. Reprinted by permission of the publisher.)

A. Asexual reproduction (especially sporangia) by bread/etc. molds (Zygomycota)

This division includes various saprobic and parasitic fungi, all of which have coenocytic hyphae. Some Zygomycota are weak parasites of flowers and fruits.

A1. Rhizopus (bread mold)

See Fig. 12–5. This common saprobe grows on bread, soft fruit, dung, or other organic matter.

●●● See Fig. 12–5. Then examine the DEMO culture of *Rhizopus stolonifer* and note the transparent coenocytic hyphae spreading over the surface of the agar and ramifying into it as rhizoids. The rhizoids secrete digestive enzymes and absorb materials. Opposite the rhizoids are groups of long black hyphae (***sporangiophores***) that bear black sporangia containing asexual (i.e., mitotic) spores. Similar sexual sporangia result after fusion of gametangia and the resultant meiosis.

●●● Take some sporangia from the culture of *Rhizopus* and prepare a water mount slide preparation (use a cover glass). Each sporangium has a bulbous central region called a ***columella*** (singular, plural ***columellae***) that is covered by the spores or the spore-producing (sporogenous) tissue. Can you observe spore release? How does this occur? What agent disperses the spores? How does bread become contaminated with fungi (not all mold seen on bread is *Rhizopus;* much of it is the blue mold *Penicillium,* a sac fungus—see Parts III-B and III-C). What is the function of preservatives in bread?

●●● *Suggested diagram and labels: Rhizopus* (a bread mold) asexual sporangia: mycelium (rhizoids, surface hyphae), sporangiophore, sporangium, spores.

A2. Phycomyces

The sporangiophores, each a coenocytic chamber bearing the sporangia, may reach 20 cm (7.9 in.) high and are sensitive to both light and gravity; a sporangium may contain up to 100,000 spores (Scagel et al. 1982). These structures have been featured in Ripley's "Believe it or not."

●●● Examine the DEMO culture of the asexual sporangia of *Phycomyces blakesleeanus*. Do *not* make a water mount slide preparation. The sporangia are essentially gigantic *Rhizopus* sporangia.

A3. Pilobolus (cannon fungus)

The Los Angeles times said of "Pilobolus": "The Piloboli create their own gravity, establish their own vocabulary of abstraction. Complicated geometric patterns sprout and grow organically. Wondrous shapes emerge, merge, split and remerge." Reference was made to the "Pilobolus Dance Co.," but it might as well have been to the fungus, which, however, grows on cow pies or meadow muffins.

●●● The asexual sporangia of *Pilobolus* and their sporangiophores have evolved a most unusual form of dispersal. Examine the DEMO diagram and photograph (from Raven et al. 1992:214), which are juxtapositioned with DEMO pictures of and newspaper articles (from *The New York times*) on the "Pilobolus" dancers. *Pilobolus* grows on horse and cow dung. Normally the spores must be eaten by an animal and passed through its digestive tract before the spores will germinate. The sporangiophores are positively phototropic (stimulated or attracted by light). Orientation of the sporangiophores toward light actually aids the explosive discharge of the enlarged tips of the sporangiophores. The sporangia are forcibly ejected nearly two meters (6.6 ft.) at a rate of 16 m/sec (52.5 ft./sec) (Bold et al. 1987:728) and at one fifth of the acceleration of a bullet exiting from a rifle at 3,000 m/sec (9842.5 ft./sec) (Vogel 1988:291). The sporangia adhere where they land. If this is on blades of grass that are subsequently eaten, the life history begins anew.

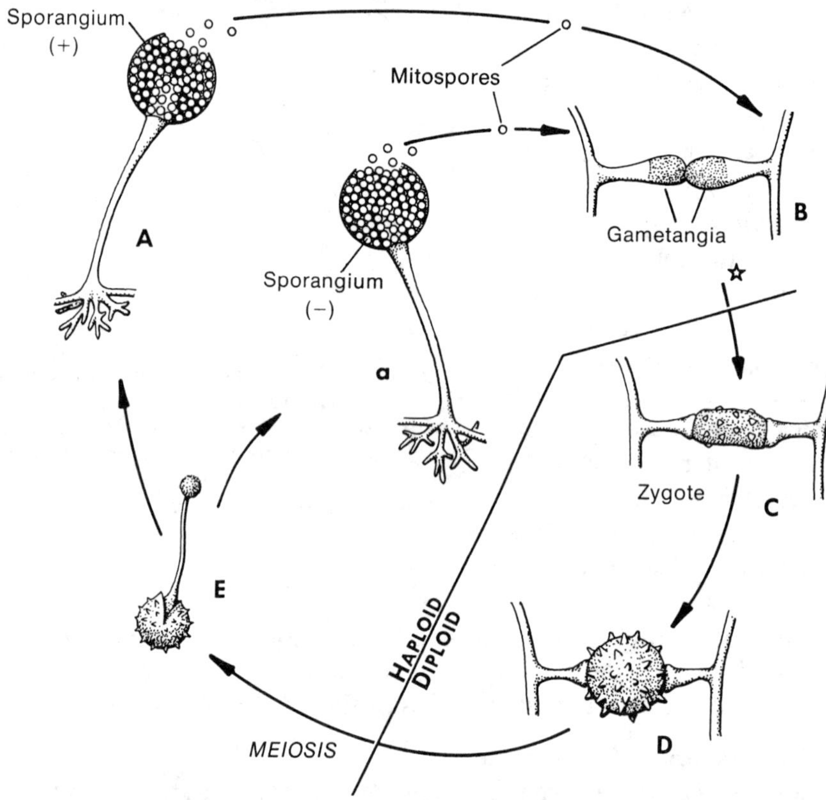

Fig. 12-5. Haploid (1*n*) life history of *Rhizopus nigricans* (black bread mold), a coenocytic (tubular, siphonous) fungus (Zygomycota). *a, A,* mycelia and sporangia of opposite mating strains (note the basal rhizoids); *B,* ☆, fusion of sex organs (gametangia) of opposite mating strains, i.e., syngamy (fertilization); *C,* 2*n* zygote; *D,* encysted zygote; *E,* germination of zygote after meiosis. (From Norstog & Long 1976:258.)

●●● Next examine the DEMO cartoon and photograph (from Carolina Biological Supply Co., 1986 advertisement, or from another available source) of the sporangial apparatus of *Pilobolus*. Note the following on the photograph:

1. the apical sporangia at the tips of the sporangiophores;
2. the enlarged, vesicles at the tips of the sporangiophores, which will explode to propel the sporangia;
3. the many water droplets on the sporangiophores.

What is morphologically inaccurate about the cartoon?

●●● Finally examine the DEMO agar cultures of *Pilobolus,* which were covered with black paper with a central pinhole. The sporangiophores will have shot their sporangia at this central pinhole bull's-eye. Squash some of these sporangia in water on a slide and verify that it is indeed a sporangium and not a spore that had been shot off.

A4. Entomophthora muscae (fly fungus)

Entomophthora muscae (fly fungus) is often found on the dead bodies of house flies on long unwashed windowpanes in garages, attics, offices, and especially university classrooms and offices (Alexopoulos & Mims 1979). A dead fly will usually be surrounded by a white, halolike zone that consists of thousands of spores (conidia—see next part) forcibly discharged by the spore-producing (sporogenous) cells growing out of the body of the fly. Such spores on contacting another fly quickly germinate and penetrate its cuticle. Infected flies usually die within a week or so after infection, and the spore discharge process is repeated. Infected flies tend to crawl upward, finally dying in a higher place (fly heaven) conducive to better spreading of spores. Spores that do not land on a suitable substratum can still germinate to form a sporophore bearing another spore at its tip. This "secondary spore" is also forcibly discharged.

●●● Examine the DEMO permanent slide of a housefly (*Musca domestica*) bearing *Entomophthora.* The soft parts of the fly have been almost entirely replaced by the fungus, which consists mostly of whitish conidia (stained greenish). The blackish hard parts of the insect are made of chitin, which is also found in the walls of most fungi.

B. Asexual reproduction (conidia)

See Fig. 12–3. There are over 100 species each of *Aspergillus* and *Penicillium.* Most of these species are imperfect fungi (Deuteromycota), that is, reproducing only asexually, although some are sac fungi because they also reproduce sexually (see Supplement 12, Part IV). Asexual reproduction is by conidia, that is, nonmotile spores produced naked at the tips or sides of hyphae (conidiophores).

Aspergillus and *Penicillium* are of great importance economically. Fermentation of *A. niger* produces citric acid for use in soda (this is a cheaper source than citric acid extracted from citrus juice). *Penicillium camemberti* and *P. roqueforti* are used to make, respectively, Camembert cheese and blue cheeses such as Roquefort, Danish blue, Wisconsin blue, and Gorgonzola. *Penicillium griseofulvum* is the source of the antifungal antibiotic griseofulvin, which is used to treat athlete's foot and other ringworm diseases. The species of penicillin fame is *P. notatum.* Bold et al. (1987:797) concluded that all these species, due to their extensive study by biologists, are probably true imperfect fungi (see Supplement 12, Part IV) and hence totally lack sexual stages.

●●● See Fig. 12–3. Then take a *small* piece of agar from a culture of *Penicillium* and examine it under the low power (10X objective) of a compound microscope *whole,* sans cover glass; that is, do *not* make a water mount because the conidia will fragment. The conidiophore of *Penicillium* is branched and brushlike, whereas that of *Aspergillus* is a large, multinucleate club-shaped vesicle, like the aspergillum used to sprinkle holy water. The blue in the DEMO of blue cheese is the fungus (*P. roqueforti*).

C. Rotten food

"Molds" are troublesome fungi well known to any person who neglects to clean out a refrigerator. The common blue or green molds growing on jams, jellies, bread, citrus fruit, and other foods are generally species of *Penicillium* and *Aspergillus. Rhizopus,* the bread mold, appeared in Part III-A1. Most of such mold seen represents asexual reproductive structures such as conidia and sporangia.

●●● Some rotting fruit or vegetables may be available on DEMO. What fungal groups seem to be represented?

D. Sac versus club fungi (Ascomycota versus Basidiomycota)

Compared to the other divisions of fungi, the sac fungi (Ascomycota) and club fungi (Basidiomycota) have the following characteristics:

- 1. predominantly terrestrial and multicellular, though the yeasts are unicellular (also rarely filamentous);
- 2. cells walls chitinous, generally lacking cellulose;
- 3. hyphae that are septate throughout the thallus (but the septa may be perforate);
- 4. flagellate cells completely lacking;
- 5. intercalation of a ***dikaryotic phase,*** that is, a $n+n$ stage (see Supplement 12, Part V), between ***cytoplasmic fusion*** (plasmogamy) and ***nuclear fusion*** (karyogamy);
- 6. asexual spores never enclosed in sporangia;
- 7. sexual spores (ascospores or basidiospores) enclosed in or on sporangia (asci or basidia);
- 8. sexual spores few and definite in number per sporangium;
- 9. sexual sporangia often borne in a definite structure, a "fruiting body" (ascocarp or basidiocarp) consisting of pseudoparenchyma.

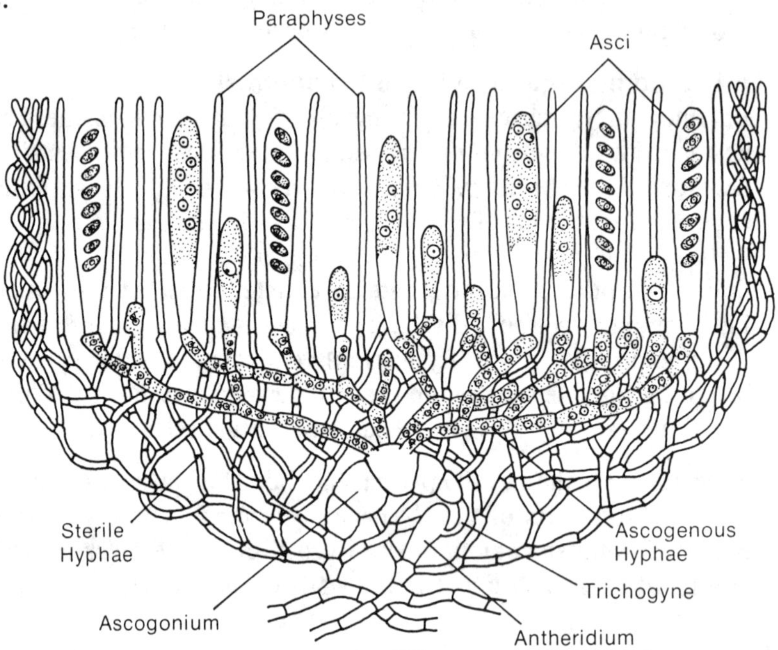

Fig. 12–6. Longisection of an ascocarp (cup-shaped type or apothecium) of a sac fungus (Ascomycota). The contents of the gametangia (female ascogonium and male antheridium) fuse via a long hairlike hyphal tube (trichogyne). Dikaryotic ($n+n$) hyphae (ascogenous hyphae) form after cytoplasmic fusion (plasmogamy). Note the relationship between the dikaryotic ($n+n$) and $1n$ sterile hyphae (respectively, shaded and unshaded) in this haploid ($1n$) and dikaryotic ($n+n$) life history. Ascospores, eight per ascus (sporangium), result from nuclear fusion (karyogamy), meiosis, and one mitotic division; $1n$ sterile hyphae (paraphyses) separate the asci. (From Norstog & Long 1976:261.)

Some of these statements clearly do not apply to the sac or club fungi such as the yeasts, which are unicellular (or also rarely filamentous) (see Part II-B). The dikaryotic phase is a very important variation in the life history of the "higher fungi" (see Supplement 12, Part V).

Items 7 to 9 contrast the sac and club fungi, namely (compare Figs. 12–6 and 12–7 with Fig. 12–8):

- *sac fungi (Ascomycota):* sexual spores (***ascospores***) borne internally within a saclike sporangium (singular ***ascus,*** plural ***asci***), typically 8 (Figs. 12–6, 12–7F), seldom 4 (Fig. 12–2E), occasionally 16 or more per ascus, the asci often aggregated into a "fruiting body" (***ascocarp***) (Fig. 12–6), the dikaryotic ($n+n$) phase brief;
- *club fungi (Basidiomycota):* sexual spores (***basidiospores***) borne externally on a clublike sporangium (singular ***basidium,*** plural ***basidia***), typically 4 per basidium (Figs. 12–8D, 13–1E), this sometimes failing to produce spores, the basidia often aggregated into a "fruiting body" (***basidiocarp***) (Fig. 12–8), the dikaryotic ($n+n$) phase long.

There are other differences between these divisions, including such specialized structures as dolipore septa and clamp collections, which are present in many club fungi but absent in sac fungi. Also, club fungi often lack asexual spores and usually lack differentiated gametangia.

E. Ascocarp structure of sac fungi (Ascomycota)

See Figs. 12–6 and 12–7. Sac fungi are not as familiar as mushrooms (see Part III-F). Morels (*Morchella*) and truffles (*Tuber,* not chocolate), which are edible and highly esteemed, are examples of sac fungi having ascocarps, as is the parasitic ergot of rye pathogen (*Claviceps purpurea*) described in Supplement 12, Part II. Most yeasts (see Part II-B) are also sac fungi, but, being mainly unicellular (or also rarely filamentous), they do not have ascocarps. There are several main types of ascocarps, including:

- ***cleistothecia,*** globose and lacking openings;
- ***perithecia,*** flask-shaped and with an opening at the top;
- ***apothecia,*** cup-shaped or other shapes, but always open, with the asci exposed (Fig. 12–6).

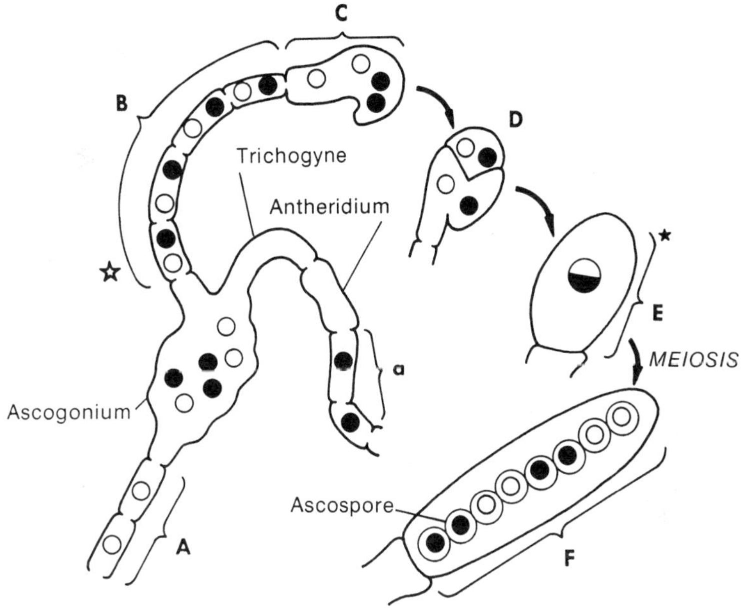

Fig. 12–7. Details of sexual reproduction in a sac fungus (Ascomycota); compare with Fig. 12–6. *a, A,* 1*n* hyphae of opposite mating strains; *B,* ☆, dikaryotic (*n*+*n*) hypha resulting from cytoplasmic fusion (plasmogamy) of syngamy (fertilization); *C, D,* early stages of ascus (crozier) formation; E, ★, nuclear fusion (karyogamy) of syngamy (fertilization); *F,* mature ascus (sporangium) with eight ascospores. (From Norstog & Long 1976:261.)

●●● See Fig. 12–6 of an apothecium. Then examine a permanent slide of a longisection of the apothecium of *Sclerotinia sclerotiorum* or a similar fungus. Note the asci (with ascospores) and sterile hyphae (paraphyses) at the top of the cup. Finally examine the DEMO slides of cultures of *Eurotium echinulatus* and *Sordaria* (wild type X gray mutants), or similar fungi. The ascocarps were crushed in preparing the slide to release ascospores from the cleistothecia (*Eurotium*) or to eject ascospores through the openings of the perithecia (*Sordaria*). How many ascospores does an ascus have?

●●● *Suggested diagram and labels: Eurotium* (a sac fungus) cleistothecium OR *Sordaria* (a sac fungus) perithecium: ascocarp (specify type), asci, ascospores. Draw crushed *and* uncrushed!

●●● Ascocarps of several species are on DEMO, including supermarket material of *Morchella esculenta* (morel). Morels are not mushrooms; each morel produces one highly convoluted apothecium.

●●● Examine the DEMO articles (from *Insight, The New York times,* and See's Candies) on truffles. Truffles are easily the most valuable fungi by weight, the Italian white variety costing as much as $1,750 a kilo (2.2 lb.). In France specially trained dogs and pigs are used to sniff out underground truffles in the wild. Truffle-scented mineral water recently went on sale for $4.00 a 285 g (10 oz.) bottle. However, other things are even more expensive, for example (closing prices on 8 Mar. 1989): saffron in Berkeley at $3.79/g (= $1,719.14/lb., $3,790.00/kg), gold in New York at $393.10/troy oz. (= $5,732.89/lb., $12,638.65/kg), cocaine on Berkeley campus (according to campus police) averaging $50.00/g (= $22,680.00/lb., $50,000.00/kg).

F. Basidiocarp (especially mushroom) structure of club fungi (Basidiomycota)

See Fig. 12–8. Mushrooms along with the related shelf (bracket) fungi (*Polyporus*), puffballs (*Calvatia, Lycoperdon*), stinkhorns (*Phallus*), earthstars (*Geastrum*), and bird's nest fungi (*Cyathus,*

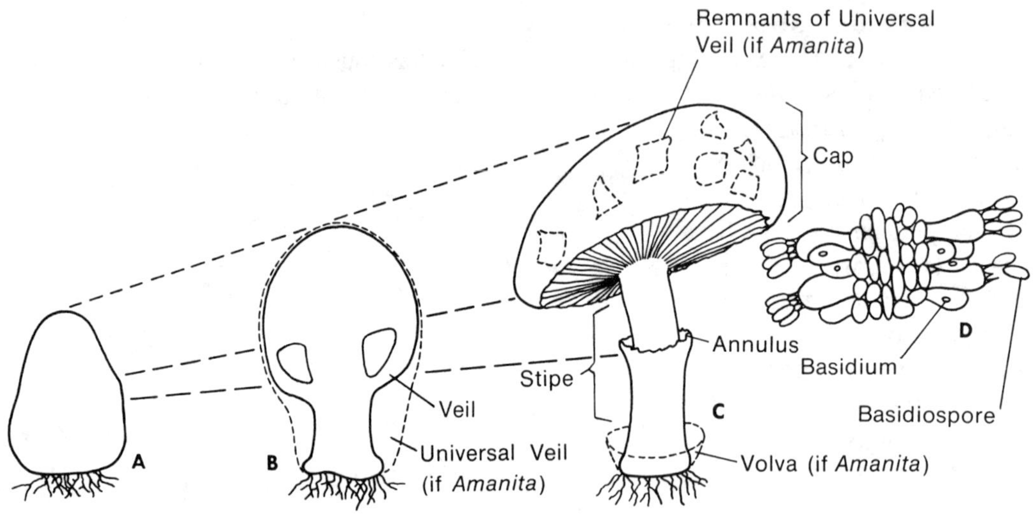

Fig. 12–8. Development of the basidiocarp of a mushroom, a club fungus (Basidiomycota). *A,* the early or button stage; *B,* later stage with partial veil; *C,* mature basidiocarp with stalk (stipe), ring (annulus), cap (pileus), and gills (lamellae); *D,* section of gill showing basidia, each bearing four basidiospores. Note the basal mycelium in *A-C.* The poisonous mushroom *Amanita* (death cap, destroying angel, etc.) has a second membrane, the universal veil (*B*); as the mushroom grows remnants of this veil are seen as a basal cup (volva) and as scales on the surface of the cap (*C*). (From Norstog & Long 1976:277.)

Nidularia) are all examples of club fungi having basidiocarps. In contrast, rusts (e.g., *Puccinia graminis,* wheat rust, Fig. 13–1) and smuts (*Ustilago maydis,* corn smut) are parasitic club fungi lacking basidiocarps. Some puffballs are very large. In spring 1992 a specimen of *Calvatia gigantea* was discovered in San Ramon, California; its circumference was 221 cm (87 in.), surpassing the old record by 25 cm (10 in.).

Mushrooms arise from a mycelium, a diffuse hyphal mass growing hidden in soil or rotting wood. The mycelium may expand evenly in all directions from a central point, up to about 30 m (101.7 ft.) in diameter or more (Raven et al. 1992:233). Mushrooms usually form at the outer edges of the mycelium, where there is the most abundant nutritive material for active growth. Mushrooms thus may appear in a circular formation, the "fairy rings" of folk legends dating to the Middle Ages. These are especially evident in lawns, golf courses, and grassy fields. The grass inside the fairy ring is noticeably greener than the grass outside because the former has been enriched by nitrogenous compounds from the dying and disintegrating mycelium. Mushrooms, that is, basidiocarps, develop very quickly, sometimes overnight, because for a long time the hyphae had been assimilating large amounts of nutrients and, concomitantly, producing new protoplasm underground. This then is available for growth of the new hyphae of the basidiocarps as they form above ground.

●●● *News flash!* The mycelium of a mushroom can achieve truly massive dimensions. Examine the DEMO article (from *Time,* 13 Apr. 1992) entitled "Humongous fungus: An underground blob may be the world's largest living creature." This organism, *Armillaria bulbosa,* occurs in northwestern Michigan and based on DNA analysis was estimated to be 1,500 years old, to extend over 15 hectares (37 acres), and to weigh between 90.7 and 907.2 metric tons (100 and 1,000 American tons).

A typical mature basidiocarp consists of the following structures (Fig. 12–8C):

- ***stalk*** (stipe);
- ***ring*** (annulus);
- ***cap*** (pileus);
- ***gills*** (lamellae) bearing the basidia, each basidium with four basidiospores.

The ring represents the remnants of the ***partial veil*** (velum), a membrane covering the gills before the basidiospores mature (Fig. 12–8B). As the mushroom matures, expansion of the cap ruptures the veil and exposes the gills and spores (Fig. 12–8C). Younger mushrooms are often more desirable from a marketing viewpoint because decay is postponed; thus most supermarket material has intact or recently ruptured veils. The ***button stage*** is the early mushroom (Fig. 12–8A). Some mushrooms, for instance, the infamous *Amanita* (death cap, destroying angel, etc.), have a second veil, the ***universal veil,*** covering the entire young mushroom (Fig. 12–8B). This veil persists in the mature mushroom as a basal ***cup*** (volva) and as distinctive ***scales*** on the cap (Fig. 12–8C).

●●● See Fig. 12–8A to C. Then examine the supermarket material of *Agaricus brunnescens* (cultivated white mushroom), the common (in the United States) edible mushroom [This is a new name, a recent change from *A. bisporus* (Bold et al. 1987)]. Cut a mushroom both longitudinally and transversely. Identify the following structures: cap (pileus), gills (lamellae), stalk (stipe), ring or partial veil of adult; button stage. There may be DEMO material of *Amanita* (identify any cup and scales) and other wild mushroom species available.

●●● *Suggested diagram and labels: Agaricus brunnescens* (cultivated white mushroom, a club fungus) basidiocarps: cap (pileus), gills (lamellae), stalk (stipe), ring or partial veil of adult; button stage. Note where the basidiospores are formed!

●●● See Fig. 12–8D. Then examine a permanent slide of a transection of the cap of *Coprinus* (inky cap mushroom) and identify the following structures (examine the oldest, largest sections on the slide): gills, club-shaped basidia, the stalks of the spores (sterigmata), and the basidiospores. How many basidiospores does each basidium bear? Also note, particularly in the central parts of the gills and the cap, that tightly woven hyphae form the structural basis for the various parts of the basidiocarp. This is ***pseudoparenchyma*** (Fig. 12–8D). Because mushrooms completely lack asexual spores, they spread exclusively via an advancing mycelium and especially via their massive production of the sexual basidiospores.

●●● *Suggested diagram and labels: Coprinus* (inky cap mushroom, a club fungus) cap x.s.: cap, gills, basidia, basidiospores.

●●● Basidiocarps of several species of club fungi are on DEMO, including supermarket material of *Lentinus edodes* (shiitake), a very edible mushroom species, and live and/or dried puffballs and shelf or bracket fungi (these are not mushrooms). Shelf fungi lack gills but instead have pores lined by their basidia. As evidenced by the DEMO drawings from Hanlin & Ulloa (1988:168–171), the order Phallales has many attractive fungi besides its most famous member *Phallus* (see Supplement 1, Part V). Incidentally, there is an angiosperm genus *Clitoria,* which is a legume.

IV. THE HAPLOID-DIPLOID LIFE HISTORY OF *ALLOMYCES,* A CHYTRID

The chytrids are a mostly aquatic group of unicellular to especially hyphal organisms. The chytrids are distinguished from other fungi by their motile cells (zoospores and gametes), each of which has a single, posterior (whiplash) flagellum. The chytrids have a variety of life histories, including the haploid-diploid ($1n$-$2n$) type (Fig. 12–9; see also Fig. Sup5–3), which is rare in fungi. As elaborated in Supplement 12, Part V, fungi with this type of life history have a $2n$ spore-producing "SPT" or sporothallus that alternates with a $1n$ gamete-producing "GPT" or gametothallus.

See Fig. 12–9 of the haploid-diploid ($1n$-$2n$) life history of *Allomyces.* Identify the sporo- and gametothallus. The female gametes secrete a hormone called sirenin that attracts the male gametes. The name of the hormone honors the Sirens of Greek mythology, who lured mariners to destruction with seductive singing. Most sources say the Sirens were part-woman and part-bird, but actually no one knows what they looked like because all persons who saw them never returned (Hamilton 1942).

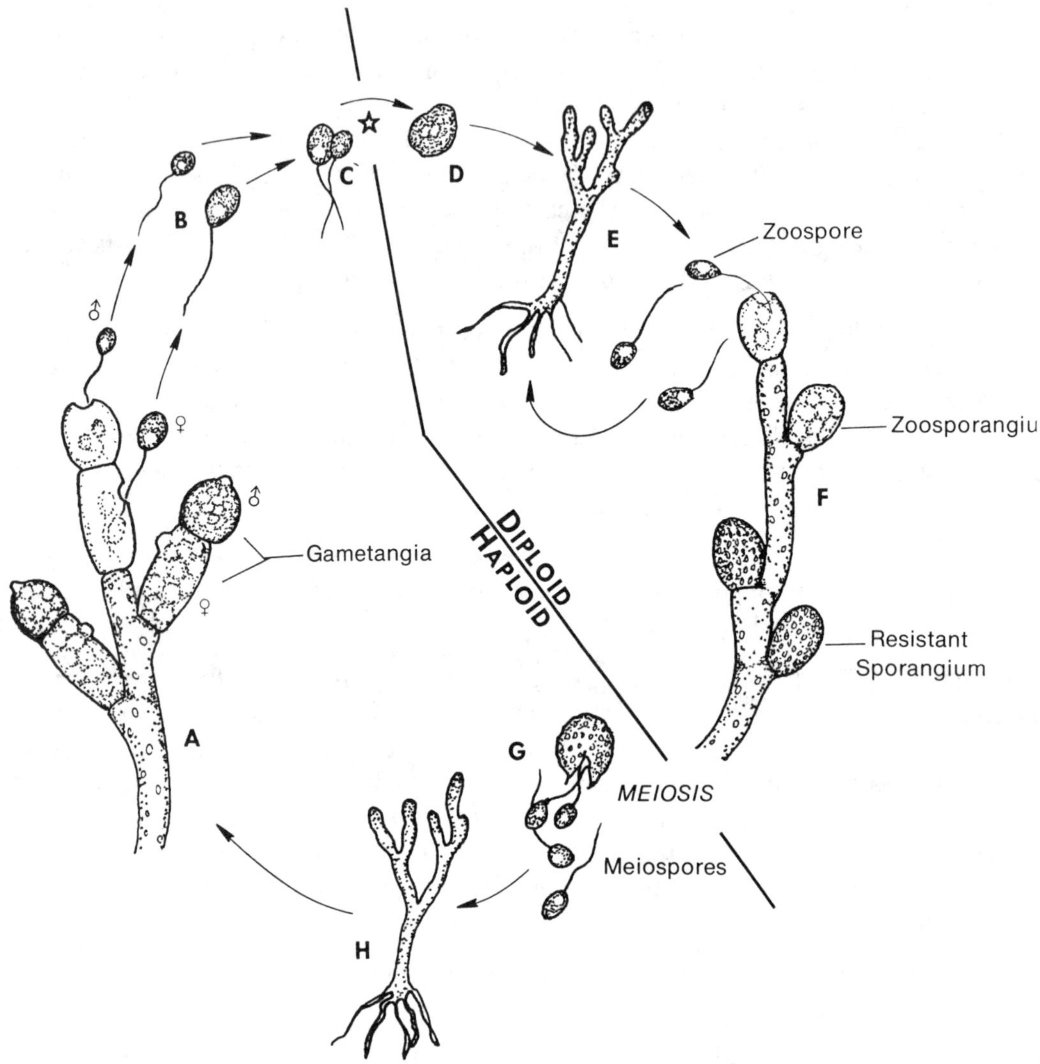

Fig. 12–9. Haploid-diploid (1*n*-2*n*) life history of the chytrid *Allomyces,* a coenocytic (tubular, siphonous) fungus (Chytridiomycota). *A,* mature 1*n* body (gametothallus) with male and female gametangia; *B,* gametes (anisogamy); *C,* ☆, syngamy (fertilization); *D,* zygote; *E,* young 2*n* body; *F,* mature 2*n* body (sporothallus) with asexual sporangia (zoosporangia, mitosporangia) and sexual resistant sporangia (meiosporangia); *G,* meiosporangium releasing zoospores (meiospores); *H,* young 1*n* body. Note the alternation of morphologically similar generations (phases). (From Norstog & Long 1976:256; redrawn and modified from R. Emerson, An experimental study of the life cycles and taxonomy of *Allomyces, Lloydia* 4:77–144, 1941.)

Important note: From the perspective of this course manual, the details of the life histories *of fungi* above and in the figures are *not* especially important. What *is* important are the effects of meiosis, syngamy (fertilization), and mitosis on the ploidy level of various structures in the life history, to wit:

- 2*n* structures ——— meiosis ———> 1*n* structures
- 1*n* structures ——— syngamy ———> 2*n* structures
- 2*n* structures ——— mitosis ———> 2*n* structures
- 1*n* structures ——— mitosis ———> 1*n* structures.

For elaboration see Supplements 4 and 5.

A stein of beer, pretzels, and me.—Rudi Schmid

NAME ______________________________ LAB SECTION # ______________ DATE ____

TITLE ______________________________

__________________ TOTAL MAG ____

TITLE ______________________________

__________________ TOTAL MAG ______

TITLE ______________________________

__________________ TOTAL MAG ____

TITLE ______________________________

__________________ TOTAL MAG ______

LAB EXERCISE **13**

Fungi II

(Fungal Interactions with Other Groups of Organisms, Especially the Wheat Rust Pathogen and Lichens)

OBJECTIVE

To examine various fungal interactions with organisms, especially (1) the complex life history of the wheat rust pathogen (*Puccinia graminis*) and (2) the structure of lichens, a fungal-algal symbiosis.

Note: Read this material *before* coming to the lab; there is insufficient time in the lab to read the material from scratch. Due to the large amount of background information presented here, a ●●● is used to mark those paragraphs calling for actual observations or diagrams in the lab. Most unmarked paragraphs reiterate information presented in the lecture. A bullet (●) signifies an important term or concept.

Suggested Lab Diagrams (see part noted for suggested labels):
Puccinia graminis, any one stage (Part I)
Sticta (a lichen) thallus x.s. (Part II-C)
Mycorrhizal fungus in *Pinus* root x.s. OR *Lycopodium* GPT l.s.

PERSPECTIVE

The fungi exhibit various symbiotic and non-symbiotic interactions (see Supplement 11) with plants, animals, protists, and other fungi. To reiterate from Supplement 11, Part III, the close, symbiotic associations include:

- ***parasitism*** (+ −), where the association is harmful to one of the organisms (e.g., ergot of rye, *Puccinia graminis* (wheat rust) parasitic on wheat and barberry—see Part I, and many other fungal parasites);
- ***mutualism*** (+ +), where the association is advantageous to both organisms (e.g., mycorrhizae—see Part III-A);
- ***commensalism*** (+ 0), where one organism is benefitted but the other is neither stimulated nor inhibited (e.g., lichens according to recent interpretations—see Part II).

In contrast, the loose, non-symbiotic associations of fungi with other organisms include carnivorous or predaceous fungi, the fungal gardens of insects, and wood-rotting fungi (see Part III).

I. THE COMPLEX LIFE HISTORY OF THE WHEAT RUST PATHOGEN (*PUCCINIA GRAMINIS*)

Puccinia graminis, the pathogen causing black stem rust of wheat, has the most complex life history of any plant or of any fungus (except other rusts, of course).

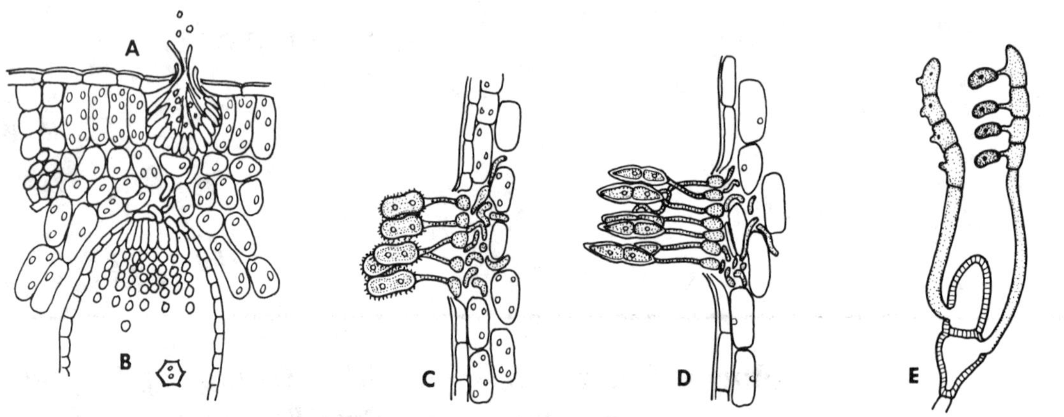

Fig. 13–1. Haploid (1*n*) and dikaryotic (*n*+*n*) life history of *Puccinia graminis* (wheat rust), a club fungus (Basidiomycota) parasitic on two hosts, the dicotyledon *Berberis vulgaris* (barberry) (*A, B,* transection of leaf), and the monocotyledon *Triticum aestivum* (wheat) (*C, D,* transections of stems). *A,* 1*n* spermagonium with chainlike spermatia; *B, n*+*n* aecium with chains of aeciospores; *C, n*+*n* uredium with uredospores; *D, n*+*n* telium with teliospores; *E,* basidia with 1*n* basidiospores. (From Norstog & Long 1976:279; redrawn from *Cryptogamic botany,* 2nd ed., vol. 1, by G. M. Smith, © 1955 by McGraw-Hill Book Co., Inc., New York. Reprinted by permission of the publisher.)

A. General

The fungus actually infects two hosts, wheat (*Triticum aestivum*) and barberry (*Berberis vulgaris* and other species). These hosts are needed to complete the life history. Wheat is a monocotyledon whereas barberry, the alternate host, is a dicotyledon. Actually, different strains of *Puccinia graminis* also parasitize other cereals (e.g., barley, oats, rye) and various species of wild grasses. Other species of rusts alternate between dicotyledonous hosts and fern or gymnospermous hosts.

●●● See Fig. 13–1 of the haploid (1*n*) and dikaryotic (*n*+*n*) life history of *Puccinia graminis*. Then examine the DEMO diagrams (from Raven et al. 1992:236–237 and Weier et al. 1982:532–533) showing in more detail the life history as you examine the material in the following sections. Some dried and live material and a poster on control of the pathogen may also be on DEMO.

●●● *Suggested diagram and labels: Puccinia graminis,* any one stage. Specify the host and label the relevant parts of the stage (for parts see Fig. 13–1 and the account below).

B. Spermagonia and aecia on barberry

See Fig. 13–1A, B. Barberry bears two stages of the fungal parasite, spermagonia (also called "pycnia," an old name) and aecia: In early spring 1*n* basidiospores are spread by wind from wheat and land and germinate on the adaxial ("upper") leaf surfaces of barberry. Within a few days the resulting hyphae proliferate in the leaf and produce flask-shaped gametangia called ***spermagonia*** (plural, singular ***spermagonium***) (Fig. 13–1A). Each spermagonium produces chains of minute uninucleate conidialike ***spermatia*** and long hyphae called ***receptive hyphae.*** The sweet liquid (improperly called "nectar," a term that should be reserved for the floral nectar of angiosperms) exuded by the spermagonia attracts flies and other insects, which carry spermatia to other spermagonia. In mid spring the spermatia will fuse with the receptive hyphae of an opposite mating strain. A dikaryon (a *n*+*n* cell) results through cytoplasmic fusion (plasmogamy); however, as in other sac and club fungi (Ascomycota and Basidiomycota), nuclear fusion (karyogamy) does not occur immediately, but is delayed. The dikaryon proliferates into a dikaryotic (*n*+*n*) hypha.

Rather quickly cuplike structures called ***aecia*** (plural, singular ***aecium***) are produced on the abaxial ("lower") leaf surfaces of barberry (Fig. 13–1B). These aecia result from the dikaryotic (*n*+*n*) hyphae arising from the aforenoted plasmogamy. How does the fungus get from the adaxial to the abaxial side of the barberry leaf, the latter surface bearing the aecia? Is there any evidence in the DEMO slide for this? The aecia produce chains of binucleate ($n+n$) ***aeciospores,*** which are conidia.

In summary, the spermagonia represent the sexual (plasmogamic) stage on barberry that produces aecia, whereas the aecia, which occur on barberry, represent the asexual infecting stage of wheat.

●●● See Fig. 13–1A, B. Then examine the DEMO permanent slide of the two fungal stages on barberry, namely, spermagonia and aecia. Identify the following structures: adaxial leaf surface, spermagonia, receptive hyphae, spermatia, abaxial leaf surface, aecia, aeciospores.

C. Uredia on wheat

See Fig. 13–1C. In late spring and early summer aeciospores spread by wind to wheat and there germinate on the stems and leaves to produce pustulelike lesions called ***uredia*** (plural, singular ***uredium;*** also uredinia, uredosori). The uredia produce rather large, spiny, binucleate ($n+n$) ***uredospores*** (urediniospores), each of which is dikaryotic ($n+n$). Wind carries these spores (i.e., more conidia) to other wheat plants. The uredospores have been called "summer spores," and the uredial stage has also been called the "red stage" of the infection because the uredospores en mass are red. In summary, the uredia represent the asexual repeating (i.e., spreading) stage on wheat.

●●● See Fig. 13–1C. Then examine the DEMO permanent slide of uredia on wheat. Identify the uredia and uredospores.

D. Telia on wheat

See Fig. 13–1D. In mid to late summer when the wheat matures, the pustules that produced the uredospores now form ***teliospores*** (teleutospores). The former uredia are now called ***telia*** (plural, singular ***telium;*** also teliosori); this is also called the "black stage" of the rust due to the black telia. Each teliospore is red-brown, large, thick-walled, and two-celled. During winter nuclear fusion (karyogamy) occurs in the teliospore. Hence each one is at first dikaryotic ($n+n$) but, after nuclear fusion, diploid ($2n$). In summary, the tilia represent the sexual (karyogamic) overwintering stage on wheat.

●●● See Fig. 13–1D. Then examine the DEMO permanent slide of telia on wheat. Identify the telia and teliospores.

E. Basidia on wheat

See Fig. 13–1E. Early in the following spring each of the two $2n$ cells of a teliospore germinates and undergoes meiosis, producing a $1n$ ***basidium*** (singular, plural ***basidia;*** also basidial apparatus, promycelium, etc.) bearing four $1n$ ***basidiospores*** that are spread to the barberry. The life history is complete. In summary, the basidia represent the sexual (meiotic) stage on wheat that produces basidiospores infecting barberry.

●●● See Fig. 13–1E. Then examine the DEMO permanent slide of teliospores germinated into basidia. Each basidium bears four basidiospores, but many basidiospores may be dislodged (respectively, right and left sides of Fig. 13–1E). Identify the teliospores, basidia, and basidiospores. If basidia of *Puccinia graminis* are unavailable, substitute material may be used, for instance, the rust *Gymnosporangium clavipes,* which alternates between *Juniperus* (juniper) and *Malus* (apple or crab apple). "Cedar apples" are the familiar reddish balls that ornament junipers and consist of galls penetrated by mycelia bearing telia.

F. Précis of the life history

The following summary of the haploid ($1n$) and dikaryotic ($n+n$) life history of the wheat rust pathogen (Fig. 13–1) nicely shows the considerable time, some eight to ten months, separating cytoplasmic fusion (plasmogamy) and nuclear fusion (karyogamy):

- cytoplasmic fusion (plasmogamy) in mid spring in the spermagonia (Fig. 13–1A);
- nuclear fusion (karyogamy) during the following winter in the tilia (Fig. 13–1D);
- meiosis early in the following spring in the tilia.

1. Two hosts—wheat (*Triticum aestivum*) and barberry (*Berberis vulgaris* and other species);
2. ***Spermagonia*** on adaxial leaf surface of barberry (in early spring);
   a. *Activity:* 1*n* basidiospores spread by wind to barberry and germinate to produce flask-shaped gametangia (***spermagonia***) that form ***spermatia*** and ***receptive hyphae;*** sweet liquid (improperly called "nectar") exuded by the spermagonia attracts insects, which carry spermatia to other spermagonia; spermatia and receptive hyphae of opposite mating strains undergo *cytoplasmic fusion (plasmogamy)*; the resultant dikaryon (a $n+n$ cell) becomes a dikaryotic ($n+n$) hypha;
   b. *Function:* the sexual (plasmogamic) stage on barberry that produces aecia;
3. ***Aecia*** on abaxial leaf surface of barberry (in late spring);
   a. *Activity:* cuplike ***aecia*** produce $n+n$ ***aeciospores*** (i.e., conidia) that spread by wind to wheat;
   b. *Function:* the asexual infecting stage of wheat via aeciospores, which produce uredia;
4. ***Uredia*** on wheat (in late spring and early summer);
   a. *Activity:* pustulelike ***uredia*** produce $n+n$ ***uredospores*** (i.e., more conidia) that spread by wind to more wheat;
   b. *Function:* the asexual repeating stage on wheat via uredospores, which produce more uredia;
5. ***Telia*** on wheat (in mid to late summer to the following early spring);
   a. *Activity:* as the wheat matures the uredia become ***telia*** and produce $n+n$ ***teliospores,*** which during winter undergo *nuclear fusion (karyogamy)* to become 2*n;*
   b. *Function:* the sexual (karyogamic) overwintering stage on wheat via teliospores, which produce basidia;
6. ***Basidia*** on wheat (in the following early spring);
   a. *Activity:* teliospores undergo *meiosis* on germination to produce 1*n* ***basidia*** (two per teliospore), each basidium bearing four 1*n* ***basidiospores;***
   b. *Function:* the sexual (meiotic) stage on wheat infecting barberry via basidiospores, which produce spermagonia.

••• *Puccinia graminis* has a haploid (1*n*) and dikaryotic ($n+n$) life history. Identify on Fig. 13–1 and the DEMO life history diagrams the 1*n*, 2*n*, and dikaryotic ($n+n$) phases of this organism. How does this type of life history differ from those of other fungi and those of plants?

G. Biological control

Wheat rust is a serious pathogen. Many efforts have been made to control the disease (Alexopoulos & Mims 1979; Bold et al. 1987). With two hosts, eradicating barberry, the less valuable host, can be very effective. French farmers recognized this and in 1660 had enacted a law decreeing eradication of barberry, but details of the life history were not clarified until 1865 by Anton de Bary, the great German mycologist. Disrupting the life history of *Puccinia graminis* by eradicating barberry works in Eurasia because severe winter weather here kills the fungus on the wheat and high east-west mountain ranges prevent northern movement of southern-based uredospores. Thus barberries are necessary for survival of the fungus in Eurasia. However, barberry eradication only lessens the severity of the disease in the United States and Canada, because uredospores overwinter in Mexico and the southern United States and in spring travel 1,600 km (1,000 mi.) or so and infect new wheat plants without needing the barberry; that is, the sexual, especially fusion (plasmogamic) stage of the life history is circumvented. In North America other means of control are the use of early-maturing varieties of wheat in the northern states and breeding rust-resistant varieties of cereals. Alas, *P. graminis* has over 200 races to which wheat varieties are differentially susceptible, and, moreover, genetic diversity of *Puccinia* can arise by nuclear exchange during the dikaryotic ($n+n$) uredinial stage.

II. LICHENS, A FUNGAL-ALGAL SYMBIOSIS

Lichens are an association between a fungus, the ***mycobiont,*** and a green alga or cyanobacterium, the ***phycobiont.*** The fungus is usually a member of the sac fungi, rarely of the club or imperfect fungi (other fungal divisions are not involved). Some lichen species have both a green algal and a blue-green bacterial (cyanobacterial) component. Therefore, lichens involve five divisions of unrelated organisms, respectively: Ascomycota, Basidiomycota, Deuteromycota, Chlorophyta, Cyanophyta. Lichens have traditionally been grouped with the fungi because (1) in most lichen species the morphology of the fungus determines the morphology of the entire lichen, and (2) a single species of phycobiont might reside in several morphologically distinct lichen thalli.

The lichens with their fungal-algal or fungal-cyanobacterial relationships have historically been considered an example of mutualism. However, recent studies cast doubt on the direct advantage of the relationship to the phycobiont (algal or cyanobacterial) component. Thus this has been regarded as "controlled parasitism" (Raven et al. 1992:228), but strictly speaking it is a case of commensalism.

There are some 18,000–20,000 species of lichens (Bold et al. 1987:801), and these occur in very diverse habitats ranging from deserts to polar regions, including a few species living in fresh and salt water with periodic exposure to air. Lichens can survive temperatures of −198° C (−388.4° F) and can photosynthesize actively at −18.5° C (−65.3° F) (Bold et al. 1987:802). Lichens living on rock are important in soil formation. In Antarctica some lichens live *in* rock just beneath the surface of the sandstone, where winter temperatures are −60° C (−140° F).

A. The secret life of Beatrix Potter

The dual organismal nature of lichens was discovered in 1867 by a Swiss botanist, Simon Schwendener, and quickly accepted as truth by German botanists. The British, however, for a long time denied this relationship, one botanist (J. M. Crombie) even dismissing it as an "unnatural union between a captive Algal damsel and a tyrant Fungal master" (Gilpatrick 1972).

Beatrix Potter (1866–1943), the famous illustrator and author of children's books, notably *The tale of Peter Rabbit* (1901), during her childhood became deeply involved in painting fungi and lichens and in this way started to make scientific observations on them. By 1894 she actually had good scientific evidence for the symbiotic nature of lichens. However, this idea and her other work on fungi was not readily accepted by the British botanists due to their prejudice against Schwendener's theory and substantial anti-feminism with regard to women in science. Eventually, with the help of her uncle Sir Henry E. Roscoe, a chemist knighted for his scientific work, Potter managed to get her work on fungal spores read (not published) by the Linnean Society. However, even here there were two ironies: The paper was read by a man, because women were not allowed at meetings of the Society, and it was read on, of all dates, April 1st 1897. Potter withdrew her paper from publication because she wanted to do further research on fungi and lichens. However, frustrated with plant biology, Potter abandoned the infamous botany for her famous bunny.

●●● A DEMO article by Gilpatrick (1972) on "the secret life of Beatrix Potter" elaborates on the above story. Note that the mycological pictures in this article are of mushrooms, *not* of lichens.

B. Growth habits

The lichen thallus exhibits four growth habits:

- ***"leafy"*** (foliose), with a leaflike organization ("'leafy' lichens");
- ***"encrusting"*** (crustose), with a crustlike organization ("'encrusting' lichens");
- ***"shrubby"*** (fruticose), with a branching, cylindrical organization ("'shrubby' lichens");
- ***"scaly"*** (squamulose), with a scalelike organization ("'scaly' lichens").

Some authors (e.g., Bold et al. 1987) do not recognize the "scaly" type.

In the large family Cladoniaceae each species has two growth habits. Initially a strictly vegetative "encrusting," "leafy," or "scaly" primary thallus forms. Then a "shrubby" secondary thallus arises vertically from the primary thallus. The secondary thallus may be simple or branched, always bears the reproductive structures (apothecia, etc.), and always characterizes the species. Two famous species of *Cladonia* are *C. cristatella,* a cosmopolitan species called "British soldiers" because of the bright red apothecia borne on the long erect ("shrubby") columns arising from a "leafy" base, and *C. rangifera* and related species, the "reindeer mosses" of northern latitudes that form the main diet of reindeer (caribou, *Rangifer tarandus*).

Special anchoring and water-absorbing hyphae called ***rhizines*** enter the substratum from the lower surface of the thallus. Rhizines, which are basically functionally like the rhizoids of yore, are complex bundles of hyphae that have anastomosed by fusion of young hyphal tips.

•• Examine the DEMO diagram (from Bold et al. 1987:806) of growth habits of lichens (this figure shows the first three types listed above). Then examine the DEMO photographs (source unknown) of lichens and the DEMO material of live and/or dried lichens. Finally, using the DEMO diagram as a guide, determine the type of growth habit of each lichen and note if any rhizines are present. *Ramalina menziesii* (= *R. reticulata;* lace lichen, California Spanish moss) is common on trees in fog zones along coastal California, especially around Monterey. *Letharia columbiana* is a spectacular, lemon-green, highly branched lichen common on tree trunks in the Sierra Nevada. In material of *Cladonia* note the aforementioned aspects of the dual growth habit.

C. Internal structure

In the lichen thallus the phycobiont and mycobiont components may be more or less evenly distributed or, usually, the phycobiont forms a distinct layer within the thallus. In the latter case three or four layers are generally distinguishable (the thallus type is pseudoparenchymatous):

- the ***upper cortex,*** the more compact outer and upper layer of the fungus;
- the ***algal layer*** (not surprisingly the photosynthesizing phycobiont occurs near the surface); this is the standard term, although the phycobiont may be either a green alga or a blue-green bacterium (cyanobacterium);
- the ***medulla,*** a loose middle layer of the fungus;
- the ***lower cortex,*** the more compact outer and lower layer of the fungus, which is often absent.

All four growth habits noted above may have such layering. In addition, rhizines may extend from the lower surface of the thallus.

••• Reexamine the DEMO diagram (from Bold et al. 1987:806) of growth habits and the anatomy of lichens, which also shows sections through rhizines. Then examine the DEMO permanent slide of the "leafy" (foliose) lichen *Sticta pulmonaria,* or a similar lichen, and identify the following parts of the thallus: upper cortex, algal layer, medulla, lower cortex, and rhizines. Species of *Sticta* have as the phycobiont either the blue-green bacterium *Nostoc* or the green alga *Trebouxia,* both of which are the most common phycobionts of all lichens. How might you tell which phycobiont is present on the slide of *S. pulmonaria?* Is the thallus of this lichen composed of pseudoparenchyma or parenchyma?

••• *Suggested diagram and labels:* *Sticta* (a lichen) thallus x.s.: upper cortex, algal layer, medulla, lower cortex, rhizines. Note that the cortex, medulla, and rhizines are fungal!

D. Reproduction

Many lichens reproduce asexually by fragmentation of the thallus. Also produced may be various types of specialized asexual structures that consist of *both* mycobiont and phycobiont. Thus microscopic masses of algal cells enveloped by fungal hyphae are dispersed by wind, animals, or splashing raindrops. In addition, the phycobiont and mycobiont components of the lichen may reproduce asexually *separately* by spores or other means.

Only the mycobiont reproduces sexually. This is by ascocarps, which are mainly apothecia and perithecia, or by basidiocarps in those few species of lichens having a club fungus (Basidiomycota) as the mycobiont. The imperfect lichens where an imperfect fungus (Deuteromycota) is involved have, of course, only asexual spore production.

●●● Reexamine the DEMO diagram (from Bold et al. 1987:806) of growth habits of lichens and note the apothecia in Fig. B of the DEMO. Then, especially with a dissecting microscope, reexamine the DEMO material of lichens and identify any ascocarps present. *Letharia* and some other species have conspicuous black apothecia. As noted above, *Cladonia cristatella* (British soldiers) has red apothecia borne on long erect columns.

III. OTHER TYPES OF FUNGAL INTERACTIONS WITH ORGANISMS

There are several other types of symbiotic and non-symbiotic associations involving fungi such as mycorrhizae, carnivorous or predaceous fungi, wood-rotting and wood-staining fungi, and the fungal gardens of insects.

A. Mycorrhizae

Mycorrhiza (singular, plural ***mycorrhizae***) is commonly defined as "a symbiotic association between certain fungi and plant roots" (Raven et al. 1992:751), but this fungal/plant association involves not only roots (Fig. 13–2) but also underground stems (rhizomes) and, in pteridophytes, underground GPTs (Fig. 5–3B). "Mycorrhizal" is usually applied to all these conditions, which exemplify mutualism. Mycorrhizae are very important to the ecology of trees; most tree species seem to have mycorrhizae.

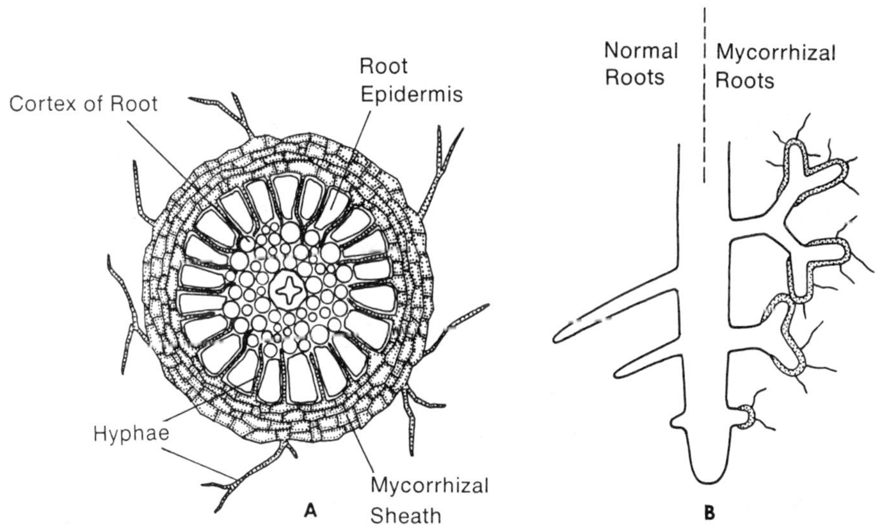

Fig. 13–2. Mycorrhizal fungi on roots of higher plants. *A,* transection of root with intercellular mycorrhiza (ectomycorrhiza); *B,* overview showing normal and mycorrhizal roots of *Pinus* (pine). In intercellular mycorrhizae the fungal cells live around and/or between the plant cells but do not penetrate them. (From Norstog & Long 1976:281; redrawn from *The biology of mycorrhizae,* 2nd ed., by J. L. Harley, © 1969 by Leonard Hill, London. Reprinted by permission of the publisher.)

Mycorrhizae are of two structural types:

- ***intercellular mycorrhizae*** (ectomycorrhizae), the fungal cells living around and/or between the plant cells, but *not* penetrating them;
- ***intracellular mycorrhizae*** (endomycorrhizae), the fungal cells living inside the plant cells and thus penetrating them.

Intercellular mycorrhizae are less common, involve many species of fungi (sac and especially club fungi—Ascomycota, Basidiomycota), and occur in roots, which are usually swollen (due to an external fungal mantle) and lack root hairs in infected regions (Fig. 13–2B). In contrast, intracellular mycorrhizae are more common, involve very few species of fungi (club fungi and mostly bread molds—Basidiomycota, Zygomycota), and occur in roots, rhizomes, and even GPTs (Fig. 5–3B).

●●● See Fig. 13–2. Then examine the DEMO live material of mycorrhizal roots of *Pinus* (pine) and note the swollen roots (Fig. 13–2B). Finally examine the DEMO permanent slide of a root transection of *Pinus* showing its intercellular mycorrhizal association. The inner part of the root is unaffected and thus reflects the normal anatomy of a root examined in Lab Exercise 7, Part I-B. Therefore, this is a good time to review basic root anatomy for a possible lab exam. In other words, the mycorrhizal root consists of two parts, an inner part unaffected by the fungus and an outer part affected by the fungus. The normal, unaffected part of the root has the following structures, as seen centrifugally (Figs. 7–2A, 13–2A):

1. central, continuous mass of primary xylem consisting of larger diameter cells centrally and several (3–5) lobes of smaller diameter cells peripherally;
2. discontinuous patches (3–5) of primary phloem located in the bays of the xylem arms;
3. single layer of thin-walled parenchyma cells comprising the pericycle;
4. endodermis with its casparian strips on the radial cell walls;
5. inner part of cortex consisting of parenchyma cells often containing much storage starch.

A pith is lacking. The outer part of the root affected by the fungus has (Fig. 13–2A)

- a peripheral ***fungal sheath*** (mantle) that sends (intercellular *not* intracellular)
- hyphae between the epidermal and cortical root cells to form a network (Hartig net) in the first outer layers of the cortex, rarely deeper (Fig. 13–2A).

The network in the intercellular spaces of the cortex appears beaded. Locate on the DEMO slide the structures noted above. To reiterate, the inner cortex and the vascular cylinder, that is, the blue-green staining tissue inside the thick-walled endodermis, lack the fungus. What anatomical feature probably prevents further penetration of the fungus into the host?

●●● See Fig. 5–3B. Then examine the DEMO permanent slide of a subterranean GPT of *Lycopodium* (club moss) showing its intracellular mycorrhizal association. Locate the dark-staining fungal region at the edge of the GPT.

B. Carnivorous or predaceous fungi

Non-symbiotic associations include the fungal gardens of insects (Batra & Batra 1967) and the carnivorous or predaceous fungi (Hauser 1984; Thorn & Barron 1984), which consume nematodes, or roundworms, found in soil and fresh and salt water. There are over 150 species of carnivorous fungi, which are mostly imperfect fungi (Deuteromycota), occasionally water molds (Oomycota), chytrids (Chytridiomycota), bread/etc. molds (Zygomycota), and club fungi (Basidiomycota) (Hauser 1984). Ten species of gill mushrooms attack and consume nematodes, perhaps to supplement low levels of nitrogen available in wood (Thorn & Barron 1984).

The trapping devices of the fungi are varied (Hauser 1984): adhesive knobs, lateral branches, sticky nets, constricting rings, and non-constricting rings. The fungus penetrates the animal by a haustorial hypha and then swells to produce a spherical vesicle (mortiferous excrescence, infection bulb) that produces hyphae to digest the animal's guts.

●●● Examine the DEMO cartoon and photograph (from Carolina Biological Supply Co., 1986 advertisement, or from another available source) and the DEMO article (Hauser 1984) of the fungus *Arthrobotrys connoides* (a member of the imperfect fungi) with its nematodic *Rhabditis* prey. Some actual culture material of this or a similar species might be available on DEMO. *Arthrobotrys* has constricting rings, each of which consists of three curved cells on a short hyphal stalk.

C. Wood-rotting and wood-staining fungi

For convenience wood-rotting fungi are included in this treatment, but certainly here neither symbiosis, disease, or even an interaction of a fungus with an organism is really involved because wood is largely non-living tissue when it is functional. In other words, dead things do not suffer diseases. However, if a wood-rotting fungus occurs on or in the non-living parts of a living tree, eventually weakens it, and causes its death by falling, then by some definitions "disease" and "parasitism" would be involved.

Fungi attacking wood cause considerable economic loss (Panshin & de Zeeuw 1980). Wood-destroying or wood-rotting fungi obtain the nourishment for their growth and reproduction by the enzymatic degradation of cell walls of wood, thus causing the breakdown (decay) of wood. Brown and white rots are caused chiefly by club fungi (Basidiomycota), whereas soft rots are caused by sac (Ascomycota) and imperfect fungi (Deuteromycota). The cell walls of wood consist of the carbohydrate cellulose, a phenolic compound lignin, and other materials (see Lab Exercise 6, Part II-D2a).

- Brown rots attack primarily cell wall carbohydrates and leave behind a network consisting of modified lignin and other materials. The affected wood becomes brown.
- In contrast, white rots digest both cellulose and lignin and leave behind a spongy or stringy mass. The affected wood becomes white, grayish white, or sometimes yellow or light brown.

In both types of decay, hyphae penetrate the cell walls through the pit-pairs or by forming boreholes (minute holes caused by enzymatic action) and by ramifying through the cavities of the wood cells. The hyphae in the cell cavities liberate enzymes that diffuse into the cell walls and degrade them.

Compared to wood-rotting fungi, ***wood-staining fungi*** obtain most of their nourishment from materials stored in cell cavities and hence have rather little disintegrating effect on the wood material (i.e., cell walls) itself. Therefore, there is here no controversy with wood-staining fungi about the occurrence of "disease" and "parasitism."

●●● Examine any DEMO material available of brown rots, white rots, or wood-staining fungi.

IV. THE KEROSENE FUNGUS, *CLADOSPORIUM RESINAE*

Cladosporium resinae is a member of the sac fungi (Ascomycota). For over 75 years this fungus has been known to live on plant resin and on various petroleum products such as creosote, kerosene, and diesel fuel. In fact, it is commonly referred to as the kerosene fungus. The fungus lives on the water in the fuel.

The fungus in its asexual conidial stage is known from Europe, North America, and Australia. The sexual stage, a cleistothecium, is known only from lab culture, but was induced in many isolates from Australia, Britain, France, and Germany. The asexual conidial stage is called *Cladosporium resinae;* the sexual stage is called *Amophotheca resinae;* the entire life history stage is called *Hormoconis resinae* (this summary is condensed from Alexopoulos & Mims 1979:289).

●●● The DEMO photographs (from Parbery 1969:356–357) show stages of sexual reproduction (ascocarps, Figs. 29 to 32) and asexual reproduction (conidia, Figs. 33 to 39). The single DEMO photograph (from Schmid, unpublished) is of its asexual conidial stage found growing in diesel fuel. The DEMO glass jar contains the fuel filter (do *not* open) from a 1977 VW Diesel Rabbit showing most of the filter clogged with the fungus. This little DEMO cost me $96 in February 1990. The VW technician thought the problem was due to an alga. How would you tell the difference between an alga and a fungus?

THEY HAD "CEILING PRICES"
ON A CUP OF BEER IN BABYLON
4000 YEARS AGO!

Beer was so popular 4000 years ago that King Hammurabi found it necessary to impose wise regulations on the price that Bit-Sikari (ancient taprooms) might charge for it.

C. SCHMIDT & SONS, INC. • IN PHILADELPHIA SINCE 1860

Schmidt's
Beer & Ale
SINCE 1860 • SINCE 1774*
*Predecessor: Rob't Smith Ale Brewery

BUY WAR BONDS FOR KEEPS

CARRYING ON THE FINEST TRADITIONS OF BREWING IN ONE OF AMERICA'S MOST MODERN BREWERIES

© 1945, C. S. & SONS, INC.

Fig. 13–3. Beer has been brewed for at least 5,000 years. Testifying to the honorable tradition of beer is this 1945 advertisement of C. Schmidt & Sons, Philadelphia, which alas is not my family business.

NAME ______________________________ LAB SECTION # ______________ DATE ____

TITLE ______________________________

__________________ TOTAL MAG _____

TITLE ______________________________

__________________ TOTAL MAG ______

TITLE ______________________________

__________________ TOTAL MAG _____

TITLE ______________________________

__________________ TOTAL MAG ______

SUPPLEMENT 13

Inside Herbs, Spices, and Drugs

(An Anatomist's View)

I. PERSPECTIVE

Plants and fungi produce various chemicals that serve as attractants, rewards, and deterrents (see definitions below). Humanity has exploited these chemicals, which occur in herbs, spices, and drugs. In many cases distinct structures, called ***secretory structures,*** actively secrete these chemicals.

II. DEFINITIONS OF "HERB," "SPICE," AND "DRUG"

First one might ask: What is an herb? What is a spice? Thus, *Origanum* (oregano) is an herb, *Syzygium aromaticum* (clove) a spice. These distinctions are often made for ***herbs*** versus ***spices:***

- *Herbs* are more common, generally temperate in origin, rather easy to grow in one's garden, whereas *spices* are more exotic, often tropical in origin, more difficult to grow. However, this distinction reflects the cultural and linguistic biases of European cultures.
- *Herbs* tend to come from herbaceous or semi-woody plants, *spices* from woody shrubs and trees.
- *Herbs* tend to involve leaves and flowers, *spices* stems, seeds, and fruits.
- *Herbs* are generally milder in flavor and action than *spices.*

Obviously, there is no strict distinction between herbs and spices, and there are many exceptions to the aforenoted generalizations. For instance, many herbal teas come from trees. From a plant anatomical viewpoint, there is also no anatomical definition of herbs and spices.

Just as there is a rather thin line between herbs and spices, there is also a rather thin line between herbs and especially spices, on the one hand, and drugs and poisonous plants, on the other hand. *Atropa* (belladonna), *Cannabis,* or *Digitalis* (foxglove) obviously are *drug plants,* that is:

- ***Drugs*** have properties or effects that may be medicinal (real or assumed), psychoactive (including hallucinogenic), and/or poisonous.

However, the spice clove is also a drug as it has various mild medicinal qualities (e.g., aromatic stimulant, antispasmodic, astringent, antiemetic, carminative, etc.), but no hallucinogenic ones.

III. SECRETION PRODUCTS AND THEIR USES

In the case of herbs and spices, the chemicals secreted are largely in the form of hydrocarbons of a complex (usually terpenoid) nature. There are basically three categories:

volatile oils, also called aromatic or essential oils;
liquid to more or less solid, amorphous products, that is, gums, resins, mucilages;
various chemicals such as alkaloids and glycosides.

These volatile oils and liquid to solid compounds can occur together, as with oleoresins, or oleogums. Drugs often derive their qualities from alkaloids and glycosides.

It seems appropriate to ask what are some uses of secretory products in general from the perspective of people, other animals, and plants:

USES BY PEOPLE:

- because of their taste, obviously as a condiment, as a flavoring agent, and historically to mask bad odors of decaying food, especially in areas lacking refrigeration;
- because of their aromatic properties, in perfumery, nowadays as an attractant, but, incidentally, historically to mask bad odors of unwashed bodies, that is, decaying people;
- because of their various effects on the body, in medicine and dentistry;
- because of their various effects on the mind, in medicine and by persons abusing drugs;
- because specific types of secretory structures and/or chemicals often characterize certain groups of plants, as taxonomic markers to characterize a plant group (see Part IV).

USES BY PLANTS AND ANIMALS:

- broadly, as attractants and rewards for pollinating agents or dispersing agents to induce an action, namely as:
  - ***attractants,*** things starting a *reaction chain* by direct or indirect action of the sensory apparatus of the visitor, that is, acting as an "advertisement" to bring useful visitors to flowers or to fruits/seeds, for example, shape, and especially odor and color;
  - ***rewards,*** things satisfying *physiological demands* of visitors such as those for food, etc., that is, acting to "reward" visitors to flowers or fruits/seeds (i.e., here the visitor usually receives some caloric benefit), for instance:

    in *pollination,* nectar, pollen, water, liquid oils, food bodies, shelter/protection, odor compounds, stigmatic exudates;

    in *dispersal,* especially food tissues.
- also as ***deterrents*** (including repellents), mechanical and chemical factors preventing exploitation of plant resources by less favorable visitors to flowers or to fruits/seeds or by assorted pests such as animal predators, fungal pathogens, etc.

> We're proud of humanity's powers
> But these potions and medians of ours
> Coffee, garlic, and spices,
> Evolved as devices
> So that insects would stop bugging flowers.

(from R. Cowen, 1990, *History of life,* 1990, p. 302)

IV. SECRETION PRODUCTS AND THEIR ANATOMICAL SOURCES (SECRETORY STRUCTURES)

Secretory structures are classified into external and internal types, depending on whether the structure is located inside or outside the plant. The table on the next page lists the types of external and internal secretory structures. For herbs, spices, and drugs most of the relevant secretory structures are internal in nature. Many of the external secretory structures listed below produce watery and/or salty products that are not especially aromatic (fatty oils are an exception).

External	*Internal*
Trichomes (glandular hairs, papillae, etc.)*	Secretory cells (= oil cells)*
Colleters	Secretory spaces
Glands	Secretory cavities*
Nectaries#	Secretory ducts (= canals)
Osmophores	Resin ducts*
Elaiophores	Gum ducts
Hydathodes	Kino veins
Stigmas*	Laticifers*
Ordinary epidermal cells*	

Key
* = Structures relevant to herbs, spices, and drugs;
# = may be internal with opening to the exterior.

Although secretory cells produce a diversity of products, there is, from a cytological and ultrastructural viewpoint, a certain commonality of structure. That is, secretory cells are active sites of secretion and hence active sites of metabolism, and this is manifested by a dense cytoplasm with many cellular organelles and a concomitant smallish vacuole system.

There is no distinction between given types of secretory structures. Thus, a type of secretory structure in one species might be virtually indistinguishable structurally from the same type of secretory structure in a very unrelated species, although usually there are chemical differences in products the structures secrete. On the other hand, different types of secretory structures may produce virtually identical chemicals (e.g., oil from the secretory cells of star anise, *Illicium verum,* is indistinguishable from oil obtained from the secretory ducts of the umbel *Pimpinella anisum*). In addition, the presence or absence of types of secretory structures and chemicals may be of taxonomic significance in the characterization of families and other large groups of plants. For instance, the clove or eucalyptus family Myrtaceae is characterized by secretory cavities.

V. SOME EXAMPLES OF SECRETORY STRUCTURES

Listed below are some examples of secretory structures. Only the most important species of a genus is noted; species names are mostly from Mabberley (1987).

SECRETORY CELLS
- bay (*Laurus nobilis, Umbellularia californica*) leaves
- cardamom (*Elettaria cardamomum*) seeds
- cinnamon (*Cinnamomum aromaticum* in China, but usually *C. culilaban*) bark
- ginger (*Zingiber officinale*) rhizomes
- lemon grass (*Cymbopogon citratus*), citronella (*C. nardus*) leaves
- linden (*Tilia cordata*) flowers
- nutmeg/mace (*Myristica fragrans*) seeds (nutmeg is the seed kernel, mace the outer fleshy aril)
- pepper (black) (*Piper nigrum*) fruits
- pepper (paprika, cayenne, chili) (*Capsicum annuum* var. *annuum*) fruits
- star anise (*Illicium verum*) fruits
- turmeric (*Curcuma longa*) rhizomes
- wintergreen (*Gaultheria procumbens*) leaves

SECRETORY CAVITIES
- allspice (*Pimenta dioica*) fruits
- citrus (*Citrus* spp.) fruit peels

clove (*Syzygium aromaticum*) flower buds (especially; rarely inflorescence axes) (see Fig. Sup13–1)

SECRETORY CANALS/DUCTS

chamomile (*Anthemis* spp., *Matricaria,* especially *M. chamomilla*) flowers

juniper (*Juniperus communis*) female cones

myrrh (*Commiphora abyssinica*) stems

tarragon (*Artemisia dracunculus*), vermouth (*A. ponica* leaves

umbel family (Umbelliferae or Apiaceae): fruits mostly, as those of anise (*Pimpinella anisum*), caraway (*Carum carvi*), fennel (*Foeniculum vulgare*), but also parsley (*Petroselinum crispum*) and chervil (*Anthriscus cereifolium*) leaves, lovage (*Levisticum officinale*) roots. *Note:* Poison hemlock (*Conium maculatum*) is in this family!

SECRETORY TRICHOMES/PAPILLAE

rose (*Rosa* spp.) petals

saffron (*Crocus sativus*) stigmas (= papillae) and styles (see the price for saffron in Lab Exercise 12, Part III-E)

SECRETORY TRICHOMES/HAIRS

hops (*Humulus lupulus*) female inflorescence bracts (see about beer-making in Lab Exercise 12, Part II-B)

marijuana, hemp (*Cannabis sativa*) leaves and flowers. *Note:* Hops and hemp are in the same family, Cannabidaceae!

mint or labiate family (Labiatae or Lamiaceae): leaves especially, as those of mint (*Mentha* spp.) and oregano (*Origanum* spp.)

vanilla (*Vanilla planifolia*) fruits

LATICIFERS

opium poppy (*Papaver somniferum*) fruits

Note: Many herbs, spices, and drugs derive from chemicals (especially alkaloids) *not* located in secretory structures, for example: cacao (*Theobroma cacao*) fruits, cinchona (*Cinchona calisaya*) bark (for quinine), coffee (*Coffea arabica*) fruits, licorice (*Glycyrrhiza glabra*) rhizomes and roots, nux vomica (*Strychnos nux-vomica*) seed endosperm (for strychnine), and tobacco (*Nicotiana tabacum*) leaves.

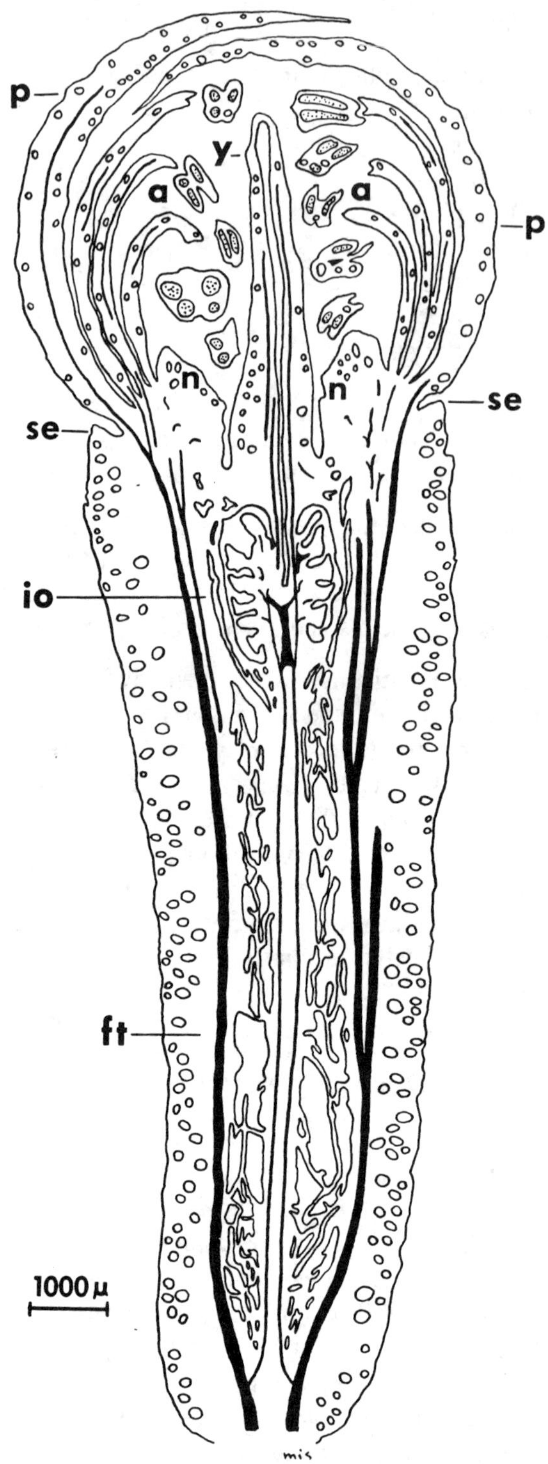

Fig. Sup13–1. Camera lucida longisection of the flower of *Syzygium aromaticum,* the spice clove. The many circles in all the floral parts represent the secretory cavities containing the essential oil (eugenol) for which clove is highly esteemed; the heavy black lines represent vascular tissue. *a,* androecium (stamens); *ft,* floral tube; *n,* nectary; *io,* inferior ovary with ovules; *p,* petal; *se,* sepal; *y,* style. (From R. Schmid, Floral anatomy of Myrtaceae. I. *Syzygium, Bot. Jahrb. Syst.* 92:433–489, 1972.)

References Cited for Plant and Fungal Diversity

Alexopoulos, C. J. & C. W. Mims. 1979. *Introductory mycology.* 3rd ed. New York: John Wiley & Sons. [Previous eds. 1952, 1962.]

Arditti, J. 1967. Factors affecting the germination of orchid seeds. *Bot. Rev.* 33:1–97.

Arnold, C. A. 1964. Mesozoic and Tertiary fern evolution and distribution. *Mem. Torrey Bot. Club* 21:58–66.

Balick, M. J. & J. M. Beitel. 1989. *Lycopodium* spores used in condom manufacture: Associated health hazards. *Econ. Bot.* 43:373–377.

Batra, S. W. T. & L. R. Batra. 1967. The fungus gardens of insects. *Sci. Amer.* 217(2):112–120.

Bhojwani, S. S. & S. P. Bhatnagar. 1978. *The embryology of angiosperms.* New Delhi: Vikas Publishing House PVT. [Previous eds. 1974, 1975.]

Bold H. C., C. J. Alexopoulos & T. Delevoryas. 1987. *Morphology of plants and fungi.* 5th ed. New York: Harper & Row, Publishers. [Previous eds. 1957, 1967, 1973, 1980.]

——— & M. J. Wynne. 1985. *Introduction to the algae: Structure and reproduction.* 2nd ed. Englewood Cliffs, New Jersey: Prentice-Hall. [Previous ed. 1981.]

Bornman, C. H. 1978. *Welwitschia: Paradox eines verdorrten Paradieses/Paradox of a parched paradise.* Cape Town: C. Struik Publishers. [With dual German-English text.]

Brodie, H. J. 1978. *Fungi: Delight of curiosity.* Toronto: University of Toronto Press.

Burns, J. M. 1975. *BioGraffiti: A natural selection.* New York: Quadrangle/The New York Times Book Co.

Canright, J. E. 1962. A little-known talent of Lester W. Sharp. *Amer. Fern J.* 52:160–162.

Chamberlain, C. J. 1935. *Gymnosperms: Structure and evolution.* Chicago: The University of Chicago Press. [Also 1966 reprint by Dover Publications, New York.]

Chastain, R. A. & J. G. Stewart. 1985. Studies on *Berkeleya hyalina* (Round & Brooks) Cox, a marine tube-forming diatom. *Phycologia* 24:83–92.

Christensen, C. M. 1965. *The molds and man: An introduction to the fungi.* 3rd ed. Minneapolis: University of Minnesota Press. [Previous eds. 1951, 1961.]

Cooke, R. C. 1977. *Fungi, man and his environment.* London: Longman.

Copeland, H. F. 1938. The kingdom of organisms. *Quart. Rev. Biol.* 13:383–420.

Corner, E. J. H. 1964. *The life of plants.* London: Weidenfeld and Nicolson.

Côté, W. A., T. E. Timell & R. A. Zabel. 1966. Distribution of lignin in compression wood of red spruce (*Picea rubens*) Sarg. *Holz als Roh- Werkstoff* 24:432–438.

Cowen, R. 1990. *History of life.* Boston: Blackwell Scientific Publications.

Dahlgren, R. 1983. General aspects of angiosperm evolution and macrosystematics. *Nordic J. Bot.* 3:119–149.

Delacrétaz, J., D. Grigoriu & G. Ducel. 1976. *Color atlas of medical mycology.* Trans. from the French by J. Poterat, J. Gunn & S. W. A. Gunn. Bern: Hans Huber Publishers.

Ehrlich, P. R. & E. O. Wilson. 1991. Biodiversity studies: Science and Policy. *Science* 253:758–762. [See also commentary: C. C. Mann, Extinction. 1991. Are ecologists crying wolf? *Science* 253:736–738.]

Emerson, R. 1969. Environments of men and molds—another look at the emperor's new clothes. *Pl. Sci. Bull.* 15(1):1–8.

———. 1973. Mycological relevance in the nineteen seventies. *Trans. Brit. Mycol. Soc.* 60:363–387.

Esau, K. 1965. *Plant anatomy.* 2nd ed. New York: John Wiley & Sons. [Previous ed. 1953.]

Fahn, A. 1990. *Plant anatomy.* 4th ed. Oxford: Pergamon Press. [Previous eds. 1967, 1974, 1982.]

Foster, A. 1949. *Practical plant anatomy.* 2nd ed. New York: D. Van Nostrand Co. [Previous ed. 1942.]

Fuller, H. J. & O. Tippo. 1954. *College botany.* Rev. ed. New York: Henry Holt and Co. [Previous ed. 1949.]

Gifford, E. M. & A. S. Foster. 1989. *Morphology and evolution of vascular plants.* 3rd ed. New York: W. H. Freeman and Co. [Previous eds. 1959, 1974, as A. S. Foster & E. M. Gifford, Jr., *Comparative morphology of vascular plants.*]

Gilpatrick, N. 1972. The secret life of Beatrix Potter. *Nat. Hist.* 81(10):38–41, 88, 90, 92–97, 109.

Haeckel, E. 1866. *Generelle Morphologie der Organismen: Allgemeine Grundzüge der organischen Formen-Wissenschaft, mechanisch begründet durch die von Charles Darwin reformirte Descendenz-Theorie.* 2 vols. Berlin: Georg Reimer.

Hamilton, E. 1942. *Mythology.* Boston: Little, Brown & Co. [Also 1969 reprint by New American Library, New York.]

Hanlin, R. T. & M. Ulloa. 1988. *Atlas of introductory mycology.* 2nd ed. Winston-Salem, North Carolina: Hunter Textbooks. [Previous ed. 1978.]

Hauser, J. T. 1984. Nematode-trapping fungi. *Carolina Tips* 47:37–39.

Hirmer, M. 1927. *Handbuch der Paläobotanik.* Vol. 1. *Thallophyta - Bryophyta - Pteridophyta.* München: R. Oldenbourg.

Hoffman, M. 1992. Yeast biology enters a surprising new phase. *Science* 255: 1510–1511.

Ingold, C. T. 1953. *Dispersal in fungi.* Oxford: Clarendon Press.

———. 1971. *Fungal spores: Their liberation and dispersal.* Oxford: Clarendon Press.

Jensen, W. A. & L. G. Kavaljian. 1958. An analysis of cell morphology and the periodicity of division in the root tip of *Allium cepa. Amer. J. Bot.* 45:365–372.

Johnson, S. A. 1982. *Mushrooms.* Minneapolis: Lerner Publications Co. [Also on other Basidiomycota.]

———. 1983. *Mosses.* Minneapolis: Lerner Publications Co. [Also on liverworts.]

Katsaros, P. 1989. *Illustrated guide to common slime molds.* Eureka, California: Mad River Press.

Kaufman, P., T. F. Carlson, P. Dayanandan, M. L. Evans, J. B. Fisher, C. Parks & J. R. Wells. 1989. *Plants: Their biology and importance.* New York: Harper & Row, Publishers.

Koponen, A. 1990. Entomophily in the Splachnaceae. *Bot. J. Linn. Soc.* 104:115–127.

Kubitzki, K. (ed.). 1990. *The families and genera of vascular plants.* Vol. 1. *Pteridophytes and gymnosperms.* Ed. by Karl Ulrich Kramer & P. S. Green, assist. by Erich Götz (illus.). Berlin: Springer-Verlag.

Large, E. C. 1940. *The advance of the fungi.* New York: Henry Holt & Co. [1962 reprint by Dover Publications, New York.]

Leedale, G. F. 1974. How many are the kingdoms of organisms? *Taxon* 23:261–270.

Lellinger, D. B. 1985. *A field manual of the ferns & fern-allies of the United States & Canada.* Washington, D.C.: Smithsonian Institution Press.

Lewin, R. A., P. A. Fransworth & G. Yamanaka. 1981. The algae of green polar bears. *Phycologia* 20:303–314.

——— & L. Cheng (eds.). 1989. *Prochloron: A microbial enigma.* New York: Chapman and Hall.

Mabberley, D. J. 1987. *The plant-book: A portable dictionary of higher plants.* Cambridge, England: Cambridge University Press.

Mat Salleh, K. 1991. *Rafflesia: Magnificent flower of Sabah.* Kota Kinabalu, Malaysia: Borneo Publishing Co.

Matossian, M. K. 1989. *Poisons of the past: Molds, epidemics, and history.* New Haven: Yale University Press.

Morey, P. R. 1973. *How trees grow.* London: Edward Arnold.

Norstog, K. & R. W. Long. 1976. *Plant biology.* Philadelphia: W. B. Saunders Co.

O'Brien, M. & C. C. O'Brien. 1972. *The story of Ireland.* New York: The Viking Press.

Oostendorp, C. 1987. The bryophytes of the Palaeozoic and the Mesozoic. *Bryophytorum Biblioth.* 34:1–112, pls. 1–49.

Ovenden, G. (ed.). 1972. *The Illustrators of Alice in Wonderland and Through the Looking Glass.* London: Academy Editions. [Also 1979 revised edition in reduced format by St. Martin's Press, New York.]

Panshin, A. J. & C. de Zeeuw. 1980. *Textbook of wood technology: Structure, identification, properties, and uses of the commercial woods of the United States and Canada.* 4th ed. New York: McGraw Hill Book Co. [Previous eds. 1949, 1964, 1970.]

Parbery, D. G. 1969. *Amorphotheca resinae* gen. nov., sp. nov., the perfect stage of *Cladosporium resinae. Aust. J. Bot.* 17:331–357.

Prance, G. T. (text) & K. B. Sandved (photographs). 1985. *Leaves: The formation, characteristics, and uses of hundreds of leaves found in all parts of the world.* New York: Crown Publishers.

Raven, P. H., R. F. Evert & S. E. Eichhorn. 1992. *Biology of plants.* 5th ed. New York: Worth Publishers. [Previous eds. 1971, 1976, 1981, 1986.]

——— & G. B. Johnson. 1992. *Biology.* 3rd ed. St. Louis: Times Mirror/Mosby College Publishing. [Previous eds. 1985, 1989.]

Ray, P. M., T. A. Steeves & S. A. Fultz. 1983. *Botany.* Philadelphia: Saunders College Publishing. [Developed from the 1971 5th ed. of *Botany* by C. L. Wilson, W. E. Loomis & T. A. Steeves, Holt, Rinehart and Winston, New York.]

Scagel, R. F., R. J. Bandoni, J. R. Maze, G. E. Rouse, W. B. Schofield & J. R. Stein. 1982. *Nonvascular plants: An evolutionary survey.* Belmont, California: Wadsworth Publishing Co.

Schmid, M. 1981. *Fleurs et plantes de Nouvelle-Calédonie.* [Papeete]: Les Éditions du Pacifique.

Schmid, R. 1967. Electron microscopy of wood of *Callixylon* and *Cordaites. Amer. J. Bot.* 54:720–729.

———. 1977. Ridding botany of its sexist terminology. *Taxon* 26:441–442. [Excerpt in Curr. Contents 1978(19):13; translated into German by Roswitha Schmid as: Sexismus in der Botanik. *Naturwiss. Rundschau.* 38:389–390 (1985).]

———. 1982. Fruit. *McGraw-Hill Encyclopedia of Science & Technology,* 5th ed., 5:740–746. New York: McGraw-Hill Book Co. [Revised and substantially expanded from versions in previous editions (1960–77) by R. M. Brooks.]

———. 1989. The debt of humour in science to Lewis Carroll. Pp. 5–8 in E. Garfield, Humor in science: The Lewis Carroll connection. *Curr. Contents* 1989(4):3–8. [Reprinted, with new appendix, from 1985 article in *Jabberwocky: J. Lewis Carroll Soc.* 14:53–59. Issued summer 1987.]

Schofield, W. B. 1985. *Introduction to bryology.* New York: Macmillan Publishing Co.

Selsam, M. E. 1986. *Mushrooms.* New York: William Morrow and Co.

Starr, C. 1991. *Biology: Concepts and applications.* 5th ed. Belmont, California: Wadsworth Publishing Co. [Previous eds. 1981, 1984, 1987, 1989.]

Stern, K. R. 1991. *Introductory plant biology.* 5th ed. Dubuque, Iowa: Wm. C. Brown Publishers. [Previous eds. 1979, 1982, 1985, 1988.]

Stewart, W. N. 1983. *Paleobotany and the evolution of plants.* Cambridge: Cambridge University Press.

Taylor, T. N. 1981. *Paleobotany: An introduction to fossil plant biology.* New York: McGraw-Hill Book Co.

Thorn, R. G. & G. L. Barron. 1984. Carnivorous mushrooms. *Science* 224:76–78.

Troughton, J. H. & L. A. Donaldson. 1972. *Probing plant structure: A scanning electron microscope study . . . .* New York: McGraw-Hill Book Co.

——— & F. B. Sampson. 1973. *Plants: A scanning electron microscope survey.* Sydney: John Wiley & Sons Australasia Pty.

Vogel, S. 1988. *Life's devices: The physical world of animals and plants.* Princeton, New Jersey: Princeton University Press.

Watts, B. 1986. *Mushroom.* Morristown, New Jersey: Silver Burdett Co.

Weier, T. E., C. R. Stocking, M. G. Barbour & T. L. Rost. 1982. *Botany: An introduction to plant biology.* 6th ed. New York: John Wiley & Sons. [Previous eds. 1950, 1957, 1964, 1970, 1974.]

Weiner, M. A. 1977. *The taster's guide to beer: Brews and breweries of the world.* New York: Macmillan Publishing Co.

Wernick, R. 1988. What were Druids like, and was Lindow Man one? *Smithsonian* 18(12):146–148, 150, 152, 154, 156, 158–160, 162, 164–166, 176.

Whittaker, R. H. 1959. On the broad classification of organisms. *Quart. Rev. Biol.* 34:210–226.

———. 1969. New concepts of kingdoms of organisms. *Science* 163:150–160.

Wood, A. 1847. *A class-book of botany, designed for colleges, academies and other seminaries.* 2nd ed. Boston: Crocker & Brewster. ["Forty-first Edition, Revised and Enlarged," but only a 2nd ed.—see: E. D. Merrill, Unlisted new names in Alphonso Wood's botanical publications, *Rhodora* 50:101–130 (1948).]

ADDITIONAL REFERENCES APPEAR IN:

Lab Exercise 1, Supplement A: Use of the compound microscope

Lab Exercise 1, Supplement B: Permanent microscope slide preparations

Lab Exercise 4, Supplement: The peel technique in paleobotany (Preparation of a peel from a coal ball)

Lab Exercise 7, Supplement: Dendrochronology (Using the growth rings in the wood of bristlecone pine)

Lab Exercise 11: Flowering plants (angiosperms) III (Some organ and whole-plant modifications)

Sample Exam Questions for Your Amusement

These questions on plant and fungal diversity represent the type that can be expected on the examinations. Some important points: (1) Although all questions will be in multiple-choice format, they fall into two categories: (a) interpretative-type questions based on some conceptual material and/or involving some degree of correlation and synthesis; (b) recall-type questions based largely on definitions, identifications, and similar factual material. (2) Questions of both types are often posed in the lab exercises; hence you should try to answer satisfactorily these questions sometime before the examinations. (3) The questions below are arranged by topics roughly in the sequence presented in the manual, with general or interdisciplinary questions at the end. (4) Unless noted to the contrary, each question has only *one* correct answer. (5) Some of these questions may well appear in identical or revised form on your examinations. (6) For accountability of terminology and common names see comments in the section "Introduction to Material on Plant and Fungal Diversity." (7) Note that not all of these questions may be relevant to the midterm examination(s). (8) *Note:* To facilitate answering multiple-choice type questions determine if *each* statement choice is true or false; the odd-man out is the correct answer (see sample question). For example (● indicates the correct answer):

From an economic viewpoint, the most important bryophytes are:
(A) Liverworts, because they are an alternate source of liverwurst.—FALSE
(B) True mosses, because there is nothing false about bryophytes.—FALSE
(C) Peat mosses, because they can dissolve organisms very well.—FALSE
●(D) Peat mosses, because of their fantastic water-holding capacity.—TRUE

Sample Exam Questions (See answer key at end.)

(1) The phase separating a cell division (mitosis and meiosis) is:
(A) Interphase
(B) Prophase
(C) Metaphase
(D) Anaphase
(E) Telophase.

(2) A beetle bores from the outside of a typical, herbaceous *root* into its middle and in so doing passes through, in sequence:
(A) Epidermis, cortex, endodermis, pericycle, phloem, xylem
(B) Epidermis, cortex, endodermis, pericycle, xylem
(C) Epidermis, cortex, endodermis, phloem, xylem
(D) Epidermis, cortex, xylem, phloem
(E) Epidermis, cortex, phloem, xylem.
Note: Two answers are possible. Why?

(3) Secondary growth in a tree does *not:*
(A) Produce cork
(B) Produce xylem
(C) Produce phloem
(D) Provide additional stability and strength to the stem
(E) Increase the length of branches.

I have answered three questions, and that is enough.—Father William

(4) Which of the following does *not* contain meristematic tissue?
(A) Periderm
(B) Axillary bud
(C) Trunk of a young tree
(D) Root tip
(E) Secondary xylem.

Note: Why isn't "secondary phloem" a valid option here?

(5) Cells dead at functional maturity are:
(A) Tracheids
(B) Parenchyma cells
(C) Sieve tube elements
(D) Collenchyma cells
(E) Epidermal cells.

(6) Cells alive at functional maturity are:
(A) Tracheids
(B) Vessel elements
(C) Cork cells
(D) Meristematic cells
(E) Fibers.

(7) Shalabiya, the magician, sticks her 3 ft.-long sword 5 ft. from ground level into the pith of ancient oak tree 80 ft. tall and 4 ft. in diameter. Over the next 30 years the tree increases its diameter 3 ft. and its height 20 ft. Then Sir Wysiwyg visits the tree. He finds the sword:
(A) Nowhere, because it is completely embedded still 5 ft. above the ground
(B) Nowhere, because it is completely embedded but 25 ft. above the ground
(C) 5 ft. above the ground, the sword handle exposed
(D) 25 ft. above the ground, the sword handle exposed
(E) Sticking in Shalabiya's skeleton.

(8) Developing ovules of gymnosperms have all of the following structures *except:*
(A) A megaspore
(B) Antheridia
(C) Archegonia
(D) Female GPT (megaGPT)
(E) Integument.

(9) Gymnosperms may produce:
(A) Flowers
(B) Fruits
(C) Fleshy cones
(D) Endosperm
(E) Antheridia.

(10) In a pine seed, the nutritive tissue for the embryo and seedling is the:
(A) Female GPT (MegaGPT)
(B) Endosperm
(C) Nucellus
(D) Embryo sac
(E) Integument.

(11) Angiosperms do *not* produce:
(A) Flowers
(B) Fruits
(C) Gametophytes
(D) Endosperm
(E) Antheridia.

(12) In an angiosperm seed, the *direct* nutrient source for the embryo is the:
(A) Endosperm
(B) Progymnosperm
(C) Sperm
(D) Embryo sac
(E) Female GPT (megaGPT).

(13) Which process does not involve fungi?
(A) Cheese production
(B) Beer production
(C) Bread production
(D) Nitrogen fixation
(E) Penicillin (an antibiotic) production.

(14) The structures bearing the asexual spores of the sac fungi are:
(A) Sporangia
(B) Conidiophores
(C) Antheridiophores
(D) Sporangiophores
(E) Stalks.

(15) Fungi are saprobes because they:
(A) Manufacture nutrients from inorganic compounds
(B) Absorb nutrients
(C) Ingest nutrients
(D) Secrete nutrients
(E) Store nutrients.

(16) Which of these structures is haploid ($1n$)?
(A) Angiosperm stem
(B) Embryo sac
(C) Liverwort spore capsule (sporangium)
(D) Gymnosperm root
(E) Sporophyll.

(17) Which of these structures is diploid ($2n$)?
(A) Sperm
(B) Embryo sac
(C) Moss GPT
(D) Pollen grain
(E) Sporocyte.

(18) The thallus is:
(A) Haploid ($1n$)
(B) Diploid ($2n$)
(C) Triploid ($3n$)
(D) Dikaryotic ($1n+1n$)
(E) Another ploidy level ($4n$, $5n$, etc.).

(19) The integument is:
(A) Haploid ($1n$)
(B) Diploid ($2n$)
(C) Triploid ($3n$)
(D) Dikaryotic ($1n+1n$)
(E) Another ploidy level ($4n$, $5n$, etc.).

(20) Which term does *not* belong with the others?
(A) Apical meristem
(B) Protoderm
(C) Provascular tissue
(D) Ground meristem
(E) Vascular cambium.

(21) Which term does *not* belong with the others?
(A) Cone
(B) Sporangium
(C) Sporophyll
(D) Spore
(E) Stem.

(22) Which term does *not* belong with the others?
(A) Epidermis
(B) Xylem
(C) Periderm
(D) Cork
(E) Protoderm.

(23) Which term does *not* belong with the others?
(A) Antheridium
(B) Archegonium
(C) Sperm
(D) Egg
(E) Sporocyte.

(24) Which term does *not* belong with the others?
(A) Cork
(B) Periderm
(C) Cork cambium
(D) Cork parenchyma
(E) Protoderm.

(25) A plant in which each pollen grain has only one aperture is most likely a:
(A) Gymnosperm
(B) Progymnosperm
(C) Dicotyledon
(D) Bryophyte
(E) Pteridophyte.

(26) Male gametes form in:
(A) Microsporangia
(B) Megasporangia
(C) Antheridia
(D) Archegonia
(E) Ovules.

(27) The *first* multicellular plants present on earth most likely were:
(A) Mosses
(B) Algae
(C) Bryophytes
(D) Fungi
(E) Pteridophytes.

(28) A group with a dominant GPTic generation (phase) (i.e, the SPT is less conspicuous) is the:
(A) Pteridophytes
(B) Bryophytes
(C) Gymnosperms
(D) Bryophytes and pteridophytes
(E) Club fungi.

(29) A group in which the SPT does *not* achieve eventual independence from the GPT is the:
(A) Pteridophytes
(B) Bryophytes
(C) Gymnosperms
(D) Bryophytes and pteridophytes
(E) Club fungi.

(30) Which group lacks antheridia?
(A) Bryophytes
(B) Pteridophytes
(C) Conifers
(D) Ferns
(E) Lycopods.

(31) Which group lacks archegonia?
(A) Bryophytes
(B) Pteridophytes
(C) Angiosperms
(D) Ferns
(E) Lycopods.

(32) The embryo does *not* begin its development in the archegonium in the:
(A) Bryophytes
(B) Pteridophytes
(C) Angiosperms
(D) Ferns
(E) Lycopods.

(33) Pollen tubes occur in:
(A) Mosses
(B) Lycopods
(C) Ferns
(D) Cycads
(E) Lycopods.

(34) Meiosis in *all* land plants occurs in the:
(A) Sporangium
(B) Gametangium
(C) Anther
(D) Zygote
(E) Seed.

(35) In the land plants, the last stage of the GPTic generation (phase) is always the:
(A) Zygote
(B) Sporocyte
(C) Spore
(D) Gamete
(E) Egg.

(36) In the land plants, the first stage of the GPTic generation (phase) is always the:
(A) Zygote
(B) Sporocyte
(C) Spore
(D) Gamete
(E) Egg.

(37) In the lab we examined heterospory in:
(A) Pines (Coniferophyta)
(B) Ferns (Pterophyta)
(C) Horsetails (Sphenophyta)
(D) Mosses (Bryophyta)
(E) Whisk ferns (Psilophyta).

(38) A group having xylem and phloem but lacking seeds is the:
(A) Ferns
(B) Algae
(C) Gymnosperms
(D) Angiosperms
(E) Bryophytes.

(39) Nuclei do *not* occur in:
(A) Ferns
(B) Algae
(C) Fungi
(D) Prokaryotes
(E) Mosses.

(40) Chloroplasts do *not* occur in:
(A) Ferns
(B) Algae
(C) Fungi
(D) Cycads
(E) Mosses.

(41) Under a dissecting microscope when you move the object to the left the image moves:
(A) To the left
(B) To the right
(C) Toward you
(D) Away from you
(E) Diagonally.

(42) Under a compound microscope when you move the object to the left the image moves:
(A) To the left
(B) To the right
(C) Toward you
(D) Away from you
(E) Diagonally.

It takes all the running you can do, to keep in the same place—The Red Queen

Note well: Specific type questions emphasizing "trivia" will not be asked, for instance (● indicates the correct answer):

(1) Filaments do *not* occur in:
- (A) Blue-green bacteria (Cyanophyta)
- (B) Green algae (Chlorophyta)
- ●(C) Euglenoids (Euglenophyta)
- (D) Brown algae (Phaeophyta)
- (E) Red algae (Rhodophyta).

Note: However, a question on the silica walls of diatoms is fair because this was noted several times.

(2) The conifers have how many species?
- (A) 750
- ●(B) 536
- (C) 137
- (D) 76
- (E) 1.

(3) Tofu and bean curd are soybeans fermented by:
- (A) Sac fungi (Ascomycota)
- (B) Club fungi (Basidiomycota)
- ●(C) Bread mold/etc. fungi (Zygomycota)
- (D) Imperfect fungi (Deuteromycota)
- (E) Water molds (Oomycota).

Note: However, a question on some economic/historical aspects of plants or fungi would be fair game if they were extensively discussed.

ANSWER KEY FOR SAMPLE EXAM QUESTIONS:

1: A	8: B	15: B	22: B	29: B	36: C
2: A,B	9: C	16: B	23: E	30: C	37: A
3: E	10: A	17: E	24: E	31: C	38: A
4: E	11: E	18: A	25: A	32: C	39: D
5: A	12: A	19: B	26: C	33: D	40: C
6: D	13: D	20: E	27: B	34: A	41: A
7: A	14: B	21: E	28: B	35: D	42: B

Note for question 2: Phloem in roots appears discontinuous in transection (it is continuous in longitudinal extent) and thus there are two tissue sequences that would appear arrayed along the radii.
Note for question 4: Secondary phloem contains meristematic tissue if periderm develops in it.
Note for questions 20–24: The format for these questions is a good one for a review but usually is too open-ended to use on actual exams.

It's time for you to answer now.—The Red Queen

STUDY NOTES FOR EXAMINATIONS